Prisoners of Reason

Is capitalism inherently predatory? Must there be winners and losers? Is public interest outdated and free-riding rational? Is consumer choice the same as self-determination? Must bargainers abandon the no-harm principle?

Prisoners of Reason recalls that classical liberal capitalism exalted the no-harm principle. Although imperfect and exclusionary, modern liberalism recognized individual human dignity alongside individuals' responsibility to respect others. Neoliberalism, by contrast, views life as a ceaseless struggle. Agents vie for scarce resources in antagonistic competition in which every individual seeks dominance. This political theory is codified in noncooperative game theory; the neoliberal citizen and consumer is the strategic rational actor. Rational choice justifies ends irrespective of means. Money becomes the medium of all value. Solidarity and good will are invalidated. Relationships are conducted on a quid pro quo basis. However, agents can freely opt out of this cynical race to the bottom by embracing a more expansive range of coherent action.

S. M. Amadae is a research affiliate at the Massachusetts Institute of Technology and has held positions as an assistant professor of political science at the Ohio State University and as an associate professor of political science at the Central European University. Amadae's first book, *Rationalizing Capitalist Democracy: The Cold War Origins of Rational Choice Liberalism* (2003), was awarded the American Political Science Association's J. David Greenstone book award for History and Politics in 2004. This thought-provoking political theorist who works on the foundations of liberalism and the philosophy of political economy has also contributed articles to the *Journal of Economic Methodology*, *History of European Ideas*, *Studies in the History and Philosophy of Science*, *Economics and Philosophy*, the *American Journal of Economics and Sociology*, *Ethics*, and *Idealistic Studies*. Amadae graduated with a PhD from the University of California, Berkeley and has held appointments at Cambridge University, the London School of Economics, the University of British Columbia Vancouver, the New School, and Harvard University's Kennedy School of Government.

Prisoners of Reason

Game Theory and Neoliberal Political Economy

S. M. AMADAE
Massachusetts Institute of Technology

CAMBRIDGE
UNIVERSITY PRESS

32 Avenue of the Americas, New York NY 10013-2473, USA

Cambridge University Press is part of the University of Cambridge.

It furthers the University's mission by disseminating knowledge in the pursuit of education, learning and research at the highest international levels of excellence.

www.cambridge.org
Information on this title: www.cambridge.org/9781107671195

First published 2015

A catalogue record for this publication is available from the British Library

Library of Congress Cataloguing in Publication data
Amadae, S. M., author.
Prisoners of reason : game theory and neoliberal political economy / S.M. Amadae.
    pages   cm
Includes bibliographical references and index.
ISBN 978-1-107-06403-4 (hbk. : alk. paper) – ISBN 978-1-107-67119-5 (pbk. : alk. paper)
1. Game theory – Political aspects.   2. International relations.   3. Neoliberalism.
4. Social choice – Political aspects.   5. Political science – Philosophy.   I. Title.
HB144.A43   2015
320.01′5193 – dc23   2015020954

ISBN 978-1-107-67119-5 Paperback

To András Rátonyi

A Hippocratic Oath for Humankind

I swear to fulfill, to the best of my ability and judgment, this covenant:

I will respect the hard-won scientific gains of those ... [researchers] in whose steps I walk, and gladly share such knowledge as is mine with those who are to follow.

I will apply, for the benefit of the ... [suffering], all measures which are required, avoiding those twin traps of overtreatment and therapeutic nihilism.

I will remember that there is art to ... [politics] as well as science, and that warmth, sympathy, and understanding may outweigh the ... [expert's judgment] or the ... [law's enforcement].

I will not be ashamed to say "I know not," nor will I fail to call in my colleague when the skills of another are needed for ... [an individual's] recovery.

I will respect the privacy of my ... [fellows], for their problems are not disclosed to me that the world may know. Most especially must I tread with care in matters of life and death. If it is given me to save a life, all thanks. But it may also be within my power to take a life; this awesome responsibility must be faced with great humbleness and awareness of my own frailty. Above all, I must not play at God.

I will remember that I do not treat a fever chart, a cancerous growth, but a[n] ... [anguished] ... human being, whose illness may affect the person's family and economic stability. My responsibility includes these related problems, if I am to care adequately for the ... [deprived].

I will prevent ... [affliction] whenever I can, for prevention is preferable to cure.

I will remember that I remain a member of society, with special obligations to all my fellow human beings, those sound of mind and body as well as the infirm.

If I do not violate this oath, may I enjoy life and art, respected while I live and remembered with affection thereafter. May I always act so as to preserve the finest traditions of my calling and may I long experience the joy of healing those who seek my help.

Quoted from, with language more general than of the medical doctor–patient relationship in the original, Louis Lasagna, "Oath," undated document, D302, Box 2, Folder 3, Courtesy of the Department of Rare Books, Special Collections and Preservation, University of Rochester River Campus Libraries. I replaced references to sickness and medicine with references to suffering and science more broadly.

Contents

Tables

Figures

Acknowledgments

I did not set out to write about the Prisoner's Dilemma and nuclear strategy. Yet the entanglement of the two with social contract theory and liberalism made it impossible to continue my exploration of the philosophical foundations and institutionalized practices of free markets and democracy without understanding how game theory has altered our vision of citizenship and sovereignty. I am grateful for the intellectual gifts offered to me by the following inspiring intellectual pioneers: Peter Galison, Jerry Green, Amartya Sen, Sheila Jasanoff, Richard Tuck, Robert Jervis, Jerry Gaus, Patrick Morgan, and Rohit Parikh. I have also benefited from generous support and feedback from numerous researchers: Nicola Giocoli, Shaun Hargreaves Heap, Nancy Rosenblum, Ken Shepsle, Nicolas Guilhot, John Mueller, Herb Weisberg, Donald Hubin, John Blackburn, David Kaiser, Randy Calvert, Shannon Stimson, David Hollinger, Robert McMahon, Jenny Mansbridge, Holger Strassheim, Margaret Schabas, Paul Erickson, Philip Mirowski, Hunter Hayek, David Ciepley, Egle Rindzeviciute, Judy Schwartzbaum, Peter Hershock, Jill Hargis, Jennifer Burns, Cricket Keating, Nicole Pallai, Christina McElderry, and the anonymous reviewers of this book.

I am grateful to those who invited me to give presentations in their departments and conferences because these opportunities were vital to developing my ideas: Robert Cavelier, Lorraine Daston, Mark Bevir, Jenny Andersson, Silja Samerski, Anna Henkel, Heidi Voskuhl, Jessica Wang, Alexei Kojevnikov, and Stefan Schwarzkopf. I benefited from the rich facilities at the University of Oldenburg; the National Center for Scientific Research, CRNS Campus, Paris; Sciences Po, Paris; the University of Arizona's Center for the Philosophy of Freedom; the Program on Science, Technology and Society, Harvard's Kennedy School of Government; Universitat Pompeu Fabra, Barcelona; the Center for International Research in the Humanities and Social Sciences, New York University; the Center for Science, Technology, Medicine and Society and Department of Political Science, University of California, Berkeley; the

Departments of the History of Science and Government, Harvard University; the Organization of American Historians; the Copenhagen Business School, Denmark; the History of Economics Society; the Social Science History Association; the Seminar in Logic and Games, CUNY; the Society for Social Studies of Science; the Political Science Department, Central European University; the American Political Science Association; the Western Political Science Association; the Jimmy Carter Presidential Library and Museum; the Hoover Institution Archive, Stanford University; University Archives, University of Rochester; and the Old Cambridge Baptist Church, Cambridge, Massachusetts. I appreciate Carnegie Mellon University Press's permission to use my article, "James M. Buchanan, John Rawls, and Democratic Governance," in *Deliberative Democracy: Theory and Practice*, ed. Robert Cavalier (Carnegie Mellon University Press, 2011), 31–52.

This project would not have been possible without the generous and exceptional assistance afforded by many distinguished colleagues: Robert Dreesen, Peter Dimock, Jonathan Cohn, John Ledger, Toni Maniaci, Diana Camella, Brittany Paris, Helen Gavel, Brianda Reyes, Jerry Udinsky, Alicia Anzivine, Philip Alexander, Linda Benson, and Melinda Wallington. I have benefited from the assistance of talented undergraduate researchers: Daniel McKay, John Adams, Halie Vilagi, James Chevako, Katie Grammenidis, Chris Urbano, Tori Mahoney, Hunter Phillips, Emilia Schrier, Lan Anh Le, and Garrett Nastarin.

I am grateful for the intangible support I received from individuals who stood as pillars of friendship and wisdom throughout the past years: Agi Risko, Diana Wear, Denise Bauman, Susan Spath, Kate Stocklin, Nancy and Wayne Hayner, Psyche North Torok, Teri McClung, Judi Stock, Jessie Labov, Kati Rátonyi, Sheryl, Olene and Jim Foley, Pauline Curley, and Ji Kwang Dae Poep Sa Nim. I appreciate my family members for their understanding of my unyielding commitment to writing and research: Sarah and Barry McLean, T Dani Adams, Paul Adams, Jim and Penny Adams, Pieter and Jacqueline Noppen, Iván Rátonyi, and Dennis Gifford. Thus, finally, I am delighted to have this opportunity to express my boundless gratitude to András Rátonyi, without whose steadfast brilliance, ingenuity, and wide-ranging problem-solving skills *Prisoners of Reason* would not have been completed.

Prologue

In this century, great advances in the most fundamental and theoretical branches of the physical sciences have created a nuclear dilemma that threatens the survival of our civilization. People seem to have learned more about how to design physical systems for exploiting radioactive materials than about how to create social systems for moderating human behavior in conflict. Thus, it may be natural to hope that advances in the most fundamental and theoretical branches of the social sciences might be able to provide the understanding that we need to match our great advances in the physical sciences. This hope is one of the motivations that has led many mathematicians and social scientists to work in game theory during the past 50 years.

—Roger B. Myerson, 1991[1]

In 1989, as the Cold War was coming to a close, Francis Fukuyama argued that liberal democracy represented the "end point of mankind's ideological evolution" and the "final form of human government." It constituted, he asserted, the "end of history."[2] With the fall of the Berlin Wall and the collapse of the Soviet Union, many Westerners concluded that there were no longer viable alternatives to capitalist democracy.[3] To its advocates, this system, manifest in a combination of consumer capitalism and thin political democracy, "resolved all of the contradictions of life for which, through the course of history, individuals have been prepared to fight."[4] Yet within a mere dozen years, the United States,

[1] Myerson was awarded the Sveriges Riksbank Prize in Economic Sciences in Memory of Alfred Nobel in 2007; quote is from *Game Theory: Analysis of Conflict* (Cambridge, MA: Harvard University Press, 1991), 1–2.

[2] Francis Fukuyama writing about his 1989 article, "The End of History?" (*National Interest*, summer 1989), in *The End of History and the Last Man* (New York: Avon Books, 1992), xi.

[3] For reference to this institution see Richard A. Posner, *Crisis of Capitalist Democracy* (Cambridge, MA: Harvard University Press, 2011); S. M. Amadae, *Rationalizing Capitalist Democracy* (Chicago: University of Chicago Press, 2003); phrase first sentence of Kenneth J. Arrow, *Social Choice and Individual Values*, 2nd ed. (Yale University Press, 1963), 1.

[4] Michael W. Doyle's description of Francis Fukuyama's argument in *Ways of War and Peace* (New York: W. W. Norton, 1997), 474–475.

which had been the world's leading proponent of this ideology, took on a leading role in a global war on terror. American presidents, pundits, and citizens confronted an uncomfortable new reality characterized by entrenched military engagement in the Middle East and a shift of economic power from the West to the Far East.

Prisoners of Reason suggests that this unsatisfactory conclusion to the Cold War in part resulted from seeds sown from within that gave rise to neoliberal capitalism. The price of winning the Cold War was not only a vast nuclear arsenal and budget deficit but also the transformation of individual autonomy and collective sovereignty. By the close of the twentieth century, the free markets and democratic governance alluded to by Fukuyama had become unmoored from their classical liberal ideals and refashioned according to the strategic rationality of game theory. The no-harm principle at the root of classical liberalism no longer, neither in theory nor practice, animates the action of rational actors who instead seek gain despite others. The concept of mutual benefit has yielded to the inevitability of winners and losers. Norm bound negotiation has given way to coercive bargaining. Financialization, risk management, and algorithmic control replace the efficient use of resources and technological innovations as the major engines for profit. Freedom, once rooted in self-determination and equality before the law, is reduced to individual choice as defined by one's willingness and ability to pay for a product or service.

The exemplary neoliberal citizen and consumer is the strategic rational actor modeled by orthodox game theory. This observation is both trivial and profound. It is trivial because rational choice theory, which dominated the social sciences and professional schools by the 1980s, makes the identical claim. It is profound because of how game theory has come to shape the unique practices of contemporary late-modern capitalism.[5] The strategic rational actor was codified in a mathematically tractable and operationalizable form particularly suited to rational deterrence: the US national security state was the world's first rational actor.[6] Given demands of nuclear security, strategists' pursuit of fail-safe policy created the appeal for a comprehensive science of decision making. In its standard form, strategic rationality is consequentialist, realist, individualistic, and amoral. Undoubtedly, this canonical rational actor is a simplification, a straw man.[7] Yet

[5] Max Weber used the phrase "instrumental rationality" to refer to bureaucratic rule-following rationality that left little room for individual judgment. He writes, "Rational calculation . . . reduces every worker to a cog in this [bureaucratic] machine and, seeing himself in this light, he will merely ask how to transform himself . . . to a somewhat bigger cog . . . The passion for bureaucratization at this meeting drives us to despair." Quote from Max Weber, *Economy and Society: An Outline of Interpretive Sociology* (Berkeley: University of California Press, 1978), lix; for reference to the "iron cage of reason," see Weber's chap. 5, "Asceticism and the Spirit of Capitalism," in *The Protestant Ethic and the Spirit of Capitalism* (London: Routledge, [1905]/1985).

[6] Jack L. Snyder, *The Soviet Strategic Culture: Implications for Limited Nuclear Options* (Santa Monica, CA: RAND, Sept. 1977), "Generic Rational Man versus Soviet Man," 4.

[7] Amartya K. Sen, "Rational Fools: A Critique of the Behavioral Foundations of Economic Theory," *Philosophy and Public Affairs* (1977) 6:4, 317–344.

within the context of the Cold War, this fictional character came to stand in for ideal rational agents throughout international relations, civil politics, and even evolutionary biology. Through its dissemination in the pedagogy of elite institutions, the neoliberal subject invented by game theory has come to animate contemporary markets and politics.[8]

Prisoners of Reason advances three arguments. First, strategic rationality, which game theorists understand to be an all-encompassing theory of rational decision making, informs an important strand of postmodern neoliberal subjectivity and agency: that operative in advanced capitalism. Rational choice theorists model complex interactions to predict collective outcomes and generate public policy, legislation, or institutional design.[9] Thus, second, their research has resulted in a canonical set of findings that characterize neoliberal theory and which this book will explore. These novel game theoretic findings directly correlate with the particular late-modern expression of capitalist democracy. Third, the political theory consistent with strategic rationality marks a distinct break with classical liberalism. In its standard form, rational choice theory rejects the rule-following normativity and logic of appropriateness embodied in classical liberal principles of no-harm, fair play, consent, and contractual commitment.[10] Game theory develops a theory of fungible value consistent with philosophical realism, which, in alliance with international relations realism, further distances this system of thought and practice from

---

[8] Studies have shown that students exposed to game theory will tend to demonstrate behavior consistent with its tenets. See Robert Frank, Thomas Gilovich, and Donald Regan, "Does Studying Economics Inhibit Cooperation?" *Journal of Economic Perspectives* (1993) 7, 159–171; Dale Miller, "The Norm of Self-Interest," *American Psychologist* (1999) 54, 1–8; for studies showing that students exposed to economics are more prone to cheating, see Donald McCabe and Linda Klebe Trevino, "Academic Dishonesty," *Journal of Higher Education* (1993) 64:5, 522–538; Donald McCabe and Linda Klebe Trevino, "Cheating among Business Students," *Journal of Management Education* (1995), 19:2, 205–218; and Donald McCabe, Kenneth Butterfield, and Linda Klebe Trevino, "Academic Dishonesty in Graduate Business Programs," *Academy of Management Learning and Education* (2006) 5:3, 294–305.

[9] Gary Becker, Sveriges Riksbank (Nobel) Prize winner in Economic Science, 1992, *The Economic Approach to Human Behavior* (Chicago: University of Chicago Press, 1978); Douglass C. North, Sveriges Riksbank (Nobel) Prize winner in Economic Science, 1993, *Institutions, Institutional Change and Economic Performance (Political Economy of Institutions and Decisions)* (Cambridge: Cambridge University Press, 1993); for a recent example, see The *Stern Review: The Economics of Climate Change*, which is available online, http://webarchive.nationalarc hives.gov.uk/+/http:/www.hm-treasury.gov.uk/sternreview_index.htm, see chap. 21, accessed July 20, 2015.

[10] My critique is not of formal modeling per se, but rather of the orthodox application of game theory. Joseph Heath, *Communicative Rationality and Rational Choice* (Cambridge, MA: MIT Press, 2003); and Gerald Gaus, *The Order of Public Reason: A Theory of Freedom and Morality in a Diverse and Bounded World* (Cambridge: Cambridge University Press, 2012); both strive to accommodate the side constraints or commitment characteristic of classical liberalism by expanding the formal framework initiated by game theory. For another approach to commitment, see Margaret Gilbert, *Joint Commitment: How We Make the Social World* (New York: Oxford University Press, 2013).

classical liberalism.[11] Orthodox game theory assumes that actors must be strategic, or individualistically competitive against others, and thus rejects joint maximization and shared intention, and reduces preference satisfaction to narrow self-interest.[12]

The association of game theory, as a systematic body of knowledge encompassing all coherent choice, with late-modern neoliberal political economy may seem jarring at first. On the one hand, "neoliberalism" is a phrase coined by leftist critics of contemporary capitalism to draw attention to its extraction of monetary value from all human relations and its erosion of the public sphere, producing unprecedented levels of inequality. What is new about this stage of capitalism, contemporary critics argue, is the financialization and commodification of *all* experiential value.[13] On the other hand, "neoliberal institutionalism" is a school of international relations theory that applies the tools of game theory to model complex interactions.[14] These theorists argue that actors in a state of nature can achieve cooperation through building regimes and institutions. From their perspective, cooperative norms can emerge which reflect stable behavioral patterns that arise when actors' preferences and choices cohere naturally or when appropriate incentives are introduced to modify choice.

On the surface, these two uses of the term "neoliberalism" are distinctly at odds with each other. Neoliberal critics of late-modern capitalism carry forth Karl Marx's dissatisfaction with laborers' plight in industrial capitalism and thus critically assess contemporary market practices, their apparent destruction of lower and middle classes, and their creation of new means of extracting surplus value.[15] Neoliberal institutionalists use game theory to show how

---

[11] See John von Neumann and Oskar Morgenstern, *Theory of Games and Economic Behavior* (Princeton, NJ: Princeton University Press), 60th anniversary edition, 2004, 617–631; on the assumption of fungible value in international relations, see Michael Doyle, *Ways of War and Peace: Realism, Liberalism, and Socialism* (New York: W. W. Norton, 1997), 47.

[12] Robert Sugden and Michael Bacharach expand individualistic maximization to permit team reasoning or joint maximization: Michael Bacharach, *Beyond Individual Choice: Teams and Frames in Game Theory*, ed. by Natalie Gold and Robert Sugden (Princeton, NJ: Princeton University Press, 2006). Joseph Heath touches on these themes in *Following the Rules: Practical Reasoning and Deontic Constraint* (New York: Oxford University Press, 2011), 12–41. Altruistic or other-regarding preferences can be admitted, see, e.g., Alvin Goldman, *Joint Ventures: Mindreading, Mirroring, and Embodied Cognition* (New York: Oxford University Press, 2012). However, because game theory tends to treat other actors as "complex objects" that can be incorporated into instrumental action, it loses the moral grounding of treating other agents as ends in themselves with human dignity; see Heath, *Following the Rules*, 2011, 41, 10–11.

[13] David Harvey, *A Brief History of Neoliberalism* (New York: Oxford University Press, 2007); Pierre Bourdieu, *Acts of Resistance: Against the Tyranny of the Market* (New York: The New Press, 1999).

[14] For an excellent overview of the neoliberal school of international relations theory, see David A. Baldwin, ed., *Neorealism and Neoliberalism* (New York: Columbia University Press, 1993). In particular, see Robert Axelrod and Robert O. Keohane, "Achieving Cooperation under Anarchy: Strategies and Institutions," Baldwin, ed., *Neorealism and Neoliberalism*, 85–115.

[15] Harvey, *A Brief History of Neoliberalism*, 2007.

classical liberal achievements, namely effective governance and economic exchange, can be attained notwithstanding the minimalist and even cynical view of human agency that rational choice accepts.[16] Surely, they reason, if mutually beneficial governance and markets can be derived from strategic rationality, proponents of an international relations realism are overly pessimistic in finding the inevitability of conflict. However, *Prisoners of Reason* argues that neoliberal market capitalism and neoliberal institutionalism share a common foundation: the assumption that strategic rationality governs all purposive agency.

Elite institutions of higher education in the West treat orthodox game theory as a canonical statement of instrumental rationality.[17] It thus seems a mere truism to observe that contemporary citizens and consumers must either conform to its dictates or risk acting "irrationally." Yet, if game theory shapes actors' subjective awareness of the meaning of their interactions; alters their behavior; and informs public policy, laws, and institutions, then we must examine the possibility that late-modern capitalism enacted by strategically rational actors is distinct from its earlier forms.[18] If indeed so, it would not be surprising that game theoretic models both analytically predict outcomes consistent with neoliberal theory and that these analytic conclusions directly correlate with contemporary empirically evident neoliberal practices.

Such a view of the transformative aspect of game theory is antithetical to its portrayal as a value-neutral tool and the perennial structure of purposive agency.[19]

---

[16] Robert Axelrod and Robert O. Keohane, "Achieving Cooperation under Anarchy: Strategies and Institutions," in Baldwin, ed., *Neorealism and Neoliberalism*, 1993, 85–115; on the cynicism inherent in rational choice explanations, see Russell Hardin, *Collective Action* (Johns Hopkins University Press, 1982), xv.

[17] For a compelling statement that expected utility theory represents our state of the art understanding of instrumental rationality, see Donald C. Hubin, "The Groundless Normativity of Instrumental Rationality," *The Journal of Philosophy* (2001) 98, 445–465; for discussion of the restrictions on expected utility theory useful in game theory (strategic rationality and the debate over whether language is primordial and strategic action parasitic, or vice versa), see also Heath, *Following the Rules*, 2008, 12–41; and Heath, *Communicative Action and Rational Choice* (Cambridge, MA: MIT Press, 2001), 49–81; Daniel M. Hausman and Michael S. McPherson, *Economic Analysis, Moral Philosophy, and Public Policy*, 2nd ed. (Cambridge: Cambridge University Press, 2006).

[18] Elinor Ostrom, *Governing the Commons* (Cambridge: Cambridge University Press, 1990), 1–28; Robert Frank, Thomas Gilovich, and Donald Regan, "Does Studying Economics Inhibit Cooperation?" *Journal of Economic Perspectives* (1993) 7, 159–171; Dale Miller, "The Norm of Self-Interest," 1999, 1–8; for studies showing that students exposed to economics are more prone to cheating, see McCabe and Klebe Trevino, "Academic Dishonesty," 1993; McCabe and Klebe Trevino, "Cheating among Business Students," 1995; and McCabe, Butterfield, and Klebe Trevino, "Academic Dishonesty in Graduate Business Programs."

[19] Herbert Gintis, *Bounds of Reason* (Oxford: Oxford University Press, 2009); Stephen Quackenbush, "The Rationality of Rational Choice Theory," *International Interactions: Empirical and Theoretical Research in International Relations* (2004) 30:2, 87–107. For the relationship between rational choice and confidence in social science, see Gary King, Robert O. Keohane, and Stanley Verba, *Designing Social Inquiry: Scientific Inference in Qualitative Research* (Princeton, NJ: Princeton University Press, 1994).

Game theory itself is a mathematical theory and thus derives its validity from analytic consistency.[20] It cannot be tested as a valid instrumental theory of rationality without establishing a consistent common means to provide a universally recognized mapping of every individual's subjective evaluation of the idiosyncratic features of outcomes and their relationship to the tangible phenomenal world governed by the laws of physics.[21] Operationalizing orthodox strategic rationality invents a particular subjectivity, either as an ideal type or as an experiential fact, insofar as individuals are taught to master and apply strategic rationality in various contexts of choice.[22] Thus, *Prisoners of Reason* analyzes what assumptions game theorists introduce about individuals' choice to make interaction contexts, or "games," susceptible to modeling and application in descriptive, normative, and prescriptive contexts.[23]

This book analyses the work of game theorists who address rational deterrence, the social contract and collective action, and the emergence of cooperation among pre-social actors in evolutionary biology. Theorists apply the same tools and models to these widely divergent fields of investigation, thus showing how game theory offers a unified methodology and results in a comprehensive understanding of agency from actors without deliberate intention, to humans and the nation state as a multiparty composite actor. Chapters focus on the prominent theorists Thomas Schelling, James R. Schlesinger, James M. Buchanan, Richard Posner, Russell Hardin, Richard Dawkins, and Robert Axelrod. Most chapters centrally address the Prisoner's Dilemma game because many of these theorists gave this model priority in their contributions.[24] Part I:

---

[20] Ken Binmore, *Game Theory and the Social Contract*, vol. 1, *Playing Fair* (Cambridge, MA: MIT Press, 1994), 95.

[21] Myerson, *Game Theory: Analysis of Conflict*, 1991, does the best job outlining this, 22–25; on the common knowledge assumption, see also Shaun Hargreaves Heap and Yanis Varoufakis, *Game Theory: A Critical Introduction*, 2nd ed. (New York: Routledge, 2004), 27; see also David Lewis, *Convention: A Philosophical Study* (Oxford: Blackwell, 2002), 52–60.

[22] For an accessible statement of orthodox game theory, see Heath, *Following the Rules*, 2011, 12–41.

[23] Mary Morgan argues that the practice of formal modeling has led to the model itself being treated as reality, rather than as a mere representation of reality: "The Curious Case of the Prisoner's Dilemma: Model Situation?" in A. N. H. Creager, E. Lunbeck, and M. N. Wise, *Science without Laws: Models Systems, Cases, Exemplary Narratives* (Durham, NC: Duke University Press, 2007), 157–85.

[24] The most prominent texts are the following: Duncan R. Luce and Howard Raiffa, *Games and Decisions: Introduction and Critical Survey* (New York: Wiley, 1957); Anatol Rapoport and Albert M. Chammah, *Prisoner's Dilemma: A Study in Conflict and Cooperation* (Ann Arbor: University of Michigan Press, 1965); Anatol Rapoport, *Fights, Games and Debates* (Ann Arbor: University of Michigan Press, 1960); Thomas Schelling, "Hockey Helmets, Concealed Weapons, and Daylight Saving: A Study of Binary Choices with Externalities," *Journal of Conflict Resolution* (1973) 17:3, 381–428; Richmond Campbell and Lanning Snowden, eds., *Paradoxes of Rationality and Cooperation: Prisoner's Dilemma and Newcomb's Problem* (Vancouver: University of British Columbia Press, 1985); Roger B. Myerson, *Game Theory: Analysis of Conflict* (Cambridge, MA: Harvard University Press, 1991); Dennis C. Mueller,

War discusses the development of nuclear deterrence theory and practice.[25] Part II: Government analyzes the implications of the novel game theoretic findings for neoliberal, as opposed to classical liberal, political theory and practice. Part III: Evolution explores how game theorists attribute strategic rationality to biological organisms in their behavioral programming. These three parts together show how game theorists suggest that strategic rationality provides a comprehensive account of purposive action operative at all levels of organization from pre-intentional action to individual's choice and international relations strategy.

Prisoners of Reason participates in the enduring Western celebration of reason and the contemporary critical discussion of rational choice. This discussion is comprehensive and directly impinges on the implications of game theory for understanding and achieving social order.[26] With respect to international relations theory, the discussion of critical assessments of rational choice mainly

Public Choice III (New York: Cambridge University Press, 2004); Hargreaves Heap and Varoufakis, *Game Theory*, 2004.

[25] The most prominent here are Herman Kahn, *On Thermonuclear* War (Princeton, NJ: Princeton University Press, 1960); Thomas Schelling, *Strategy of Conflict* (Cambridge, MA: Harvard University Press, 1960); Glenn H. Snyder, *Deterrence and Defense: Toward a Theory of National Security* (Princeton, NJ: Princeton University Press, 1961); Philip Green, *Deadly Logic: The Theory of Nuclear Deterrence* (Columbus: Ohio State University Press, 1966); Jack Snyder, *The Soviet Strategic Culture: Implications for Limited Nuclear Options* (Santa Monica, CA: RAND Corporation, Sept. 1977); Robert Jervis, *The Illogic of Nuclear Strategy* (Cornell University Press, 1984); Lawrence Freedman, *The Evolution of Nuclear Strategy* (New York: St. Martin's Press, 1981); Fred M. Kaplan, *The Wizards of Armageddon* (New York: Simon & Schuster, 1983); Gregg Herken, *Counsels of War* (New York: Alfred A. Knopf, 1985); Douglas P. Lackey, "The American Debate on Nuclear Weapons Policy: A Review of the Literature 1945–85," *Analyse and Kritik* 9 (1987), 7–46; Joseph S. Nye Jr., *Nuclear Ethics* (London: Free Press, 1986); Steven J. Brams, *Superpower Games: Applying Game Theory to Superpower Conflict* (Yale University Press, 1985).

[26] With respect to the large and general critique of game theory, key texts are Sen, "Rational Fools," 1990; Jane Mansbridge, ed., *Beyond Self Interest* (Chicago: University of Chicago Press, 1990); Jon Elster, *Sour Grapes: Studies in the Subversion of Rationality* (New York: Cambridge University Press, 1985), and *Ulysses and the Sirens: Studies in Rationality and Irrationality* (New York: Cambridge University Press, 1979); Donald P. Green and Ian Shapiro, *Pathologies of Rational Choice Theory: A Critique of Applications in Political Science* (Yale University Press, 1994); Kristen Renwick Munroe, *The Economic Approach to Politics: A Critical Reassessment of the Theory of Rational Action* (New York: HarperCollins, 1991). With respect to the particular texts analyzing the implications of strategic rationality for social contract theory and collective action, see James M. Buchanan, *Limits of Liberty: Between Anarchy and Leviathan* (Chicago: Chicago University Press, 1975); Michael Taylor, *Anarchy and Cooperation* (New York: Wiley, 1976), and *The Possibility of Cooperation* (New York: Cambridge University Press, 1987); Russell Hardin, *Collective Action* (Johns Hopkins University Press, 1982); David Gauthier, *Morals by Agreement* (Oxford: Clarendon Press, 1986); Martin Hollis, *Trust within Reason* (Cambridge: Cambridge University Press, 1998); Ken Binmore, *Playing* Fair, 1994, and *Game Theory and the Social Contract*, vol. 2, *Just Playing* (Cambridge, MA: MIT Press, 1998); Joseph Heath, *Communication Action and Rational Choice* (Cambridge, MA: MIT Press, 2001); Gintis, *Bounds of Reason*, 2009; and Gaus, *Order of Public Reason*, 2012.

follows the contours of the debate between the schools of realism and neoliberal institutionalism.[27] This book intersects with researchers investigating the historical roots of game theory in the Cold War.[28] *Prisoners of Reason* also relates to the burgeoning critical engagement with neoliberal capitalism.[29]

The central argument of *Prisoners of Reason* builds on recent analyses of modern liberal political theory by Michael Doyle, Richard Tuck, and Elaine Scarry. Doyle, in his *Ways of War and Peace* (1997), provides a discussion of theories of international relations, including classical realism and classical liberalism, to pave the way toward revitalizing a commitment to liberal international relations theory. He identifies theoretical principles of classical liberalism that ground individual freedom, private property, and equality of opportunity and extend to the domain of relations among

[27] Robert Jervis wrote an early paper addressing the relationship between realism and game theory: "Realism, Game Theory, and Cooperation," *World Politics* (1988) 40, 317–349; see also Jervis, "Realism, Neoliberalism, and Cooperation: Understanding the Debate," *International Security* (1999), 24, 42–63. Baldwin's edited collection sets forth the debate as it developed in the 1980s and makes clear that advocates of realism promote strategic rationality and a commitment to underlying objective sources of power, and that neoliberal institutionalists concede this framework but argue that even under this most limited set of assumptions, cooperation in international institutions and regimes is still attainable: Baldwin, *Neorealism and Neoliberalism*, 1993. Michael Doyle's *Ways of War and Peace* (1997) discusses the debate between a contemporary renewed commitment to the classical liberal tradition in international relations and a realist approach. *Prisoners of Reason* dovetails with Doyle's *Ways of War and Peace* in drawing attention to the assumptions and structure of classical liberalism vis-à-vis contemporary realism and instrumentalism consistent with orthodox game theory. On rational deterrence and rational decision theory (game theory), see Keith Kraus, "Rationality and Deterrence in Theory and Practice," in *Contemporary Security and Strategy*, ed. by Craig A. Snyder (New York: Routledge, 1999), 120–149.

[28] Key texts are Philip Mirowski, *Machine Dreams: Economics Becomes a Cyborg Science* (New York: Cambridge University Press, 2002; S. M. Amadae, *Rationalizing Capitalist Democracy* (Chicago: University of Chicago Press, 2003); Robert Leonard, *Von Neumann, Morgenstern, and the Creation of Game Theory: From Chess to Social Science, 1900–1960* (New York: Cambridge University Press, 2010); Paul Erikson et al., *How Reason Almost Lost Its Mind: The Strange Career of Cold War Rationality* (Chicago: University of Chicago Press, 2013); William Thomas, *Rational Action: The Sciences of Policy in Britain and America, 1940–1960* (Cambridge, MA: MIT Press, 2015); Paul Erickson, *The World the Game Theorists Made* (Chicago: University of Chicago Press, 2015). On Cold War social science more generally see Jamie Cohen-Cole, *The Open Mind: Cold War Politics and the Sciences of Human Nature* (Chicago: Chicago University Press, 2014); and Mark Solovey, *Shaky Foundations: The Politics-Patronage-Social Science Nexus in Cold War America* (New Brunswick, NJ: Rutgers University Press, 2013).

[29] Crucial texts in this investigation are Harvey, *Brief History of Neoliberalism*, 2007; Philip Mirowski and Dieter Plehwe, eds., *The Road from Mount Pelerin: The Making of the Neoliberal Thought Collective* (Cambridge, MA: Harvard University Press, 2009); Manfred B. Steger and Ravi K. Roy, *Neoliberalism: A Very Short Introduction* (New York: Oxford University Press, 2010); Michel Foucault, *Birth of Biopolitics: Lectures at the Collège de France, 1978–1979* (London: Picador Reprint ed., 2010); Angus Burgin, *The Great Persuasion: Reinventing Free Markets since the Depression* (Cambridge, MA: Harvard University Press, 2012).

nations. For liberal political theory, individual freedom is premised on "the right to be treated and the duty to treat others as ethical subjects, not as objects or means only."[30] Although the "will to subjugate" may be an ever-present concern, this need not compromise an actor's intention to seek peace and build "the mutual confidence and respect that establishing a true peace will require."[31]

Tuck's *Free Riding* (2008) provides a template for *Prisoners of Reason* in combining historical insights with philosophical analysis to focus attention on how contemporary rational choice theory inverts commonsense understanding of the causal efficacy of collective intention. Tuck provides examples from modern European political thought to demonstrate that theorists, including David Hume and John Stuart Mill, stressed the rationality of collaboration. Thus, throughout the modern era, the state had to introduce legally backed sanctions to prevent collusion among firms or collaboration among individuals pursuing collective bargaining rights.[32] Tuck argues that large-scale market competition defies the logic of strategic competition because no individual actor has the causal power to make any appreciable difference on collective outcomes. The foremost concern over the failure of collective action is thus the worry about individuals' lack of causal efficacy in large-scale undertakings, and not the Prisoner's Dilemma concern that rational individuals seek to free ride on others' efforts. Like *Free Riding*, *Prisoners of Reason* highlights how rational choice introduces a novel approach to coherent action that displaces earlier conventional wisdom.[33] Specifically, neoliberalism jettisons the commonsense understanding that civil society depends on mutual respect and the good will to make at least one person better off and no one worse off in every interaction.

Scarry's *Thermonuclear Monarchy: Choosing between Democracy and Doom* (2014) draws attention to how the invention of nuclear weapons altered the US practice of sovereignty from republican democracy with the military under civilian authority, to a system of command and control with its own prerogatives and little respect for either citizens' participation or the no-harm principle. Scarry contrasts modern liberal political theory, which stresses the inviolability of corporeal persons, with the post–World War II reliance on secrecy and disregard for embodied persons. Scarry's discussion of the social contract points out the ways in which the exercise of sovereignty through the

---

[30] Doyle, *Ways of War and Peace*, 1997, 207. Richard Tuck's *The Rights of War and Peace: Political Thought and the International Order from Grotius to Kant* (New York: Oxford University Press, 2001) argues that the classical liberal subject was first established with respect to nations before it was articulated in civil political theory.

[31] Doyle, *Ways of War and Peace*, 1997, 255–256.

[32] Amadae, *Rationalizing Capitalist Democracy* (2003) argues that the theory of liberalism consistent with rational choice theory proved that cooperation in various guises is irrational.

[33] Russell Hardin, "The Free Rider Problem," *Stanford Encyclopedia of Philosophy*, May 21, 2003, available online at http://plato.stanford.edu/entries/free-rider/, accessed June 16, 2015 provides an example of how the argument that it is rational to free ride on others' efforts goes across the grain of conventional wisdom.

threat to destroy peoples and civilizations with weapons of mass destruction interrupts the classical liberal commitment to a contract among the ruled and those who rule. Her chapter on "Consent and the Body" reminds readers how throughout modern political philosophy, giving consent was an embodied practice with direct implications for the well-being of those voicing a willingness to participate. Scarry's analysis in *Thermonuclear Monarchy* complements *Prisoners of Reason*'s argument that the development of game theory and its integration into law and public policy facilitated the exercise of nuclear sovereignty. Game theory views individuals as abstracted sets of preference rankings free from corporeal embodiment, and it empties the practice of consent of meaning because actors' preferences exist outside of time and thus can be artificially incorporated into public decision making without the direct participation of citizens.

Finally, this book incorporates methodological insights from Sheila Jasanoff's *States of Knowledge: The Co-Production of Science and the Social Order* (2004). Jasanoff contributes to the methodology of science, technology, and society studies and argues that scientific innovation and its incorporation into social life unfold in a process of co-production.[34] Applied to the US exercise of superpower sovereignty through deterrent threats of catastrophic harm and the manifestation of neoliberal subjectivity in the form of deferring to rational choice theory as the ultimate science of purposive decision making, Jasanoff's perspective invites us to investigate into how individual autonomy and collective agency coevolve with the decision technologies that rationalize action. Such a perspective helps show how game theory developed together with the "nuclear dilemma." John von Neumann, a cardinal architect of neoliberal economics, formalized game theory, axiomatized quantum thermodynamics, and contributed to the Manhattan Project. Von Neumann views the social world in terms of relentless competition over scarce resources that everyone alike seeks, and he inscribes this worldview into the mathematical foundations of his social science. *Prisoners of Reason* excavates these original foundational assumptions and reveals their implications for instantiating late-modern political economy on their basis.

OUTLINE OF THE BOOK

Prisoners of Reason starts with two preliminary chapters that set the stage for its central argument that game theory provides a singular approach to purposive agency and collective action that informs late-modern neoliberal descriptive and predictive models. Chapter 1, "Neoliberalism," discusses how classical liberalism differs from neoliberalism, primarily in the latter's rejection of the Archimedean no-harm principle that grounds the reciprocal respect of

---

[34] *Thermonuclear Monarchy: Between Democracy and Doom* (New York: Routledge, 2014). I am grateful to Sheila Jasanoff who invented the title *Prisoners of Reason*.

individuals' rights to their persons, property, and contracts. This chapter addresses the apparent gap between the uses of the term "neoliberal" by critics of late-modern capitalism and the international relations school of neoliberal institutionalism. Critics of neoliberal political economy point to its propagation of gross disparities in wealth distribution, its tendency to socialize debt and privatize profit, and its treatment of all experiences as potential means to extract surplus value. In contrast, neoliberal institutionalism strives to rescue the classical liberal vision of mutual benefit while accepting that only strategic rationality governs purposive conduct.

Given the neoliberal institutionalists' commitment to employing game theoretic models to make the case that cooperation can emerge among egoists even in a state of anarchy in which no other actor can be trusted to cooperate, not even oneself, the association of this school with neoliberal economic practices may seem off-putting or unjustified. However, the goal of this book is to identify how, despite researchers' best intentions, strategic rationality condones predatory behavior as a feasible and even necessary means to secure individual survival and success. Rational choice theorists acknowledge they are "generally concerned with pushing cynical explanations to their limit" to address the worst possible case and come up with sound remedies.[35] Yet, as the pages ahead document, rather than finding such remedies, this analytic style is more likely to lead to the validation of cynical *modus operandi* and dismissal of alternatives. Thus, because strategic rationality is recognized as the gold standard of reason, actors are rewarded for, or pressured into, following its rules for choice. Idiosyncratic neoliberal economic practices coincidentally conform to the predictions of the game theoretic models.

The second preliminary chapter, "Prisoner's Dilemma," follows game theorists in their identification of the particular prominence of this recalcitrant game in which individuals' self-interest is mutually destructive, rather than mutually beneficial.[36] The Prisoner's Dilemma belies the classical liberal argument that self-interest results in mutual prosperity. The chapter demonstrates how the pedagogy of the Prisoner's Dilemma (PD) relies on a set of tacit assumptions that must be accepted to master this game. Mastering the Prisoner's Dilemma, and therein embracing the assumptions underlying its standard operationalization, makes it difficult to resist the powerful logic of strategic rationality and inevitably leads actors to assimilate and accommodate neoliberal subjectivity.

Part I: War follows the progress of game theory in its entanglement with nuclear strategy. Chapter 3, "Assurance," centers on the economist and strategist Thomas Schelling. This future Nobel Memorial Prize recipient uses game

[35] Hardin, *Collective Action*, 1982, xv.
[36] E.g., ibid.; Campbell and Sowden, *Paradoxes of Rationality and Cooperation*, 1985; Ostrom, *Governing the Commons*, 1990; Hargreaves Heap and Varoufakis, *Game Theory*. For a more contemporary discussion of the Prisoner's Dilemma's central role in modeling the neoliberal challenge of achieving social order, see Heath, *Following the Rules*, 2011, 35–40.

theory to develop the mentality of strategic rationality, which he first used to analyze nuclear deterrence before applying it more broadly to economics, politics, and conflict resolution. Although his research precedes the neoliberal institutionalist school, his work is consistent with those researchers' aims. Schelling uses strategic rationality to identify and resolve the worst-case scenario to build a nuclear peace resembling the classical liberal commitment to mutual respect in the form of reciprocal deterrence. The accompanying discussion of nuclear strategy and arms development shows how game theory informed and made possible the integration of nuclear planning into US sovereignty using rational decision technologies. Game theorists and game theory were indispensable for wielding nuclear weapons and exercising nuclear sovereignty. Strategists altered sovereignty so that it resided less in the judgments of an individual commander in chief and more in the office itself, seamlessly integrated with algorithmic judgment afforded by strategic rationality.

This chapter also introduces the consequential nuclear security debate between the two positions of mutual assured destruction (MAD), supported by Schelling, and nuclear utilization targeting selection (NUTS), supported by the defense analyst Herman Kahn. Given the outlay of hundreds of billions of dollars during the Cold War on nuclear hardware, and the additional $900 billion spent on command and control, the nuclear security debate is not only central to US history but also reflective of the application of game theory to render intelligible nuclear deterrence.[37] Although the average US citizen may have been unaware of NUTS, this aggressive strategy dedicated to preparing to fight and prevail in protracted nuclear war won the Cold War nuclear security debate by the end of Jimmy Carter's presidency in 1980. Rational deterrence theory armed American strategists with the confidence and clout to mobilize these weapons of mass destruction in the service of American security and sovereignty.

Chapter 4, "Deterrence," follows the conclusion of the debate between the liberal stance of MAD and the offensive realist policy of NUTS. The champions of neoliberal agency firmly suppose that in facing the worst-case scenario, mutual cooperation resembling classical liberalism can still be achieved. And yet, as is evident in the logical capitulation of MAD to NUTS, the inherent cynicism of supposing that every individual pursues ends despite other agents without regard for the dignity of personhood or the legitimacy of principles of conduct ends up ceding the classical liberal promise of mutual benefit to the neoliberal realism of coercive bargaining, predatory gain, and asymmetric deterrence. Chapter 4 follows nuclear strategy under President Jimmy Carter to show how neoliberal principles consistent with strategic rationality grew out of an ultimately futile effort to retain classical liberal practices of mutual security and exchange in a non-classical world.

[37] Daniel Volmar, PhD dissertation, The Power of the Atom: US Nuclear Command, Control, and Communications, 1945–1965, Harvard University, forthcoming 2016, p. 13 ($926 billion on Command and Control, in 1996 US dollars).

Part II: Government follows the application of game theory to modeling interactions constitutive of markets and governance. The themes in Part II will be familiar to anyone who knows game theory: the Prisoner's Dilemma analysis of Hobbesian anarchy, the Prisoner's Dilemma account of the social contract, the role of unanimous agreement, the role of consent, and the problem of collective action. These chapters discuss the core game theoretic findings in these areas of research and pinpoint what is unique to neoliberal political philosophy predicated on strategic rational action, specifically insofar as it deviates from classical liberalism.

Chapter 5, "Hobbesian Anarchy," discusses how game theorists interpret Hobbes's state of nature to be a Prisoner's Dilemma game and see in his authoritarian *Leviathan* a solution: maintaining social order at the point of a sword. However, Hobbes viewed the problem of achieving social order akin to an Assurance Game, in which actors who prefer to cooperate end up seeking individual gain despite others because of a lack of trust, not a Prisoner's Dilemma. Members of a commonwealth choose to cooperate once assured of others' like preferment or commitment; their first preference is not to defect. Moreover, Hobbes's solution is achieved by individuals laying down the right to all things. Any mode of action resembling strategic rationality is laid aside in favor of adopting side constraints as a precondition for harmonious living in civil society.

Chapter 6, "Social Contract," follows James M. Buchanan's application of the Prisoner's Dilemma to social contract theory. Buchanan's *Limits of Liberty: Between Liberty and Leviathan* exemplifies neoliberal political theory. Buchanan uses the Prisoner's Dilemma to argue that as long as the social contract offers more than the state of nature – for him defined by the worst every individual can threaten on every other actor – citizens will conform to its division of goods. Whereas John Rawls believes that stability under the social contract is achieved when actors are motivated by tacit consent under the veil of ignorance, Buchanan argues that actors will accommodate any social contract backed by sufficient coercive force. Therefore, he calls for force rather than legitimacy to address the civil unrest characterizing the late 1960s and early 1970s.

Chapter 7, "Unanimity," follows Buchanan further in his attempt to apply a key insight of the neoclassical economists who identified in the principle of unanimous agreement a solid ground for government and collective decisions. Surely, if every member of a community agrees that one outcome is superior to all the rest, then the leadership and citizens can confidently enact it as law. However, counterintuitively, in neoliberal political theory, unanimous agreement has no intrinsic mobilizing force for collective action. Since individuals each relentlessly maximize expected utility in competition with one another, their agreement to an outcome or a law can only ever be strategic. At any moment when spoils are on the table to be shared, regardless of any prior unanimous agreement on divisions, renegotiations are inevitable, with actors forming unstable coalitions, engaging in coercive bargaining, and preferring to sucker others.

Chapter 8, "Consent," examines Richard Posner's treatment of this titular topic throughout his school of law and economics, which defines justice as wealth maximization. Posner relies on rational choice theory's stipulation that rational individuals have preference rankings over all outcomes independent of time to conclude that consent prior to an exchange and consent after an exchange are equivalent. Thus, consent to terms no longer connotes deliberate action or conscious judgment. Posner's approach justifies the state's arbitrary redistribution of rights on the basis of wealth maximization: if one party has a greater willingness and ability to pay for a good or a better prospect of reaping profit from a resource than its current owner, the state can forcibly transfer the property rights accordingly.

Chapter 9, "Collective Action," examines the standard game theoretic extension of the two-person Prisoner's Dilemma to a large number of individuals. Game theorists argue that voluntary cooperation must fail because all actors prefer to free ride on others' contributions. This chapter contrasts this neoliberal PD logic with Mancur Olson's *Logic of Collective Action* (1965), which argues that group size dictates a group's ability to cooperate. Olson proposes that in large-scale collective ventures, agents fail to cooperate for the same reason that perfect competition requires that no single actor can alter the price of a product: in large group settings, no single individual has the wherewithal to causally impact the outcome for the group or any one of its members. Individuals' failure to cooperate, then, stems from a sense of causal impotence rather than a desire to free ride.

Part III, "Evolution," explores how strategic rationality has been used to model organisms in a state of evolutionary natural selection. Chapter 10, "Selfish Gene," follows the application of game theory to evolutionary biology with the implication that all life forms are deduced to follow the laws of strategic rationality to survive the demands of natural selection. Game theory provides algorithms that may have been programmed into organisms' behavior and assumes no conscious choice. Evolutionary game theorists hold that organisms compete over a source of objective fitness value and that each organism's survival depends on gaining more than others. Richard Dawkins uses these analytic models to argue that humans must have evolved to carry a gene for selfish behavior that conforms to the principles of noncooperative game theory.

Chapter 11, "Tit for Tat," discusses Robert Axelrod's argument that cooperation can emerge among the egoistic utility maximizers modeled by noncooperative game theory. Axelrod uses a repeated Prisoner's Dilemma game played by two actors to argue that reciprocal altruism along the line of the golden rule would be a successful strategy in this setting. Axelrod's argument is pivotal for the neoliberal institutionalists' hopes that cooperation can emerge even among actors who abide by the principles of noncooperative game theory. Hence, according to this analysis, markets and institutions reminiscent of classical liberalism can be sustained notwithstanding the pessimistic assumptions about the character of actors. However, as this chapter shows, Tit for Tat cooperation is an idealized solution that must be supplemented by coercive sanctions and vigilante punishment for cooperation to emerge in settings beyond those with two individuals interacting indefinitely with perfect recall who play their strategies without error and value the future as much as the present.

The final chapter, *Pax Americana*, restates the unique findings of neoliberal theory derived from rational choice. Neoliberal subjects are those stymied by the Prisoner's Dilemmas theorized to exist throughout life's experiences. However, acting in accordance with the strategic imperatives of noncooperative game theory is optional. One can instead choose to actualize the classical liberal disposition to acknowledge others' right to exist, a more comprehensive view of value that transcends fixed-sum resources, and permits both joint maximization and the classic liberal imperfect duty of altruism. The Prisoner's Dilemma is a loaded trap that seems perplexing precisely because its historical development has superimposed the will to dominate others on top of the motive to act cautiously and independently to protect one's interests. The bright-line test of whether actors are really caught in a PD game, and whether they qualify as neoliberal subjects, is whether these individuals would choose to cooperate after the other actor has already chosen to do so in situations classified as tangible resource dilemmas in which everyone seeks to secure scarce goods. Classic liberal society is constructed on the premise that if others cooperate in market exchange and provision of public goods, then one will voluntarily do so as well. Neoliberal theory asserts that every actor will likely cheat, free ride, and seek self-gain by threatening harm on others even if others cooperate. It may seem that strategic actors can simply incorporate a predilection for cooperation into their utility functions. However, in its orthodox form, noncooperative game theory permits only the consideration of consequences, views joint maximization as unsound, and stresses fungible rewards directly associated with actors' chances of instrumental success. Rather than regarding norms and rule following as deliberate choices, game theory views norms as behavioral patterns that emerge solely as a consequence of individualistic preference satisfaction.

PRELIMINARIES

I

Neoliberalism

The political effects of nuclear weapons are also a serious consideration. American values have been affected. Our central government is larger, and the executive branch plays a larger role in foreign affairs. Interaction between strategic adversaries involves secrecy, and secrecy is difficult to reconcile with democracy. Many of these changes began before 1945, but enormous life-and-death decisions are nevertheless delegated to the president or his successors, and the circumstances may not permit congressional involvement. Knowledge of the details of nuclear targeting plans tends to be restricted to the military, and there have been cases in the past where a significant gap existed between military plans and what elected officials thought to be policy.

Joseph S. Nye Jr., 1986[1]

The distinction between acting parametrically on a passive world and acting non-parametrically on a world [of game theory] that tries to act in anticipation of these actions is fundamental. If you wish to kick a rock down a hill, you need only concern yourself with the rock's mass relative to the force of your blow. . . . By contrast, if you wish to kick a person down the hill, then unless the person is unconscious, bound or otherwise incapacitated, you will likely not succeed unless you can disguise your plans until it's too late for him to take either evasive or forestalling action.

Don Ross, 2006[2]

The resultant dominant ideology is founded on the illusion that observed inequality is not to be explained in terms of the social power of one class or group over the other but, instead, is the result of different abilities, work ethic, etc. . . . Indeed, mainstream economics, and by association game theory, may be thought of as the highest form of this ideology . . .

Our world may have never before been so ruthlessly divided along the lines of extractive power between those with and those without access to productive means. And yet never before has the dominant ideology been so successful at

---

[1] Joseph S. Nye Jr., *Nuclear Ethics* (New York: Free Press, 1986), 78.
[2] Don Ross, "Game Theory," *Stanford Encyclopedia of Philosophy*, published 1997, substantially revised 2014, available on line, http://plato.stanford.edu/entries/game-theory (accessed September 1, 2015).

convincing most people that there are no systematic social divisions; that the poor
are mostly undeserving and that talent and application is all the weak need in order
to grow socially powerful.

Shaun Hargreaves Heap and Yanis Varoufakis, 2004[3]

After September 11, 2001, it became obvious that Francis Fukuyama's predicted
End of History failed to materialize in prosperous global markets and inclusive
democratic governance. This disappointing outcome to the Cold War, and the
attendant ongoing deprivation faced by billions worldwide, encourages contemporary critics of capitalism to investigate the unique features of late-modern
political economy. They point out that both the reach of the market and differentials in access to resources are wholly unprecedented. Keeping company with
the varied efforts to explain the unique features of postmodern market discipline
and government rule, *Prisoners of Reason* advances the simple thesis that neoliberalism reflects both the mentality and institutions consistent with orthodox
game theory. To some, this might seem like targeting the straw man strategic
actor, much as Thorstein Veblen ridiculed the ideal *Homo economicus* of neoclassical economics.[4] However, this *Homo strategicus* remains at the epicenter of
standard game theory pedagogy and continues to represent the most readily
operationalized agency modeled by the theory.[5] It is difficult to imagine teaching
game theory without introducing new students to the canonical rational actor
who only considers outcomes in terms of direct personal advantage and renders
all decision making commensurable on a single scale translatable into money.[6]

This chapter briefly discusses contemporary authors' efforts to identify the
unique features of neoliberal political economy. This scholarly engagement
with capitalism veers from blurring the distinctions between classical liberal
and late-modern capitalism on the one hand to pointing out all of the oppressive
features of contemporary economic practice on the other. The first section
advances the thesis that neoliberal economics and politics are best understood
as the result of modeling agency and designing institutions according to the

---

[3] Shaun Hargreaves Heap and Yanis Varoufakis, *Game Theory*, 2nd ed. (New York: Routledge, 2004), 262–263.
[4] Thorstein Veblen, "Why Is Economics Not an Evolutionary Science?" reprinted in *The Place of Science in Modern Civilization* (New York: Viking, 1919), 73.
[5] Note, for example, that the Nash equilibrium solution concept of "mutual-best-reply" is only guaranteed when mixed (randomized) strategies relying on expected utility theory are permitted; furthermore, in many useful applications, interpersonally transferable sources are assumed (hence permitting that players can offer side payments to other players). For discussion, see Duncan Luce and Howard Raiffa, *Games and Decisions* (New York: Wiley, 1957), 88–113.
[6] See how Roger B. Myerson introduces game theory with the idea that money serves as a useful means to model interactions: *Game Theory* (Cambridge, MA: Harvard University Press, 1991), 3. See also Ken Binmore's treatment, *Game Theory: A Very Short Introduction* (New York: Oxford University Press, 2007). The *Economist* magazine makes clear that to use the tool of cost-benefit analysis, common in neoliberal economic practice, all decision making must be made as if out-comes have prices: "Economic Focus: Never the Twain Shall Meet," *Economist*, February 2, 2002, 82.

principles of orthodox game theory. The second section explores how the political philosophy consistent with rational choice theory departs from the founding principles of classical liberal markets and government. Neoliberal theory negates side constraints, deontic commitments, and due process. Classical liberalism depends on the no-harm principle and mutual respect for its defense of free markets and a minimal state. Neoliberalism cannot define harm; even if actors agreed on a standard of harm, still they would advance their self-interest at the expense of others, breaking agreements and free riding whenever possible.

This chapter's final section provides an overview of how neoliberal theory starts with the minimal assumptions of strategic rationality and attempts to derive a basis for a social order that roughly resembles modern free markets and democratic governance. Numerous rational choice theorists have viewed the challenge of achieving peace out of anarchy as a version of Prisoner's Dilemma under the assumption that every individual will seek self-benefit at a cost to others. Their solution is the introduction of incentives to mobilize individuals' compliance with laws. Although neoliberalism seems to offer a means to achieve stability, it differs from classical liberalism by normalizing the pursuit of gain at the expense of others and the implementation of governance through coercive sanctions, leaving little room for legitimacy and voluntary compliance.

The arc of *Prisoners of Reason* begins with the international relations nuclear security dilemma, then proceeds to the identical security dilemma proposed to underlie civil government, and ends with the similarly formulated question of how evolving life forms individually survive and achieve stable, resource efficient, equilibria. In each case, theorists have found that the Prisoner's Dilemma game represented a key decision problem confronting actors. This chapter's final section, "Neoliberalism and Nuclearism," articulates how orthodox game theory resonated with the *realpolitik* approach to international relations popular during the Cold War period. Exploring this overlap between game theory and international relations realism helps clarify the implications of applying the same assumptions to the structure of civil society.

DEFINING NEOLIBERALISM

The term *neoliberalism* is currently in vogue.[7] At least since David Harvey published *A Brief History of Neoliberalism* (2007), there has been a mushrooming

---

[7] Harvey's *Neoliberalism* was published by Oxford University Press, 2007; see also Henry A. Giroux, *The Terror of Neoliberalism: Authoritarianism and the Eclipse of Democracy* (London: Paradigm Publishers, 2004), on Carter, see 22–23; for a more recent overview of the field of inquiry, see Mathieu Hilgers, "The Three Anthropological Approaches to Neoliberalism," *International Social Science Journal* (2010) 61:202, 351–364, and his "The Historicity of the Neoliberal State," *Social Anthropology* (2012) 20:1, 80–94. Oxford University Press published *Neoliberalism: A Very Brief Introduction* by Manfred B. Steger and Ravi K. Roy, 2010. See also Pierre Bourdieu, *Acts of Resistance: Against the Tyranny of the Market* (New York: New Press, 1999); and Michel

inquiry into the nature of the practices constituting late-modern political economy, often dated to the late 1970s when President Jimmy Carter appointed Paul Volcker to chair the Federal Reserve Bank and actively promoted policies of deregulation. Following the planned economy of the New Deal and World War II, and the ensuing fiscal Keynesianism, public choice theorists supported a renewed enthusiasm for the power of markets to solve social problems and generate prosperity.[8]

This late twentieth-century incarnation of capitalism, associated with Margaret Thatcher, Ronald Reagan, Deng Xaioping, and later John Williamson's 1989 Washington Consensus, has come to be referred to as neoliberalism.[9] Harvey argues that neoliberalism, as a mode of discourse, has become hegemonic, producing "pervasive effects on ways of thought to the point where it has become incorporated into the common-sense way many of us interpret, live in, and understand the world."[10] Neoliberalism, he explains, argues that "the social good will be maximized by maximizing the reach and frequency of market transactions" and thus "seeks to bring all human action into the domain of the market."[11] Michael Sandel draws attention to similar concerns in his recently published book *What Money Can't Buy: The Moral Limits of Markets*.[12] His argument, that the contemporary practice of monetizing all value displaces moral and other-regarding dimensions of action,

Foucault, *The Birth of Biopolitics: Lectures at the Collège de France*, 1978–1979 (Lectures at the Collège de France, Picador Reprint Edition, 2010). For a theoretical analysis of the distinction between a classic liberal and neoliberal author, see Javier Avanzadi, *Liberalism against Liberalism: Theoretical Analysis of the Writings of Ludwig von Mises and Gary Becker* (London: Routledge, 2006). On the contrast between neoliberalism and neoconservatism, see Wendy Brown, "American Nightmare Neoliberalism, Neoconservatism, and De-Democratization," *Political Theory* (2006), 34:6, 690–714. For a skeptical position on neoliberalism, see Rajesh Venugopal, "Neoliberalism as a Concept," *Economy and Society* (2015) 44:2, 165–187.

[8] Some observers view the return to a pro-market orientation in opposition to the legacy of the New Deal and Keynesian economics as neoliberalism and point to Friedrich Hayek and Milton Friedman as paving the way. See Philip Mirowski and Dieter Plehwe, eds., *The Road from Mount Pelerin: The Making of the Neoliberal Thought Collective* (Cambridge, MA: Harvard University Press, 2009); see also Angus Burgin, *The Great Persuasion: Reinventing Free Markets since the Depression* (Cambridge, MA: Harvard University Press, 2012). My analysis associates the unification of *Wall Street Journal* Republicanism and traditional religion with neoconservatism and reserves the term "neoliberalism" for a secular approach to political economy and security; the term "*Wall Street Journal* Republican" is taken from Francis Fukuyama, *End of History* (New York: Free Press, 2006); Hayek is committed to the rule of law, which connotes a legitimacy distinct from the mere positive treatment of law consistent with the rational choice approach to liberalism. On public choice theory, see S. M. Amadae, *Rationalizing Capitalist Democracy* (Chicago: University of Chicago Press, 2003), 133–155.

[9] John Williamson, "Democracy and the 'Washington Consensus,'" *World Development* (1991) 21:8, 1329–1336.

[10] Harvey, *A Brief History of Neoliberalism*, 2007, 3.

[11] Ibid.

[12] Michael Sandel, *What Money Can't Buy: The Moral Limits of Markets* (New York: Farrar, Straus, Giroux Reprint edition, 2013); for commentary, see Thomas L. Friedman, "This Column Is Not Sponsored by Anyone," *New York Times*, May 13, 2012, SR13.

resonates with the thesis explored in *Prisoners of Reason*: that the exhaustive application of game theory and rational expectations as our orthodox understanding of rational action effectively distills out ethical action, other-regarding considerations, and the ability to voluntarily cooperate in groups.[13]

Neoliberalism has a number of agreed-upon facets. All value is commodified and financialized.[14] Work and gradual wealth accumulation are replaced with speculation, risk management, and casino finance.[15] Elite institutions spread the ethos of neoliberal agency and public policy.[16] Citizens experience an increasing disparity in access to resources, income, and wealth. Consumers accept the inevitability that there are winners and losers, counter to the belief that markets will bring progressively improving living conditions for everyone.[17] Experts denounce the possibility for collective action and meaningful democratic will formation, or even the existence of a public interest.[18] Government and business incentivize compliance with performance metrics and regulations formulated to achieve social order.[19] Individuals experience responsibility in terms of pay-as-you-go access to conditions necessary to sustain life.[20] Entrepreneurs accept predatory practices to promote profit, circumventing mutual exchange.[21] New practices of coercive bargaining are resolved through binding arbitration and debt bondage instead of public courts of justice and normative conduct oriented toward mutual exchange and reciprocal respect.[22]

[13] For a complementary treatment, see Daniel T. Rodgers, "Rediscovery of the Market," in *Age of Fracture* (Cambridge, MA: Belknap Press, 2012), 41–76. See also Debra Satz, *Why Some Things Should Not Be for Sale: The Moral Limits of Markets* (New York: Oxford University Press, 2010).
[14] Discussed by Hilgers, "Historicity of the Neoliberal State," 2012, 84–85; Harvey, *Neoliberalism*, 2005, 3, 33, 165–172. For recent and comprehensive treatment of neoliberalism that makes similar points, see Wendy Brown, *Undoing the Demos: Neoliberalism's Stealth Revolution* (New York: Zone Books, 2015); on financialization of all value, see 63–78.
[15] Discussed by Hilgers, "Three Anthropological Approaches," 2011, 353–354.
[16] Ibid., 359–360.
[17] A central theme in studies of neoliberalism is the increasing disparity of wealth in addition to the politics of wealth accumulation through dispossession. See Harvey, *Neoliberalism*, 2007, 31–35.
[18] Theme of Giroux, *Terror of Neoliberalism*, 2004; see also Harvey, *Neoliberalism*, 2007, 66.
[19] See Hilgers, "Three Anthropological Approaches," 2011, 356; there is an acknowledgment that rules of law are the product of interests, but there is also a sense of the inevitability of positive law and the power underlying the promotion of self-interest, e.g. Harvey, *Neoliberalism*, 2007, 77, 203. See also Brown, *Undoing the Demos*, 2015, 115–150.
[20] This is consistent with a neo-Darwinist approach to social policy that resonates with Richard Dawkins's *Selfish Gene* (Oxford: Oxford University Press, 1976). See also Hilgers, "Three Anthropological Approaches," 2011, 356; Hilgers, "Historicity of the Neoliberal State," 2012, 85–86; and Brown, *Undoing the Demos*, 2015, 131–134.
[21] See, e.g., Harvey, *Neoliberalism*, 2005, 36; see also G. A. Cohen, *Why Not Socialism?* (Princeton, NJ: Princeton University Press, 2009). See also Charles H. Ferguson, *Predator Nation: Corporate Criminals, Political Corruption, and the Hijacking of America* (New York: Crown Business, 2012).
[22] On the acceptance of encumbered debt that is essentially permanently unresolvable, see Hilgers, "Historicity of the Neoliberal State," 2012, 83–84; on the rise of binding arbitration, see the documentary, *Hot Coffee* (2011), directed by Susan Saladoff.

Some researchers have traced the origins of neoliberalism to Friedrich Hayek, the Mont Pèlerin Society, and the Chicago school of economics.[23] Both critics and proponents of the turn to privatization and free market solutions seem unclear about whether there is anything new in contemporary forms of capitalism. Consider, for example, Harvey's definition that "the neoliberal state should favour strong individual property rights, the rule of law, and the institution of freely functioning markets and free trade."[24] There seems to be little to differentiate this contemporary confidence in the power of markets to empower individuals through choice from the classical liberal idea of self-determination through the exercise of private property rights.[25] Harvey's definition leaves vague what makes neoliberalism distinct from Adam Smith's original *laissez-faire*.[26]

The two current extremes in addressing neoliberal political economy are either to define it in such a way that its core principles seem indistinct from eighteenth– or nineteenth-century *laissez-faire*, as Harvey does or to use the term as a multi-purpose critique for varying forms of oppression. Mathieu Hilgers voices such a caution, noting that "the term 'neoliberalism' has no single definition on which all agree ... it is used by alterglobalisation activists, within political debates and also as a scientific term by some academic researchers."[27] Hilgers contrasts the use of "neoliberalism" as a critique of runaway market enthusiasm with the neoliberal institutionalist school of international relations, which has its own distinct methodology and terminology.

Prisoners of Reason argues that a key feature unifying neoliberal theories, institutions, and practices is game theory, first articulated by John von Neumann and Oskar Morgenstern in their now famous *Theory of Games and Economic Behavior* (1944). Gesturing toward this thesis, Daniel T. Rodgers writes in his recent *Age of Fracture*, "Market ideas moved out of economics departments to become the new standard currency of the social sciences. Certain game theory

---

[23] See the edited collection by Mirowski and Plehwe, *The Road from Mont Pelerin*, 2009. See also Burgin, *The Great Persuasion*, 2012. For a counterargument that the center of gravity of the neoliberal movement is the Virginia School rather than the Mont Pelerin Society or the Chicago school, see Nancy MacLean, *Chaining Leviathan: The Decades-Long Plan of the Radical Right to Shackle Democracy* (New York: Viking, forthcoming).

[24] Harvey, *Brief History of Neoliberalism*, 2007, 64.

[25] For an analysis of how markets have come to be associated with freedom, see Eric MacGilvray, *Market Freedom* (Cambridge: Cambridge University Press, 2010).

[26] Adam Smith, *An Inquiry into the Nature and Causes of the Wealth of Nations*, 2 vols., ed. by R. H. Campbell and A. S. Skinner (Indianapolis: Liberty Fund, 1976); for commentary on variations between the rational choice approach to liberalism and Adam Smith's, see Amadae, *Rationalizing Capitalist Democracy* 2003, 193–219; and S. M. Amadae, "Utility, Universality, and Impartiality in Adam Smith's Jurisprudence," *Adam Smith Review* (2008) 4, 238–246.

[27] Hilgers, "The Three Anthropological Approaches," 2010; and his "The Historicity of the Neoliberal State," 2012, 20, 80–94; quote is from Hilgers, "Three Anthropological Approaches," 352. On the term's plural applications with an extensive literature review, see Rajesh Venugopal, "Neoliberalism as a Concept," *Economy and Society* (2015) 44:2, 165–187.

set-pieces – the free-rider problem, the prisoner's dilemma, the tragedy of the commons – became fixtures of common sense."[28] Rodgers points out how game theory offers the intellectual infrastructure for contemporary economic analysis. Rational choice theory assumes that actors maximize expected gain and compete with one another strategically.[29] The integration of game theory and rational expectations into economic science marks a new period in the history of economic thought sequential to both the initial classical period from Adam Smith's *Inquiry into the Nature and Causes of the Wealth of Nations* (1776) to Karl Marx's *Das Capital* (1867–94) and to the subsequent neoclassical period encompassing William Stanley Jevon's *Theory of Political Economy* (1871), Vilfredo Pareto's *Manuel of Political Economy* (1906), and Lionel Robbins's *Nature and Scope of Economic Science* (1932).[30]

The ideas of game theory are historically unprecedented and have justified, and rendered plausible or even inevitable, the iconoclastic features of neoliberalism.[31] The financialization of all value is consistent with game theory but not with classical or neoclassical economics.[32] Collective action, public

[28] Daniel T. Rodgers, *Age of Fracture* (Cambridge, MA: Harvard University Press, 2011), 10. Amadae, *Rationalizing Capitalist Democracy*, 2003, argues that the belief that rational choice theory emerged from within the discipline of economics is a common misunderstanding. Richard Tuck, *Free Riding* (Cambridge, MA: Harvard University Press, 2008) argues that free riding is, indeed, a relatively new "commonsense" intuition not sustained in earlier periods of thought.

[29] Von Neumann and Morgenstern articulate the basic mathematical arguments of game theory and mainly study two-person zero-sum games in which the total value gained is conserved. Their extension of game theory to non-zero-sum games assumes the same concept of value as that developed for zero-sum games: *Theory of Games and Economic Behavior*, 60th anniversary edition (Princeton, NJ: Princeton University Press, 2007); note the subtitle to Roger B. Myerson's *Game Theory: Analysis of Conflict*, 1991.

[30] For analysis of the paradigmatic shift from neoclassical economics to game theory, see Amadae, *Rationalizing Capitalist Democracy*, 2003, 220–250; Philip Mirowski, *Machine Dreams: Economics Becomes a Cyborg Science* (Cambridge: Cambridge University Press, 2002).

[31] The pioneers of game theory were acutely aware that they were contributing a new method for economics and social science. See von Neumann and Morgenstern, *Theory of Games*, 2007, 1–45; Luce and Raiffa, *Games and Decisions*, 1958, 1–11.

[32] To see this, consider how Lionel Robbins argued that economics is concerned with scarce resources that serve as means to ends and that money can only ever be a means and not an end, *Essay on the Nature and Significance of Economic Science*, 2nd ed. (London: Macmillan, 1962). Classical political economists viewed the cost of subsistence as the basis from which to evaluate profit and surplus value. See Istvan Hont and Michael Ignatief, "Needs and Justice in *The Wealth of Nations*," in their edited collection *Wealth and Virtue: The Shaping of Political Economy in the Scottish Enlightenment* (Cambridge: Cambridge University Press, 1983), 1–44. By contrast, according to contemporary economics, "Trade-offs can be struck between competing ends ... choices must be made. Even if environmentalists ruled the world, difficult choices would have to be confronted – and, working backwards from those choices [using revealed preference theory], made according to whatever criteria, it will always be possible to calculate economic values ... trade-offs, measurable in dollar terms, had in fact been struck" ("Economic Focus," 2002, 82). Expected utility theory, with its requirement that all outcomes be ranked on a single scale, and its application to empirical phenomena via the identification of a salient tangible property, often introduces a monetary metric as the default against which all value is ascertained; see, e.g., Myerson, *Game Theory*, 1991, 3–25.

interest, voluntary cooperation, trades unions, social solidarity, and even voting
are all irrational according to rational choice theory.[33] Consent is rendered
superfluous because knowledge of an individual's preferences over all possible
outcomes makes it possible to deduce what that individual would choose to do
"in every situation which may conceivably arise."[34] The invisible hand of
classical political economy and general equilibrium theory becomes an old-
school myth that must be countered by the backside of the invisible hand via
coercive sanctions.[35] Given the way that incentive schemes function in game
theory such that *everything* of value to an agent must be reflected in a common
metric, such as money, food calories, energy, time, and information, non-scarce
resources such as integrity and trust are treated as though they were costly and
finite.[36] This results in a mentality whereby every decision is evaluated on a cost-
benefit analysis basis of how it promotes individual interest in accordance to a
fungible rewards scheme. It thus becomes rational to cheat if one can do so
without getting caught.[37] Instead of one market-clearing, public price that

[33] This is proven by Kenneth J. Arrow's *Social Choice and Individual Values*, 2nd ed. (Yale
University Press, 1970). For discussion, see Amadae, *Rationalizing Capitalist Democracy*,
2003, 83–154; for technical discussion, see Tuck, *Free Riding*, 2008; for a critique internal to
the rational choice paradigm, see Elinor Ostrom, *Governing the Commons* (New York:
Cambridge University Press, 1990).

[34] Von Neumann and Morgenstern, *Theory of Games*, 2004, 31; this permits Richard Posner's
equation of *ex ante* and *ex post* consent, see *Prisoners of Reason*, Chapter 8.

[35] For this analysis and vocabulary, see Russell Hardin "The Back Side of the Invisible Hand,"
Collective Action (Resources for the Future, 1982), 6–15; see also Geoffrey Brennan and Philip
Pettit, "Hands Tangible and Intangible," *Synthese* (1993) 94, 191–225.

[36] "A *prize* [payoff] in our sense could be any commodity bundle or resource allocation. We are
assuming that the prizes in X [total set of possible outcomes or payoffs] have to be defined so that
they are mutually exclusive and exhaust the possible consequences of the decision-maker's decisions. Furthermore, *we assume that each prize in X represents a complete specification of all
aspects that the decision-maker cares about in the situation resulting from his decision*" [emphasis
added]; realizable utility is finite and bounded, see Myerson, *Game Theory*, 1991, 7–8, 13. This
covers the requirements for each individual's single scale ranking over all conceivable outcomes;
for this metric to be common across actors (although not assuming that a certain number of units
on this metric have equivalent experiential utility for various actors), agents must base evaluation
on a common unit of measurement, and "for many situations money serves this purpose," Luce
and Raiffa, *Games and Decisions*, 1958, 145; see also Myerson, *Game Theory*, 1991, 3; for
discussion, see Anatol Rapoport, *Fights, Games, and Debates* (Ann Arbor: University of Michigan
Press, 1960): "The real difficulty is to define and estimate 'utilities,' which game theory simply
takes for granted. Except where money or some other easily measurable commodity can be taken
to be equivalent to utiles or at least related to utiles in an ascertainable manner, the determination
of utilities or even a proof of their existence is a most difficult matter," 164, 180–194; and von
Neumann and Morgenstern, *Theory of Games*, 2004, 20–24.

[37] In rational choice scholarship, the principles of commitment, honesty, and integrity can only be
evaluated on a cost-benefit analysis basis, thereby losing intrinsic significance. See Amartya K.
Sen, "Rational Fools: A Critique of the Behavioral Foundations of Economic Theory,"
Philosophy and Public Affairs (1977) 6:4, 317–344. See also Daniel Hausman and Michael
McPherson, *Economic Analysis, Moral Philosophy and Public Policy*, 2nd ed. (Cambridge:
Cambridge University Press, 2006), 72. For an experimental critique of the rational incentives

forms an equilibrium of supply and demand, pricing becomes private and invisible to general scrutiny, finely honed to each individual's personal willingness and ability to pay, knowable through comprehensive, non-anonymous, asymmetrically leveraged data mining.[38] New means of systematically exploiting surplus value by finding ways to charge individuals scarce cash value for positive-sum, inherently unlimited resources create new opportunities for profit.[39]

Rational decision theory, first solidified as rational deterrence theory in the 1960s, was integrated into evolutionary biology and analyses of political economy in the 1970s and finally became mainstream in economics in the 1980s. John Harsanyi, Reinhart Selten, and Robert Aumann contributed game theory analyses to *Models of Gradual Reduction of Arms* (1967) before going on to win Nobel Prizes in economics in 1994, 1994, and 2005, respectively. Similarly, Thomas Schelling was preeminent for his work on nuclear deterrence and arms control in the 1960s before becoming a Nobel Laureate in economic science in 2005.[40] Strategic rationality can help build descriptive models, but more importantly, it offers a normative and prescriptive understanding of rationality that may inform decision making and structure subjective appraisals over appropriate action choices. It exists as a powerful pedagogy that can revise actors' interpretations of valid and effective courses of action.[41] Given that the ideas examined in *Prisoners of Reason* received recognition at the same time as the onset of neoliberal institutions and practices, we may hypothesize that this congruence is not coincidental or accidental. Certainly, the core ideas that inform neoliberal governance and market discipline are structured in accordance with game theory and are markedly distinct from the body of ideas defining classical liberalism.

RECALLING CLASSICAL LIBERALISM

The shift in orientation from liberalism to neoliberalism is sufficiently stark that it calls for a reexamination of the bedrock theoretical commitments underlying

approach to ethical conduct, see Nina Mazar and Daniel Ariely, "Dishonesty in Everyday Life and Its Policy Implications," *Journal of Public Policy and Marketing* (2006) 25:1, 117–126.

[38] For an analysis of aspects of this phenomenon permitting the privatization of pricing see *Assessing the Impact of Online Personalization on Algorithmic Culture*, a project of inquiry at Northeastern University, Boston, MA: www.northeastern.edu/nulab/personalization-research -northeastern/, accessed January 30, 2015.

[39] Sam Han, "American Cultural Theory," *Routledge Handbook of Social and Cultural Theory* (New York: Routledge, 2013), 239–256.

[40] Many post–World War II economists whose contributions have been acknowledged with a Sveriges Riksbank Prize in Economic Sciences in memory of Alfred Nobel contributed to or used game theory: Kenneth J. Arrow, James M. Buchanan, Gary S. Becker, John C. Harsanyi, John F. Nash Jr., Reinhard Selton, Douglass C. North, Roger B. Myerson, Robert J. Aumann, Daniel Kahneman, and two theorists whose work is critical of the paradigm – Amartya K. Sen and Elinor Ostrom.

[41] Myerson acknowledges this point, *Game Theory*, 1991, at 22.

the family of liberal political philosophies.[42] We may perceive of distinct classical, progressive, and welfarist liberalisms, which are articulated by John Locke, Adam Smith, and Immanuel Kant; T. H. Green, John Stuart Mill, and John Dewey; and John Rawls, respectively.[43] We could also include the mid-century renewed faith in free markets, pointing to Friedrich Hayek and Milton Friedman, as well as Robert Nozick's libertarianism.[44] We might also incorporate the instrumentalist approach to liberalism, encompassing Hugo Grotius, Thomas Hobbes, and David Gauthier.[45] We could see liberalism based on an empirical pragmatist approach, best characterizing Adam Smith's method, or on a deontological ethics consistent with Kant's work.[46] Classical liberalism can also be viewed in terms of achieving accord among nations.[47] Liberalism is obviously a multifaceted approach to understanding individual freedom and the proper relationship between government and private citizens.[48]

---

[42] For contemporary commentary, see Michael J. Sandel, *Liberalism and the Limits of Justice*, 2nd ed. (Cambridge: Cambridge University Press, 1998); Gerald Gaus, "Liberalism," *Stanford Encyclopedia of Philosophy*, online, substantially revised, 2010; Alan Ryan, *The Making of Modern Liberalism* (Princeton, NJ: Princeton University Press, 2014).

[43] John Locke, *Second Treatise of Government*, ed. by C.B. Macpherson (Indianapolis: Hackett, 1980); Adam Smith, *Theory of Moral Sentiments*, ed. by D. D. Raphael and A. L. Macfie (Indianapolis: Liberty Fund, 1982); Immanuel Kant, *Groundwork of the Metaphysic of Morals*, trans. by H. J. Paton (New York: Harper and Row, 1964); Immanuel Kant, *The Metaphysical Elements of Justice*, trans. by John Ladd (Indianapolis: Bobbs-Merrill, 1964); T. H. Green, *Lectures on the Principles of Political Obligation* (Kitchener, ONT: Batoche Books, 1999); John Stuart Mill, *J. S. Mill* On Liberty *and Other Writings*, ed. by Stefan Collini (Cambridge: Cambridge University Press, 1989); John Dewey, *Public and Its Problems* (Chicago: Swallow Press, 1954); John Rawls, *Theory of Justice*, revised ed. (Cambridge, MA: Belknap Press, 1999).

[44] Friedrich Hayek, *Constitution of Liberty*, ed. by Ronald Hamowy (Chicago: University of Chicago Press, 2011); Milton Friedman, *Capitalism and Freedom*, Fortieth Anniversary Edition (Chicago: University of Chicago Press, 2002); Robert Nozick, *Anarchy, State, and Utopia*, 2nd ed. (New York: Basic Books, 2013).

[45] Hugo Grotius, *The Rights of War and* Peace, 3 vols., ed. by Richard Tuck (Liberty Fund, 2005); Thomas Hobbes, *Leviathan*, ed. by Richard Tuck (Cambridge: Cambridge University Press, 1996); David Gauthier, *Morals by Agreement* (Oxford: Oxford University Press, 1987), although this latter work is pivotal in the shift from classical to neoliberalism.

[46] Smith, *Theory of Moral Sentiments*, 1982; Kant, *Groundwork*, 1964.

[47] Richard Tuck argues that classical liberalism was first defined in terms of international relations theory during early modernity: *The Rights of War and Peace: Political Thought and the International Order from Grotius to Kant* (Oxford: Oxford University Press, 2001); for discussion of the classical liberal approach in international relations theory, see also Michael Doyle, *Ways of War and Peace* (New York: W. W. Norton, 1997), 205–314.

[48] Duncan Bell argues that "liberalism" refers to such a variety of political theories that the term lacks specificity and furthermore was used in the mid-twentieth century in the Cold War effort to ground democracy; "What Is Liberalism?" *Political Theory* (2014) 42:6, 682–715. Russell Hardin argues that there was so much disagreement by the end of the twentieth century among the select theorists who actively recognized a common philosophical tradition (including John Rawls, Robert Nozick, and James M. Buchanan) that the sheer fact of such disagreement demonstrates the lack of theoretical cogency to "liberalism"; "Contractarianism: Wistful Thinking," *Constitutional Political Economy* (1990) 1:2, 35–52; Gerald Gaus identifies

In looking for a common denominator underlying the myriad liberalisms coexisting as a family of political theories, freedom from interference or freedom from arbitrary power stands prominent because even the progressive, welfarist, and so-called positive versions of liberalism rest on this elementary foundation. Kant formulates the understanding of classical liberal freedom thus: "Freedom (independence from the constraint of another's will), insofar as it is compatible with the freedom of everyone else in accordance with a universal law, is the one sole and original right that belongs to every human being by virtue of his humanity."[49] Rawls similarly states this essential observation: "Each person is to have an equal right to the most extensive system of equal basic liberty compatible with a similar system for all."[50] For this common ground to transcend a state of natural anarchy in which all individuals please themselves at will regardless of their impact on others, liberal philosophy must stipulate how any single individual's sphere of free action can be compatible with others' similar spheres.[51] This is an important point because neoliberal practices of political economy are divorced from this common liberal understanding and, at best, adopt a concept of freedom of choice defined by each particular individual's preferences only constrained by available opportunities.

Traditional liberalism, consistent with a minimal state whose function is restricted to maintaining commutative justice and security, rests on the Archimedean reference point that individuals' claim to rights is concomitant with their obligation to respect the rights of others.[52] This point of reference is self-evident to reasoned reflection.[53] Actors may test the appropriateness of actions according to criteria of impartiality and universalizability so that courses of action are mutually consistent in principle, no matter who originates them. This willingness to respect others' right to exist, as commensurate with one's own human dignity and like right, is a prerequisite for identifying the content of rights and respecting them.

---

liberalism as a coherent family of political orders, "The Diversity of Comprehensive Liberalisms"; see also Eric Mack and Gerald Gaus, "Classical Liberalism and Libertarianism: The Liberal Tradition," both essays in Gerald F. Gaus and Chandran Kukathas, eds., *Handbook of Political Theory* (London: Sage, 2004), 100–114 and 115–130.

[49] Kant, *Metaphysical Elements of Justice*, 1965, 43–44.

[50] Rawls, *Theory of Justice*, 1999, 220.

[51] For example, Amartya K. Sen defines the "minimal liberal" condition to permit every individual a choice among two outcomes, which hence limits another individual's freedom of choice over that pair of outcomes, and yet Sen discovers a contradiction between this condition and that of Pareto optimality, holding that if all members of a community prefer state *a* to state *b*, then the group as a whole must also prefer state *a* to state *b*. He therefore concludes that there is no way to uphold the classical liberal concept of freedom of choice and also respect the Pareto conditions of market efficiency: "Impossibility of a Paretian Liberal," in his *Choice, Welfare and Measurement* (Cambridge, MA: Harvard University Press, 1982), 285–290.

[52] See, e.g., Nozick's concept of side constraints, *Anarchy, State, and Utopia*, 2013, 28–32.

[53] As an example, see Locke, *Second Treatise of Government*, 1988, section 7.

The key insight of the classical liberal approach is that state intervention can be kept to a minimum because individuals by and large gratuitously respect one another's rights. Citizens recognize this commitment to be entailed by their own assertion of liberty. Richard Tuck elucidates the insights of Hugo Grotius:

All men would agree that everyone has a fundamental *right* to preserve themselves, and that wanton or unnecessary injury to another person is unjustifiable. No social life was possible if the members of a society denied either of these two propositions, but no other principles were necessary for social existence, at least on a rudimentary level.[54]

The classic liberal derivation of voluntary self-constraint thus follows from extending one's own right to self-preservation to all human actors. Self-preservation is a natural right coextensive with the reality that agents have an innate drive to protect themselves. Classical liberal civil society emerges from individuals' recognition that acknowledging and respecting others' right to exist, and the conditions for their liberty with respect to the sanctity of their persons and personal possessions, is the basis for peaceful coexistence. For classical liberal theorists, even though positive law may have differing means of legitimation, from Hobbesian authoritarianism to Lockean representative democracy, there is still a common understanding *prior to codified law* in individuals' tacit acknowledgment that the liberty of person possible in a civil society relies on individuals' voluntarily yielding their right to all things, including one another's persons.[55]

Consequently, in classical liberal theory, all individuals' rights exist as a function of respecting others' like rights as a matter of first principle, fleshed out with respect to the specific content of rights. This point is so essential and incontrovertible that the words of Locke bear recalling:

The *Freedom* then of Man and Liberty of acting according to his own Will, is *grounded on* his having *Reason*, which is able to instruct him in that Law he is to govern himself by, and make him know how far he is left to the freedom of his own will ... And reason, which is that Law, teaches all Mankind, who will but consult it, that being all equal and independent, no one ought to harm another in his Life, Health, Liberty, or Possessions.[56]

Individuals can only meaningfully exercise their own right to life, property, and the pursuit of happiness insofar as they of their own accord cede to others the liberty of person, possession, and pursuit of happiness.

Two theoretical commitments are entailed in this classical liberal approach to achieving social order, each of which neoliberalism rejects. First, individuals must recognize others' right to exist. Second, individuals must integrate this

---

[54] Richard Tuck, *Hobbes: A Very Short Introduction* (Oxford: Oxford University Press, 2002), 26; emphasis in original for all quotations unless otherwise noted.

[55] Gerald Gaus argues that various grounding principles for a liberal political order can obtain, as long as they satisfy some sufficient conditions, *The Order of Public Reason* (Cambridge: Cambridge University Press, 2011), pp. 321–33; see also his "Public Reason Liberalism," in *The Cambridge Companion to Liberalism*, edited by Steven Wall (Cambridge: Cambridge University Press, 2015), 112–40.

[56] John Locke, *Second Treatise on Government*, 1980 (subsection 63, "Of Civil Government").

respect for others into their own decisions for action. The classical liberal political process assumes these two orientations and has the task of reinforcing them in state-sponsored legislation. In contemporary parlance, this limitation on action may be viewed as each individual voluntarily complying with the Pareto principle that every choice must make at least one individual better off and no one worse off.

To differing degrees, Kant, Nozick, Rawls, and Locke reflect a deontological approach to liberal political philosophy that justifies norms of conduct in accordance with the application of reasoned reflection.[57] However, even self-claimed proponents of minimal classical liberalism who look to tradition, custom, or convention reach the same conclusion that the claim to liberty is dependent on acquiescence to guidelines of conduct that respect others' pursuit of freedom. This effective self-governance underwrites the minimal state. As Hayek explains, "It is indeed a truth, which all of the great apostles of freedom . . . have never tired of emphasizing, that freedom has never worked without deeply ingrained moral beliefs and that coercion can be reduced to a minimum only where individuals can be expected as a rule to conform voluntarily to certain principles."[58] For Hayek, the enactment of a system of mutual liberty follows from individuals' recognition of moral obligation and their voluntary compliance with these moral guidelines that inform action yet are prior to positive law.

Isaiah Berlin, who also eschews the need to supply metaphysical or deontological justifications for liberty, concurs with this elementary position. In his support of the "sanctity of person,"[59] he observes,

I must establish a society in which there must be some frontiers of freedom which nobody should be permitted to cross. Different names or natures may be given to the rules that determine these frontiers. They may be called natural rights, the word of God, natural law, the demands of utility, or the "permanent interests of man"; I may believe them to be valid a priori or assert them to be my own ultimate ends or those of my society or culture.[60]

Berlin clarifies further, "What these rules or commandments will have in common is that they are accepted so widely, and are grounded so deeply in the actual nature of men as they have developed through history, as to be, by now, an essential part of what we mean by being a normal human being."[61] He is adamant that "genuine belief in the inviolability of a minimum extent of individual liberty entails some such absolute stand."[62]

Berlin upholds the tradition of classical liberalism that gradually narrowed its claims from (1) Smith's confident "System of Natural Liberty" buttressed by an

---

[57] See Sandel, *Liberalism*, 2nd ed., 1998, 1–13.
[58] Hayek, *Constitution of Liberty*, 2011, 123.
[59] Isaiah Berlin, *Four Essays on Liberty* (New York: Oxford University Press, 2002), 69.
[60] Ibid., 210.
[61] Ibid.
[62] Ibid.

invisible hand that guarantees mutual prosperity, to (2) Mill's soft paternalistic encouragement of individual development within the framework of a no-harm principle, and then to (3) a final minimalist stand. Yet, as Berlin himself acknowledges, this stand is so pertinent to liberalism that it bears restating:

> No society is free unless it is governed by at any rate two interrelated principles: first, that no power, but only rights can be regarded as absolute, so that all men, whatever power governs them, have an absolute right to refuse to behave inhumanly; and, second, that there are frontiers, not artificially drawn, within which men should be inviolable, these frontiers being defined in terms of rules so long and widely accepted that their observance has entered into the very conception of what it is to be a normal human being.[63]

From the vantage of liberal tradition, which defines individual liberty as freedom from interference, this agreement on the boundaries that define the sanctity of persons is sufficiently self-evident so that it informs individual action such that, for the most part, individuals avoid harming one another.[64] Classical liberals are confident that individuals can distinguish between harmful and benign acts. Moreover, they accept that the rationale and motive force underlying the prohibition on committing harmful acts is sufficiently self-evident, as a result of the conditions for mutual recognition, treating every agent as an end in him- or herself, or agentive autonomy.[65]

The role of the state is thereby minimal because it need only apportion police force and judicial oversight to those pathological individuals who do not agree to recognize or uphold the sanctity of persons, or to anomalous cases defying ready judgment in accordance with precedents. Actors' recognition that liberty is dependent on their accepting the responsibility for avoiding harming other persons, or interfering with their basic rights, is essential to any form of liberalism. Otherwise, maintaining social order would rely on police force and private vigilantism. Thus, in classical liberalism and the family of liberal political philosophies that it inspired, actors are inclined to uphold a principle of noninterference. Even if, admittedly, harm may need to be legislatively defined, agents are motivated to refrain from harming others and to recognize other individuals as ends in themselves, not mere means or active opponents whom one must strategically dominate to attain personal satisfaction.[66] Actors enter into normative bargains in which they keep agreements made by their own volition, thereby obviating the coercive bargaining characteristic of noncooperative game

---

[63] Ibid., 211.

[64] Philip Pettit focuses on noninterference to characterize the classic liberal tradition, *Republicanism: A Theory of Freedom and Government* (Oxford: Oxford University Press, 1997), 9.

[65] Brown, *Undoing the Demos*, 2015, makes similar points referencing Adam Smith in contrast to contemporary neoliberal practice, 79–99.

[66] Rapoport describes how "game theory extends these methods [of strategic rationality] to apply to situations where an intelligent and usually malevolent opponent is operating" (163–164) discusses the importance of leveraging threats of harm in bargaining (171–173), and contrasts prudential and contra-prudential (threatening the worst on the opponent) strategy (192–212), *Fights, Games, and Debates*, 1960.

theory and governance via the calibrated threat of sanctions consistent with the rational choice account of action.

NEOLIBERALISM AND NUCLEARIZED SOVEREIGNTY

Classical liberalism, as explained earlier, is premised on individual freedom, typically conceptualized in terms of sanctity of personhood and private property, sustained by the negative virtue commitment to avoid harming others. Self-determination and individual initiative sustain voluntary exchange, efficient production, the gradual accumulation of wealth, and mutual prosperity. Free market forces of supply and demand alleviate scarcity and lower the cost of living and are justified for this reason.[67]

By contrast, in neoliberal political economy, individuals are identified by their preferences and opportunities. Freedom becomes the prerogative to make *any available* choice and thus conveys more of a tautological rather than normative imperative.[68] Agents profit through effective risk management or the creation of "externalities," that is, self-gain at a cost to another party.[69] Intelligence is algorithmic, bargaining is coercive, and norm-following behavior, if it arises, is an equilibrium outcome of individualistic utility maximization. The role of government is to improve social equilibria through monitoring behavior and threatening sanctions. Mass incarceration is tolerated if it costs less to house prison inmates than to guarantee employment opportunities. There is no characteristic distinction between a citizen and a criminal, because all rational actors would break the law if the benefits outweigh the costs.[70]

The attributes of classical liberalism are consistent with its founding association of liberty with the sanctity of personhood from violation by the state or individual actors. The practices characterizing neoliberalism are consistent with strategic rationality according to which common knowledge of other actors' preferences replaces reciprocal acknowledgment of one another's right to exist.[71] This new interpretation of political economy delineated using game theory is intimately connected with the changing practice of market exchange from

---

[67] On Adam Smith's system of natural liberty predicated on negative virtue and promising mutual prosperity, see Amadae, *Rationalizing Capitalist Democracy*, 2003, 193–219.

[68] The only way to prevent choice from automatically being that which the agent prefers is to identify a salient property of the decision environment enabling a criterion for consistency to be applied: Myerson, *Game Theory*, 1997, 25.

[69] Thomas Schelling employs this vocabulary in "Hockey Helmets, Concealed Weapons, and Daylight Saving: A Study of Binary Choices with Externalities," *Journal of Conflict Resolution* (1973) 17:3, 381–428.

[70] Gary Becker, *The Economic Approach to Human Behavior* (Chicago: University of Chicago Press, 1978).

[71] On common knowledge, see David Lewis, *Convention: A Philosophical Investigation* (Oxford: Blackwell, 2002), 52–60.

modernity to postmodernity.[72] The modern human sciences yield to rational choice theory and behavioral economics.[73] Double entry bookkeeping that views values as ratios of exchange gives way to market accounting that pegs value to anticipated future sales and tracks a single criterion, namely dollars.[74]

This shift in the understanding of the agents who inhabit civil society, from individuals who voluntarily respect others' human dignity and right to exist as a condition of their own, to those who pursue self-interest without regard – or even with active disregard – for their impact on others, undermines a classical liberal commonwealth. *Prisoners of Reason* identifies both the novelty of this new approach to human behavior and its coevolution with nuclear deterrence and the exercise of nuclearized sovereignty. The maintenance of nuclear deterrence seemed to require that the United States stand ready to deliver on its promise to destroy the Soviet Union should its leaders initiate nuclear war against Americans. However, strategists viewed this threat to harm as incredible because it relied on engaging in an instrumentally pointless act of atrocious destruction *after* deterrence had already failed. The bipolar Cold War world replete with an overkill supply of thermonuclear warheads guaranteed that mutual assured destruction was an existential reality and not a strategic plan. Seeking to make nuclear weapons effective deterrent threats, strategists looked to rational deterrence theory, which is theoretically equivalent to noncooperative game theory, to inform US nuclear policy. Orthodox game theory stood outside of moral reasoning, which only seemed to weaken US strategic initiatives and the credibility of deterrent threats.[75]

Orthodox game theory is particularly suited to realism in international relations, the school that flourished after World War II.[76] Four assumptions

---

[72] This is a theme that Michel Foucault worked to articulate as he investigated neoliberalism in his final lectures, *The Birth of Biopolitics*, 2010, distinguishing it from modern governmentality described in *Discipline and Punish* (New York: Vintage, 1979). A key distinction toward which Foucault gestures is that modern liberalism permits the internalization of the disciplining gaze so that actors voluntarily abide by norms of efficient capitalist productivity. In neoliberalism, actors are only governed by external sanctions, which are either the official product of a state or the informal product of shaming. See Becker, *The Economic Approach to Human Behavior*, 1978.

[73] Hausman and McPherson, *Economic Analysis, Moral Philosophy and Public Policy*, 2nd ed., 2006; Richard H. Thaler and Cass R. Sunstein, *Nudge: Improving Decisions about Health, Wealth, and Happiness*, revised and expanded (New York: Penguin, 2009). Rational choice public policy introduces institutions that compensate for individuals' inability to achieve collectively rational outcomes on their own volition, and for systematically revealed cognitive deficits preventing individuals from acting rationally.

[74] On the relationship of modern capitalism to double-entry bookkeeping, see Jacob Soll, *The Reckoning: Financial Accountability and the Rise and Fall of Nations* (New York: Basic Books, 2014); for discussion of late-modern accounting practices, see Sebastian Botzem, *The Politics of Accounting Regulation: Organizing Transnational Standard Setting in Financial Reporting* (Cheltenham, UK: Edward Elgar, 2012).

[75] This point is amply discussed by Philip Green in *Deadly Logic: The Theory of Nuclear Deterrence* (Columbus: Ohio State University Press, 1966).

[76] David A. Baldwin, "Neorealism, Neoliberalism, and World Politics" (1993) makes this point, 12, in Baldwin, ed., *Neorealism and Neoliberalism: The Contemporary Debate* (New York:

that ground the standard application of game theory are consistent with realism. First, only outcomes matter, not the means by which they are achieved.[77] Second, the source of value or power is fungible: it is divisible and transferable, like money.[78] Third, rational actors must act independently and individualistically; they cannot act jointly or with solidarity.[79] Fourth, gratuitous altruism, imperfect duty, and other-regarding preferences are irrational. By contrast, in classical liberalism, actors have the perfect and therefore justiciable duty to refrain from harming others. They also have the imperfect duty of personal conscience to consider contributing to those in need, but because every specific choice is the product of private judgment and represents one possible use of scarce resources rather than a precise prohibition, such actions are not subject to legislation.[80] Although it may appear merely coincidental that the progression of the national security debate over nuclear deterrence occurred at the same time as the development of game theory, the two fields were, in fact, intimately connected. Orthodox game theory, articulated in von Neumann and Morgenstern's *Theory of Games and Economic Behavior* as an exhaustive and comprehensive science of decision making, adopts a stance of *realpolitik* that offers such a thoroughgoing instrumentalist approach that other actors are treated as complex objects or malevolent opponents.[81] In the game

Columbia University Press, 1993), 3–28, as does Joseph M. Grieco, "Anarchy and the Limits of Cooperation: A Realist Critique of the Newest Liberal Institutionalism," in Baldwin, ed., *Neorealism and Neoliberalism*, 1993, 116–142, at 116. See also Kenneth N. Waltz, *Man, the State, and War: A Theoretical Analysis* (New York: Columbia University Press, 1959) is a key text in post–World War II realist theory.

[77] In *Nuclear Ethics* (London: Free Press, 1986), Joseph S. Nye Jr. observes that deterrence theory is consistent with a thoroughgoing instrumentalism, that some have argued that having nuclear weapons necessarily defeats the modern Western commitment to just war theory (44), and that merely having and by extension intending to use such weapons of mass destruction in itself is an immoral action (50–51).

[78] Doyle, *Ways of War and Peace*, 1997, points out the consistency between the assumption of fungible sources of power and realism, 47; Baldwin points out how Robert Axelrod's Tit-for-Tat solution of the indefinitely repeated Prisoner's Dilemma game often used to model anarchy in international relations assumes that actors compete over fungible sources of value, "Neoliberalism, Neorealism, and World Politics," 1993, 20; Thomas Schelling makes clear the importance of fungible value in repeating, multiparty Prisoner Dilemma games: "Hockey Helmets, Concealed Weapons, and Daylight Saving," 1973.

[79] The need to go it alone and pursue interests independently is favored in international realism. For discussion, see Art Stein, "Coordination and Collaboration: Regimes in an Anarchic World," in Baldwin, ed., *Neorealism and Neoliberalism*, 1993, 29–59, at 31. Although it is conceivable that solidarity, joint maximization, and team reasoning could be modeled using game theoretic tools, this method and assumption counters the game theoretic orthodoxy that actors maximize independently from one another. See Michael Bacharach, Natalie Gold, and Robert Sugden, *Beyond Individual Choice* (Princeton, NJ: Princeton University Press, 2006).

[80] Amadae, *Rationalizing Capitalist Democracy*, 2003, 205–212; on this theme, see Amartya K. Sen, "Elements of a Theory of Human Rights," *Philosophy and Public Affairs* (2004) 32:4, 315–356.

[81] Rapoport, *Fights, Games, and Debates*, 1960, 163; Heath, *Following the Rules*, 2011, 41.

of life, actors are ceaselessly jockeying for resources and status against one another.[82]

The agency normalized by strategic game theory reinforces the understanding of agency presumed by international relations realists. *Prisoners of Reason* follows the nuclear security debate between proponents of mutual assured destruction (MAD), who accepted the classical liberal position that deterrence should be restricted to bilateral or multilateral self-defense, and the advocates of nuclear utilization targeting selection (NUTS), who argued that the pursuit of strategic dominance was part of self-defense. Game theory intersected with this debate because nuclear strategy was a nonempirical science appropriate for abstract and analytic formal modeling. Although, as a theory of rational action, game theory appeared solely to provide rigorous argumentation devoid of content, the assumptions buried within its original formalization helped secure the analytic victory of NUTS over MAD. Whereas strategic rationality could be deemed one logic of action among many, its construction presupposes that it provides a comprehensive treatment of rational action that necessarily subsumes all other considerations for action under its auspices.[83] Thus, rather than permit logics of appropriateness, solidarity, and imperfect duty to coexist either alongside or within strategic rationality, game theory instead negates these alternative means of attributing intelligibility to action.

Nuclear strategy appears to be a world apart from markets and democratic governance. However, understanding the nuclear security debate is crucial for grasping how game theory came to characterize all purposive agency during and beyond the Cold War era.[84] In its orthodox form, game theory asserts a purely consequentialist evaluation of outcomes, a realist single criterion metric for value, and individualistic combat. Whereas self-determination and reciprocal respect of human dignity were the starting points for classical liberal theory, neoliberal theory offers the freedom of individual choice to do as he or she pleases constrained only by feasible options. Under neoliberal political philosophy, even affording to others rights of personhood and human dignity is regarded as a weakness and moralism at odds with strategic conduct.

In 1986, Joseph S. Nye Jr. argued that it was "a serious exaggeration to say that 'nuclearism' [the exercise of national sovereignty by projecting power through nuclear deterrence] has caused a cultural, as well as a political and

---

[82] "We cannot fail to recognize that people are constantly jockeying to better their lot in a manner which is quite analogous to playing in an extremely complicated many-person game," Luce and Raiffa, *Games and Decisions*, 1958, 105.

[83] Von Neumann and Morgenstern present an exhaustive theory of decisions that views all individual decisions and social interactions as games formalized in their *Theory of Games and Economic Decisions*, [1944]/2004. See especially 31–34: "The immediate concept of a solution is plausibly a set of rules for each participant which tell him how to behave in every situation which may conceivably arise" (at 31).

[84] Texts that make these points quickly accessible are Heath, *Following the Rules*, 2011; Rapoport, *Fights, Games, and Debates*, 1960; and Daniel Ellsberg, "Theory of the Reluctant Duelist," *The American Economic Review* (1956) 46:5, 909–923.

constitutional breakdown."[85] However, assessing the entanglement of rational deterrence theory, or game theory, with civil political theory and practice leads one to the opposite conclusion. The type of reasoning that strategists found useful for buttressing nuclear deterrence was specifically that of orthodox game theory, which openly broke with classical liberal mutual self-regard.[86] Game theory coevolved with US nuclear deterrence, which by 1980 had moved from a stance consistent with classical liberal reciprocal respect of the right to self-preservation to the neoliberal, or offensive realist, posture that credibly sustaining deterrent threats relies on demonstrating the intention and capability to prevail in all levels of conflict by leveraging asymmetric power.[87]

Nye's *Nuclear Ethics* provides an overview of the Cold War strategic debate in which those who supported NUTS "attacked MAD in the 1970s in order to urge the development of new, prompt counter-silo weapons [and] failed to distinguish between the *doctrine* of assured destruction targeting and the *condition* of ultimate vulnerability that remains even when the doctrine is changed."[88] *Nuclear Ethics* makes clear that nuclear deterrence theory was predicated on the irrationality of ever using nuclear weapons but also dedicated to establishing "credible options for the use of nuclear weapons that encourage prudence in the calculation of a rational opponent."[89] With respect to the former, which Nye

---

[85] Joseph S. Nye Jr., *Nuclear Ethics* (New York: Free Press, 1986, 78; Nye is quoting Robert Jay Lifton and Richard A. Falk, *Indefensible Weapons* (New York: Basic Books, 1983), 262. Elaine Scarry's *Thermonuclear Monarchy* (New York: W. W. Norton, 2014) concurs with Lifton and Falk.

[86] This much is evident in the common knowledge assumption, and in the ready applicability of game theoretic modeling to the evolution of prehuman ancestors. On the common knowledge assumption, see Shaun Hargreaves Heap and Yanis Varoufakis, *Game Theory*, 2nd ed. (London: Routledge, 2004), 60–78. On dissolving the distinction between humans and other life forms, see Robert Axelrod, *Evolution of Cooperation* (New York: Basic Books, 1984).

[87] The abnegation of the "no first use" pledge is consistent with the Schlesinger Doctrine's flexible response and slips into treating nuclear weapons as conventional weapons; Nye discusses the strategic rationality for rejecting a no first use pledge, *Nuclear Ethics*, 1986, 49–58, specifically 52; Thomas Schelling speaks at length of the US strategic advantage in maintaining the threat to use nuclear weapons on a first use basis only balanced by the negative evaluation this policy may evoke from other nations:

The most critical question about nuclear weapons for the United States Government is whether the widespread taboo against nuclear weapons and its inhibition on their use is in our favor or against us. If it is in the American interest, as I believe obvious, advertising a continued dependence on nuclear weapons, i.e. a U.S. readiness to use them, a U.S. need for new nuclear capabilities (and new nuclear tests) – let alone ever using them against an enemy – has to be weighed against the corrosive effect on a nearly universal attitude that has been cultivated through universal abstinence of sixty years.

"An Astonishing Sixty Years: The Legacy of Hiroshima," Prize Lecture, December 8, 2005, available at www.nobelprize.org/nobel_prizes/economicsciences/laureates/2005/schelling-lecture.pdf, accessed December 31, 2014.

[88] Nye, *Nuclear Ethics*, 1986, 110–111.

[89] Ibid., 107.

refers to as the "hard core" of nuclear deterrence, actors worldwide are, or should be, united in pursuing the avoidance of nuclear war, which risks ending advanced civilization and even exterminating the human species.

Nye's *Nuclear Ethics* explains how ethical reasoning must overshadow nuclear deterrence for the threat to deploy nuclear weapons to avoid becoming a quest for superiority, and to uphold the classical liberal commitment to "certain minimal obligations of common humanity."[90] Nye argues that even international relations realists, for whom "prudence [is] the supreme virtue of politics" will acknowledge this point, at least to a limited extent, insofar as they generally accept that "we do not kill [members of common humanity] for food or pleasure."[91] The question is how to blend ethical considerations consistent with a realist-cosmopolitan hybrid into the practice of brandishing nuclear weapons.[92] Nye supposes that we must work within a framework that accepts our common humanity but respects that every nation is autonomous, that there is no common definition of "the good," and that preserving peace and order in a nuclear world is a top priority.[93] Using the familiar language of classical liberalism, he explains, "When we recognize each other as part of common humanity despite national differences, we admit negative duties not to kill, enslave, or destroy the autonomy of other peoples as part of our definition of the term 'humanity.'"[94] He further adds that we should take responsibility for how our actions affect foreigners and accord with the positive duty of charity and good samaritanism consistent with not making fellow citizens "significantly worse off."[95]

In discussing the ethics of exercising US sovereignty through wielding the threat to engage in nuclear combat, Nye classifies considerations into intentions, means, and ends. He argues that the *intention* to use nuclear weapons should be limited to self-defense, understanding that meaningfully protecting oneself must include defending the core values for which one stands, and that self-defense cannot be defined so broadly as to encompass disproportionate interests. Prudence and morality can be aligned in the threat to use nuclear weapons only to maintain crucial interests. When it comes to *means*, it is difficult to morally justify any threat to use nuclear weapons, especially because there can be no guarantee that escalation will stop short of mutual annihilation. Herein lies the crux of the nuclear deterrence paradox: "If there is absolutely no possibility of the use of nuclear weapons, or if that is believed to be the case, they will have no deterrent effects."[96] It thus seems impossible to wield the deterrent

---

[90] Ibid., 98, 34.
[91] Ibid., 29, 36.
[92] Nye introduces the "realist-cosmopolitan" hybrid because realism alone, he suggests, does not incorporate core ethical values essential to the type of national self-identity he observes is consistent with the US and the Western tradition of universal and impartial values that encompass *legitimacy* as a sought after value not reducible to simply achieving order, ibid., 31–41.
[93] Ibid., 38.
[94] Ibid., 39.
[95] Ibid., 40.
[96] Ibid., 52.

threat and maintain just war theory because, at a minimum, deterrence relies on making a credible threat to harm innocent lives, and, at a maximum, any actor credibly threatening deterrence must acknowledge the lack of any guarantee for the survival of the human species.[97]

Nye identifies five maxims of nuclear ethics derived from extending the just war tradition to address nuclear deterrence. First, with respect to motives, self-defense is a just but limited motive. Second and third, with respect to means, "never treat nuclear weapons as normal weapons" and "minimize harm to innocent people."[98] And fourth and fifth, with respect to ends, reduce the risk of nuclear war in the short term and the reliance on nuclear weapons in the long term.[99] Given Nye's acknowledgment that deterrence has one foot in the sphere of strategic rationality and one foot in the domain of core values, and his acknowledgment that mutual assured destruction is an existential fact, he suggests that not much is necessary to deter.[100]

However, in following the contours of the debate between the limited deterrence option of mutual assured destruction and the pro–nuclear use war-fighting school, moral constraints on deterrent threats stood in the way of maintaining credible, and hence effective, deterrence. Although the classical liberal framework grounded the modern era with its negative duty to avoid harm and its positive duty to engage in charity when possible, strategic rationality broke with this tradition and buttressed nuclear deterrence with a hard-nosed realism exclusive of respect for side constraints and the recognition of common human dignity. Side constraints on action consistent with the no-harm principle, such as those Nye recommends, have no role in either nuclear deterrence or in the mathematical formalism of game theory. Orthodox game theory therefore breaks with the classical liberal tradition because it has no provision for respecting human dignity or the negative virtue of avoiding injuring people. By accepting that national security depends on wielding deterrent threats to wage nuclear war, game theory offers an abstract formal means to model the security dilemma and evaluate the credibility of threats. Subsequently, after first offering guidance in the form of rational deterrence theory, strategic rationality soon became recognized as the state of the art theory for capturing prudence throughout international relations and soon thereafter the behavioral standard for reasoned judgment for all human relationships.[101]

---

[97] Ibid., 52–57, 45.

[98] Schelling reports how it may well be in the interest of the United States to threaten the use of nuclear war as though nuclear weapons were indistinguishable from conventional weapons, "Astonishing Sixty Years," 2005.

[99] Nye, *Nuclear Ethics*, 1986, 99.

[100] Ibid., 107.

[101] See, e.g., most recently Steven J. Brams's advocacy of using game theory to understand the humanities, *Game Theory and the Humanities: Bridging Two Worlds* (Cambridge, MA: MIT Press, 2012). Brams first wrote *Superpower Games: Applying Game Theory to Superpower Conflict* (Yale University Press, 1985).

2

Prisoner's Dilemma

Puzzles with the structure of the prisoner's dilemma were devised and discussed by Merrill Flood and Melvin Dresher in 1950, as part of the Rand Corporation's investigations into game theory (*which Rand pursued because of possible applications to global nuclear strategy*). The title "prisoner's dilemma" and the version with prison sentences as payoffs are due to Albert Tucker, who wanted to make Flood and Dresher's ideas more accessible to an audience of Stanford psychologists. Although Flood and Dresher didn't themselves rush to publicize their ideas in external journal articles, the puzzle attracted widespread attention in a variety of disciplines. Christian Donninger reports that "more than a thousand articles" about it were published in the sixties and seventies. A bibliography (Axelrod and D'Ambrosio) of writings between 1988 and 1994 that pertain to Robert Axelrod's research on the subject lists 209 entries. A Google Scholar search for "prisoner's dilemma" in 2014 returns 106,000 results.

Steven Kuhn, 2014[1]

The *Prisoner's Dilemma* turned out to be one of game theory's great advertisements. The elucidation of this paradox, and the demonstration of how each player brings about a collectively self-defeating outcome, because she is *rational* in pursuing her own interests, was one of game theory's early achievements which established its reputation among the social scientists.

Shaun Hargreaves Heap and Yanis Varoufakis, 2004[2]

---

[1] Steven Kuhn, "Prisoner's Dilemma," *Stanford Encyclopedia of Philosophy*, available online at http://plato.stanford.edu/entries/prisoner-dilemma/, dated 2007, revised 2014, 1–2/49, accessed June 30, 2015; italics added to show how this nuclear security claim for significance is an ongoing feature of presenting game theory; Christian Donninger, "Is It Always Efficient to Be Nice?" in *Paradoxical Effects of Social Behavior*, ed. by Anatol Rapoport, Andreas Diekmann, and Peter Mitter (Heidelberg: Physica Verlag, 1986), 123–134; Robert Axelrod and Lisa D'Ambrosio, "Bibliography for the Evolution of Cooperation," *Journal of Conflict Resolution* (1995) 39, 190. For another statement of the wide attention the Prisoner's Dilemma game received, see Elinor Ostrom, *Governing the Commons* (New York: Cambridge University Press, 1990), 1–8.

[2] Shaun Hargreaves Heap and Yanis Varoufakis, *Game Theory*, 2nd ed. (London: Routledge, 2004), 37–38.

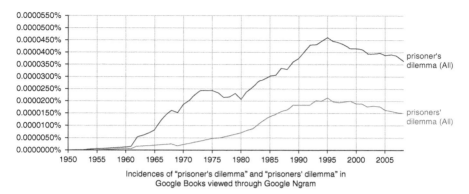

FIGURE 1. Ongoing Engagement with Prisoner's Dilemma, 1950–2010
This figure was made by running the Google Ngram function.

As these opening quotes acknowledge, the Prisoner's Dilemma (PD) represents a core puzzle within the formal mathematics of game theory.[3] Its prominence is evident in the steady rise of incidences of the phrase's use from 1960 to 1995, thereafter remaining stable into the present. This famous two-person "game" has a stock narrative cast in terms of two prisoners who each independently must choose whether to remain silent or confess implicating the other. Each advances self-interest at the expense of the other and thereby achieves a mutually suboptimal outcome, miring any social interaction it is applied to into perplexity. The logic of this game proves the inverse of Adam Smith's invisible hand: individuals acting on self-interest will achieve a mutually suboptimal outcome. However, as this chapter illuminates, the assumptions underlying game theory drive this conclusion.

The Prisoner's Dilemma is not only a core problem at the heart of analytic game theory, but it has also been applied to model and explain numerous phenomena throughout politics and economics.[4] The nuclear security dilemma, subject of Chapters 3, "Assurance," and 4, "Deterrence," was the first concrete problem for which game theorists found the Prisoner's Dilemma apt. In the same vein, by the end of the 1960s, game theorists found the Prisoner's Dilemma model useful for analyzing arms control and bargaining over weapons reduction. In the 1970s,

---

[3] The best collection of essays on the analytic puzzle of the PD is Richmond Campbell and Lanning Sowden, eds., *Paradoxes of Rationality and Cooperation* (Vancouver: University of British Columbia Press, 1985); for a discussion spanning the analytic game theory and empirical applications, see Anatol Rapoport and Albert M. Chammah, *Prisoner's Dilemma* (Ann Arbor: University of Michigan Press, 1970). Figure 2.1 is made using Google's Ngram function with the vertical axis reflecting the percentage of all the two word phrases in the English corpus searchable by Google represented by "Prisoner's Dilemma" and "Prisoners' Dilemma" between 1950 and 2010.

[4] Hargreaves Heap and Varoufakis, *Game Theory*, 175–178; Elinor Ostrom, *Governing the Commons* (New York: Cambridge University Press, 1990), 1–20.

theorists developed a treatment of bargaining in the context of market exchange in terms of the PD game. By extending the model to an exactly repeating scenario, and also by extending it to encompass any number of individuals, theorists modeled the problem of achieving a social contract as a multiple-person, indefinitely repeating, Prisoner's Dilemma.[5] Theorists also found the PD model well suited to model market failure, collective action, free riding, and public goods and to analyze the general rationale for government.[6] Some theorists have analyzed voting as a many-agent PD.[7] Climate change, pollution, individuals' decisions to get vaccinated or to stand up at sporting events are also studied with this model.[8] It is difficult to overemphasize the amount of attention the PD has received, and the numerous social interactions that have been modeled with it.[9] Finally, survival situations such as famine or the competition for nutritional value under conditions of natural selection have been modeled with the PD.[10]

This chapter lays the groundwork for understanding the recent conceptual movement from the classical liberal social contract of mutual prosperity to the neoliberal social contract of conjoint depletion. In brief, this transformation in approach follows the game theoretical dismissal of the classical liberal view that actors will respect others' right to exist and, when assured that others will do likewise, are inclined to keep the agreements they voluntarily made. In its place, game theory holds that rational actors will forge agreements premised on their ability to harm others, and will moreover break their word with impunity, even after others have kept theirs. Game theory does not acknowledge that side constraints on action, the logic of appropriateness, commitment, promising, or fair play provide valid motives for action. It generally replaces norm bound agreement and voluntary compliance with coercive bargaining and leveraged enforcement. The pages ahead show how the specific means of tracking value, in terms of the expected utility of outcomes, necessary in game theory render it imperative that these classical liberal modes of action, encompassing perfect and imperfect duties, as well as solidarity, lose their coherence.

---

[5] On the *n*-person Prisoner's Dilemma and collective action, see Russell Hardin, *Collective Action* (Resources for the Future, 1982).

[6] See, e.g., how Dennis C. Mueller's *Public Choice III* (Cambridge: Cambridge University Press, 2003) begins by the Prisoner's Dilemma model as providing the motive underlying the "origins of the state," 9–14.

[7] For example, Joachim I. Krueger and Melissa Acevedo, "A Game-Theoretic View of Voting," *Journal of Social Issues* (2008), 64:3, 467–485.

[8] Hargreaves Heap and Varoufakis, *Game Theory*, 2004, 175–178.

[9] Searching "Google Books" yields more than 4.5 million hits for "Prisoner's Dilemma."

[10] Partha Dasgupta, *Inquiry into Well-Being and Destitution* (Oxford: Oxford University Press, 1993), on the use of bargaining theory in terms of nutrition, see 324–336. The clearest statement of the relationship between the Nash Bargaining Solution and noncooperative game theory is Ken Binmore's introduction to John Forbes Nash, *Essays on Game Theory* (New York: Edward Elgar, 1997), ix–x.

Thus, in learning game theory, the Prisoner's Dilemma game, and its extensive applications to mundane problems throughout politics and economics, students who master this material will learn to limit their horizons regarding legitimate action as they conform to the tacit assumptions underlying strategic rationality. These assumptions not only rule out the classical liberal family of perfect duties but also contradict unbounded realms of experiential value, the ethos of solidarity and joint maximization, in addition to the classical liberal imperfect duties of charity and beneficence. This chapter renders explicit these latent assumptions that are evident on inspection but are not typically directly discussed in either teaching or applying the Prisoner's Dilemma game or non-cooperative game theory more generally.

The first section provides a discursive introduction to the logical structure of the Prisoner's Dilemma. This discussion is wholly didactic and cannot do justice to the formal apparatus required to specify the game. The second section introduces the means of assessing value, or expected utility, in game theory, and the third section presents the standard means of teaching the PD. The fourth section discusses the relationship between bargaining and the Prisoner's Dilemma, which was originally articulated within the context of nuclear arms control, and lays the groundwork for introducing the PD model of the social contract.

This chapter on the Prisoner's Dilemma directly addresses only the single play game and leaves discussion of the formalized many-person PD to Chapter 9, "Collective Action," and the indefinitely repeating PD game for Chapter 11, "Tit for Tat." I isolate treatment of the single-play PD because it has sufficient theoretical complexity that it warrants focus.[11] Moreover, the iterated PD, as game theorists refer to the repeated scenario, and the multi-agent PD amplify the underlying assumptions of game theory because they strictly relies on, if not interpersonally transferable utility in many contexts, then certainly at a minimum, expected utility theory. Even though many regard the indefinitely repeated PD and Robert Axelrod's Tit for Tat solution as magic bullets to demonstrate that cooperation can emerge under the limited assumptions of strategic rationality and narrow self-interest, this solution depends on perfectly repeating play with little significance for large-scale, multiple-agent political economy.[12] This is because on the one hand, a mutually beneficial solution to

---

[11] This concurs with the judgment of Campbell and Sowden, eds., *Paradoxes of Rationality*, 1985, 16 of 20 papers discuss single-play PDs.

[12] On the limitations of the cooperative solution to the indefinitely repeated PD, see Russell Hardin, "Individual Sanctions, Collective Benefits," in Campbell and Sowden, eds., *Paradoxes of Rationality*, 1985, 339–354; Daniel M. Hausman and Michael S. McPherson, *Economic Analysis, Moral Philosophy, and Public Policy*, 2nd ed. (New York: Cambridge University Press, 2006), 243–245, note in these authors' treatments how rapidly discussion of social justice and the repeated PD moves into discussion of evolutionary biology and the characteristics of successful invaders of groups whose members exhibit behavioral tactics conforming to cooperation (at 244); Ken Binmore also moves swiftly to the repeated PD within an evolutionary context, glancing on the utility assumptions required for this treatment, *Natural Justice* (Oxford: Oxford University Press, 2005), 72–92. For further discussion see Chapter 11, "Tit for Tat."

the repeated PD requires exact repetition with the same two actors over an indefinite yet potentially lengthy time horizon; on the other hand, guaranteed solutions for two-person non-zero-sum noncooperative games require randomized strategies, and are not limited to a single solution.[13]

THE STANDARD NARRATIVE[14]

You and your coconspirator have been captured by the authorities. You are separated and each given the choice between confessing and remaining silent. One of four possible outcomes will occur. If you confess while your partner remains silent, you go free. If you both remain silent, you each receive one year in prison. If you both confess, you each receive a five-year sentence. If you remain silent while your partner confesses, you face a ten-year sentence while your partner goes free. What do you do?

There are different ways to reason through which action to take. Let us consider them each in turn.

Commando: I need to remain silent to protect my partner, my country, and my honor. I have been trained to remain silent, and whatever the price may be, I will remain silent.

Team Member: As a coconspirator, I identify myself as part of a team. Although I may personally do better if I confess, we do better as a team by remaining silent. Thus, it is obvious that I should remain silent, and I choose to remain silent.[15]

Platonic Reasoner: Reason is universal. I know myself and my preferences as well as my coconspirator and his preferences. All must reason alike in like circumstances, so we must choose the same act. We will either converge on both confessing or both remaining silent. Obviously, the latter is superior. Therefore, I remain silent, confident in my partner's identical reasoning capability.[16]

Assurance Seeker: If I knew my partner would remain silent, I would too. But I am afraid that under the pressure of confrontation with the authorities, my partner will not have the wherewithal to remain silent. Although I *would*

---

[13] With respect to the former point, a clear treatment is found in the "14 Indefinite Iteration," entry on "Prisoner's Dilemma," *Stanford Encyclopedia of Philosophy*, Stephen Kuhn, 2014, accessed January 5, 2015; on the second point, see Anatol Rapoport, *Fights, Games, and Debates* (Ann Arbor: University of Michigan Press, 1960), 184–185; the Prisoner's Dilemma game does not necessarily have a symmetric payoff.

[14] See, e.g., R. Duncan Luce and Howard Raiffa, *Games and Decisions* (New York: Wiley, 1958), 94–95; Luce and Raiffa wrote the early definitive text on game theory, and it retains its insightfulness today; see also Hargreaves Heap and Varoufakis, *Game Theory*, 2004, 172–173.

[15] Michael Bacharach investigates this manner of reasoning, which is distinct from the premise of individualistic maximization assumed in orthodox game theory, *Beyond Individual Choice: Teams and Frames in Game Theory*, ed. by Natalie Gold and Robert Sugden (Princeton, NJ: Princeton University Press, 2006).

[16] Reasoning by symmetry is widely dismissed by game theorists, see Lawrence H. Davis, "Is the Symmetry Argument Valid?" in Campbell and Sowden, eds., *Paradoxes of Rationality*, 1985, 255–263; Ken Binmore views Kant's categorical imperative as a variant on symmetrical reasoning, which he refers to as magical thinking, *Natural Justice*, 2005, 63.

definitely stay silent if assured my partner would, my fear of being left alone in prison for ten years is so great that I choose to confess to protect myself from this terrible outcome.

Homo strategicus: The strategy of confessing over remaining silent is better for me, regardless of what my partner chooses. If I confess and my partner does too, then I will just get five years instead of ten. If I confess and my partner refuses to talk, then I will walk away scot-free. Unfortunately, we'll probably both end up with five-year sentences, and not one-year sentences, but this is the logical outcome of being rational.[17]

Most game theorists endorse only this last solution to the Prisoner's Dilemma game: each actor chooses to confess ("defect"), regardless of what the other does.[18] The importance of this result for modern decision theory cannot be exaggerated. *Each actor faces no dilemma of choice because each still chooses to defect, even if fully guaranteed that the other will cooperate.*[19] The larger collective social dilemma arises because individuals' maximization of expected gain results in mutual impoverishment. In the world construed as a Prisoner's Dilemma, every actor most prefers to sucker everyone else.

Game theorists have formalized this narrative and corresponding quandary into the game called the Prisoner's Dilemma. It has become a familiar conceptual artifact of expertise in strategic rationality and represents a familiar pattern of two-person moves and countermoves like tic-tac-toe, although with simultaneous rather than sequential play. The PD is typically represented in a simple form: one individual's choice is represented by rows, and the other individual's choice by columns. Each actor has the choice of remaining silent (cooperating) or confessing (defecting). Given that each person can choose one of two acts, a total of four combinations are possible. Table 1 is a normal presentation of this iconic game.[20] Given the reward structure of this "game," if I am Prisoner 2, I can choose between confessing and not confessing. Regardless of what Prisoner 1 opts to do, I am better off by confessing. If Prisoner 1 chooses to confess, then Prisoner 2 is better off confessing and getting eight years rather than ten; if Prisoner's 1 chooses to cooperate, then Prisoner 2 is better off confessing and getting three months rather than one year. Clearly, the prisoners only care about a mutually evident, salient feature of their decision environment: personal jail

[17] Game theorists concur that this is the dominant or only rational strategy: each is better off defecting whatever choice the other takes; Luce and Raiffa, *Games and Decisions*, 1958, 94–97.

[18] A subaltern position on the Prisoner's Dilemma game reflects the idea that reason should be universal, reflected by the previously mentioned Platonic Reasoner scenario; Davis, "Is the Symmetry Argument Valid," 1985. However, mainstream game theory assumes that each individual must reason independently and maximize gain independently. The only solution is thus the one in which one gains the most regardless of what the other agent decides to do. See Hargreaves Heap and Varoufakis, *Game Theory*, 2004, 184; see also Luce and Raiffa, *Games and Decisions*, 1958, 95–102. For more on the subaltern position on the Prisoner's Dilemma game, see Paul Erickson, *The World the Game Theorists Made* (Chicago: University of Chicago Press, 2015).

[19] Binmore is adamant on this point, *Natural Justice*, 2005, 63–64.

[20] Taken from Luce and Raiffa, *Games and Decisions*, 1958, 95; note that games can also be expressed as decision trees, which is referred to as the extensive form of the game.

TABLE I. *Matrix Representation of Prisoner's Dilemma*

		Prisoner 2	
		Not Confess	Confess
Prisoner 1	Not Confess	1 year each	10 years for Prisoner 1, 3 months for Prisoner 2
	Confess	3 months for Prisoner 1 10 years for Prisoner 2	8 years each

Player 1's choices reflected in rows; Player 2's choices reflected in columns

time, and this serves as each person's criterion of judgment.[21] The prisoners only evaluate personal rewards and do not contemplate how they are brought about, that is, by making the other individual worse off.[22] The prisoners are, most game theorists presume, unable to maximize as a team and thereby mutually achieve one year of jail time each instead of eight years.[23] Finally, neither has any motive to contribute to the greater good or to seek to benefit the other.[24] These assumptions ignore whether empirical actors actually view their behavior as bound by these rules and whether actors' subjective assessment of the significance of their decision environment reflects other features of the choice environment, such as the quality of actors' intentions or means by which the outcome is achieved. Thus, game theoretic analysis cannot encompass non-consequentialist motives derived from commitment, principle, or deontic constraint.[25] Neither does the standard PD treatment permit the

[21] Defining the pure PD game requires stating for certain that both actors would prefer a repeating situation of mutual cooperation rather than a repeating alternation between being the unilateral winner and the sucker; for discussion, see section 6, "Cardinal Payoffs," Kuhn, "Prisoner's Dilemma," 2014. This can be done by adding up each individual's rewards separately, without comparing their intensity across individuals, yet this lends itself to each player having an observable metric of value correlating to each choice (hence a mathematically precise payoff), which reinforces the tendency to permit a concrete measure of success, such as cash value, to stand in for subjective value (again permitting an affine transformation).

[22] This assumption is breathtaking for shifting the significance of the intelligibility of meaning from shared understanding developed by interaction to the view that "each player is to behave independently, without any collaboration or communication, with other players"; see Nicola Giocoli, "Nash Equilibrium," *History of Political Economy* (2004) 36:4, 639–666, at 645.

[23] Hausman and McPherson, *Economic Analysis*, 2006, underscores this point at 250.

[24] Ken Binmore argues that if actors were concerned about others' payoffs, then this information could be directly added into the decision maker's expected utility function; *Natural Justice*, 2006, 63–64; however, in fact, estimating how much additional welfare an individual gets from enhancing another actor's expected utility is not straightforward and deviates from the fundamental assumption that payoffs track salient features of outcomes; see Hausman and McPherson, *Economic Analysis*, 2006, 250.

[25] Hausman and McPherson, *Economic Analysis*, 2006, 210–11; see also Joseph Heath, *Following the Rules* (New York: Oxford University Press, 2011), 35–41 (although Heath seeks to extend formal modeling to incorporate deontic constraints).

consideration of non-fungible value, joint maximization, or gratuitous altruism. If agents reject any one of these assumptions, the inexorable logic of the Prisoner's Dilemma game may be dispelled.

It may seem that if only the two prisoners were able to talk and reach an agreement, they would both remain silent.[26] This, however, is a logical impossibility in the game because all players' preference for less jail time over more jail time is assumed to reflect the only pertinent information they use to weigh their choices.[27] Therefore, even after agreeing to cooperate when back in their cells, both reach the same conclusion as before: confessing is superior, no matter what the other does. Even though the name of the game suggests some sort of angst in decision making, both agents are resolute in their dominant strategy of defecting independent of any consideration of what the other might do. Neither actor faces any moral or prudential quandary of choice.[28] The jointly suboptimal outcome results when the players follow the rules of conduct standardized throughout most operationalized game theory.[29] Confusingly, if the PD were derived from an Assurance Dilemma in which both actors doubt the other's intentions, as in the "Assurance Seeker" vignette (discussed on pages 28–29 above), each actor would cooperate if assured of the other's like intent and only defects due to misgivings about the other's preferences. Yet, standard game theory does not disambiguate this crucial possibility because it fails to emphasize the bright-line test that assurance seekers always cooperate once guaranteed the other's cooperation. Instead, orthodox game theory holds that when confronted by a Prisoner's Dilemma, often demarcated by salient fungible payoffs, it is rational for every actor to defect, regardless of whether or not the other agent defects or cooperates.

A MORE FORMAL PRESENTATION

As related earlier, the standard introduction of the Prisoner's Dilemma presents its characteristic payoff matrix and assumes that every actor solely acts individualistically to maximize his or her instrumentally salient rewards, therefore making defection the only rational choice. One of the difficulties in discussing game theory generally, and the Prisoner's Dilemma specifically, is that the PD can be introduced as though it were as simple as tic-tac-toe. This section introduces the concept of expected utility theory that was first articulated by von Neumann and Morgenstern in a technical appendix to *Theory of Games and Economic Behavior*.[30] This formal treatment of actors' anticipated satisfaction is necessary

[26] See discussion in Hargreaves Heap and Varoufakis, *Game Theory*, 2004, 174.
[27] This is a primary feature of noncooperative game theory; for discussion, see Rapoport and Chammah, *Prisoner's Dilemma*, 1970, 25.
[28] Binmore, *Natural Justice*, 2006, 64.
[29] Heath, *Following the Rules*, 2011, 39.
[30] John von Neumann and Oskar Morgenstern, *Theory of Games and Economic Behavior* (Princeton, NJ: Princeton University Press, 1944), Section 3, "Theory of Games and Economic

to solve most games and limits what can count in their subjective evaluations of worth. Actors' preferences over outcomes are referred to as expected utility functions that must obey restricted formal rules. In addition, actors must follow a decision rule, which typically prescribes a form of individualistic maximization. Maximizing average expected utility, maximizing the greatest possibility of gain, or maximizing the worst-possible outcome (referred to as "maximin") are all possible decision rules in noncooperative game theory.[31]

Game theory is densely mathematical and impeccable as an abstract analytic system. Creating formal models that meet the rarified axioms governing game theory and yet can be applied to social circumstances requires the introduction of simplifying assumptions.[32] These simplifying suppositions, discussed ahead, are introduced to ensure that the social world can be subject to rigorous mathematical analysis. Game theorists strive to identify a solution concept or a determinate outcome of a game that is referred to as an equilibrium. Von Neumann developed the "minimax" equilibrium concept, which is unique in every zero-sum game, in which each player maximizes his best worst-possible outcome and simultaneously minimizes his opponent's best-possible outcome, irrespective of the opponent's choice. John Forbes Nash Jr.'s alternative equilibrium concept of mutual-best-reply, which also applies more generally to non-zero-sum games, identifies a set of players' strategies that are mutually reinforcing because no single actor could improve his outcome by having selected a different strategy.[33] Many non-zero-sum games have no single determinate solution, regardless of how the equilibrium concept is defined. However, in the Prisoner's Dilemma game, both von Neumann's and Nash's equilibria are definitive and identical: both actors select to defect.

When theorists apply the Prisoner's Dilemma game to diverse situations throughout civil society, political economy, and international relations, they must simplify the world of social interaction to fit within game theory. This necessarily compromises the existential richness of individuals' experience.[34] Mathematical tractability, or the demands of applying the theory, entails making specific assumptions about payoffs, or value.[35] Von Neumann established

Behavior"; for discussion, see Nicola Giocoli, "Do Prudent Agents Play Lotteries: Von Neumann's Contributions to the Theory of Rational Behavior," *Journal for the History of Economic Thought* (March 2006) 28:1, 95–102, 101–102.

[31] Nicola Giocoli presents the clearest distinction between rules governing the rationality (or consistency) of preferences in expected utility theory and rules governing the rationality of action choice. See "Do Prudent Agents Play Lotteries."

[32] Luce and Raiffa, *Games and Decisions*, 1958, acknowledge this point directly, 26.

[33] On the Nash Equilibrium, see Giocoli, "Nash Equilibrium," 2004.

[34] Luce and Raiffa openly acknowledge this point: "Although not 'all life is a game,' at least not in our sense, we cannot fail to recognize that people are constantly jockeying to better their lot in a manner which is quite analogous to playing in an extremely complicated many-person game," *Game and Decisions*, 1958, 105.

[35] Rapoport acknowledges this point, and the prescription nature of game theory, in *Fights, Games, and Debates*, 1970, 164, 182.

the precedent, still the default, of directly associating the tangible payoffs that are convenient for observation and measurement with agents' subjective utility rankings.[36] This radical move promotes the belief that the Prisoner's Dilemma is not just a logical construction but also a phenomenon that inheres in the world anytime fungible rewards can be construed as reflecting its payoff matrix. Even though rational choice theory states that everything an individual values can be reflected in individuals' preference rankings (expected utility functions), the means of tracking value in applied game theory categorically restricts to varying degrees the considerations strategic rational actors can incorporate into judgment.[37] Hence, this seemingly encompassing treatment of value actually operates as an imperative to limit what features of the decision environment can count in rational actors' decisions. Therefore, those operationalizing strategic rationality in concrete circumstances may not even see how this practice legitimizes conduct that only maximizes fungible rewards on an individualistic basis and negates normative, shared, or other-regarding conduct.[38] Hence, game theory favors consequentialism and excludes the logic of appropriateness, usually assumes an interpersonally transferable source of value, emphasizes individualistic maximization, and dismisses charitable actions without some tangible benefit to the benefactor.

Game theory relies on specific guidelines for tracking value. It is possible to stipulate a rudimentary game simply by using numbers to indicate actors' preferences over outcomes.[39] Table 2 depicts the iconic Cold War arms race, with higher numbers reflecting more preferable states.[40] This game theoretic payoff matrix has the characteristic Prisoner's Dilemma form. Note that the preferences may seem to pertain only to features of the world impacting that agent. This is not the case because even though individuals' payoffs are the primary basis for individual choice, outcomes are causally interdependent. The US is not indifferent between two states of being armed. The US most prefers

---

[36] Giocoli directly addresses this important point in "Do Prudent Agents Play Lotteries," 2006, 102–103.

[37] This point is crucial, although subtle: in parametric (single-agent) decision theory, incorporating subjective sentiment about processes by which ends are achieved into expected utility functions may be possible; however, this is impossible in game theory because (1) actors only appraise outcomes independently of how they arise and (2) this appraisal focuses on salient instrumentally relevant features of the decision environment, i.e., outcomes could equally well arise by the roll of a die or by deliberate choice. Myerson makes the imperative claim that expected utility functions incorporate all considerations of value in expected utility functions with the implication of ruling out of consideration experiential elements not subject to this type of appraisal; Roger B. Myerson, *Game Theory: Analysis of Conflict* (Cambridge, MA: Harvard University Press, 1991), 7–8; on parametric decision theory see Donald C. Hubin, "The Groundless Normativity of Instrumental Rationality," *Journal of Philosophy* (2001) 98:9, 445–468.

[38] For a lengthy discussion see Martin Hollis, *Trust within Reason* (Cambridge: Cambridge University Press, 1998); one current trend is to identify dispositions that are not motives associated with utility maximization; for discussion, see Hausman and McPherson, *Economic Analysis*, 2006, 210.

[39] For discussion of ordinal preference rankings, without numeric intensities of desire, see Kuhn "Symmetric 2x2 PDs with Ordinal Payoffs," entry on "Prisoner's Dilemma," 2014.

[40] Hargreaves Heap and Varoufakis, *Game Theory*, 2004, 37.

TABLE 2. *Iconic Cold War Arms Race Prisoner's Dilemma Game*

		USSR	
		Disarm	Arm
US	Disarm	2, 2	0, 3
	Arm	3, 0	1, 1

US is the row player; USSR is the column player.
Payoffs reflect the desirability of the outcome for each player; higher numbers are more desirable.
The first number in each pair reflects the US (row) payoffs; the second number in each pair reflect the
   USSR (column) payoffs.

TABLE 3. *Cold War Arms Race Modeled as Stag Hunt*

		USSR	
		Disarm	Arm
US	Disarm	3, 3	0, 2
	Arm	2, 0	1, 1

Payoff matrix conventions are the same as in Table 3.

itself to be armed and the USSR to be disarmed. In the Prisoner's Dilemma game, each actor can only realize his most preferred state by debilitating the other. If, in the Cold War, every actor had preferred mutual disarmament over unilateral armament, then the game would have been an Assurance Game, or Stag Hunt, instead (Table 3). This payoff matrix with numeric utilities reflects that both actors most prefer mutual disarmament, both have the second choice of unilaterally arming, both have the third choice of mutually arming, and each least prefers being the only nation to disarm.

To be sufficiently useful to solve most games, the numbers specifying the payoff matrices must permit the evaluation of what ratio mix of most and least preferred outcome is equivalent to a midrange outcome.[41] For example, in the perfectly defined PD game, players must know that always cooperating yields a superior outcome to alternating between unilateral victory and unilateral ignominy *as though they were engaging in* indefinitely repeated play.[42] Thus, the formal definition of PD relies on expected utility and its treatment of value.

---

[41] Rapoport, *Fights, Games, and Debates,* 1970, 180–194; Luce and Raiffa, *Games and Decisions,* 1958, 106–109.

[42] Kuhn, "Prisoner's Dilemma," has a clear discussion; a pure PD, defined in terms that agents prefer always cooperating more than alternating between unilateral defection and unilateral cooperation, is specified by the formula that CC ≥ ½ (DC + CD) [formula refers to the payoffs accruing to either player for mutual cooperation (CC), row's sole defection (DC), and column's sole defection (CD)]; on the difficulty of interpreting this requirement in single play games, see Rapoport, *Fights, Games, and Debates,* 1970, 162.

Furthermore, as will be increasingly evident ahead, the payoff matrix numbers in much applied game theory directly correlate to a measurable and observable salient feature of the decision environment and take into consideration the fully specified causal state of the outcome that simultaneously specifies the other actor's outcome.[43] Every player's outcome is physically inseparable from what the other achieves. In other words, game theory payoff matrices reflect causally interdependent states.[44] This discussion makes more sense when one understands, first, how game theory originated as an analysis of zero-sum competitions and, second, how most games of relevance to international relations, political economy, governance, and evolutionary biology rely on mathematically formalized expected utility theory that incorporates the assumption of interpersonal transferability of utility.[45] In a zero-sum game, two contestants wrestle over a fixed amount of a good (or property), so that what one individual obtains inversely correlates to what the other gets. As one player maximizes her expected gain, this player simultaneously minimizes her opponent's expected gain with mathematical precision. Operationalizing strategic rationality makes it difficult for an individual to prefer most an outcome in which both players share a fixed-sum good equally, because then these preferences over outcomes do not elegantly map onto the observable and measurable resource that characterizes the game's outcomes.[46]

Von Neumann and Morgenstern's *Theory of Games and Economic Decisions* focuses on two-person, zero-sum games in which the players wrangle over a finite and fixed amount of a utility-affording property. Von Neumann added the appendix on expected utility theory at the prompting of Morgenstern, who wanted to make their theory friendlier for economists.[47] Economists in the 1950s and 1960s were more interested in the treatment of expected utilities, which provided continuity with Daniel Bernoulli's invention of the concept to solve the St. Petersburg gambling paradox in the eighteenth century.[48] The concept of "expected utility," as opposed to straightforward "utility," provided

---

[43] Myerson, *Game Theory*, 1991, 22–26.

[44] Von Neumann and Morgenstern were acutely aware that their theory provided a mathematical formalism for complex interactions. See the introduction to *Theory of Games and Economic Decisions*, 1944, 8–15.

[45] Myerson, *Game Theory*, 1991, 3. Von Neumann and Morgenstern introduced their expected utility theory as an appendix to *Theory of Games and Decisions*, and it relies on an objective understanding of probability. L. J. Savage subsequently introduced an alternative theory that relies on a Bayesian, or learned, account of probability, *The Foundations of Statistics* (New York: Wiley, 1954). On Savage's subjective utility theory, see Nicola Giocoli, "Savage vs. Wald: Was Bayesian Decision Theory the Only Available Alternative for Postwar Economics?" June 2004, available at SSRN: http://ssrn.com/abstract=910916 or http://dx.doi.org/10.2139/ssrn.910916.

[46] Some researchers have worked to incorporate attitudes toward fairness, but this development is a distinct subfield of inquiry that is not integrated in most game theoretic presentations of or experiments with the Prisoner's Dilemma game; see, e.g., Hargreaves Heap and Varoufakis, *Game Theory*, 2004, 162–163; Binmore, *Natural Justice*, 2005, 66–67.

[47] Giocoli, "Do Prudent Agents Play Lotteries," 2006, 102.

[48] For a brief discussion, see Luce and Raiffa, *Games and Decisions*, 1957, 19–21.

the latitude to acknowledge that individuals have differing attitudes toward probabilistic outcomes.[49] One individual may readily purchase a $10 lottery ticket for a 1/11 chance to win $100; yet another individual may prefer to keep $10 for sure to a 1/9 chance to win $100. Expected utility theory allows the incorporation of individuals' attitudes toward risk into their assessment of outcomes.[50]

Expected utility theory also permits simplifying the act of choice in situations with uncertainty (unknown odds) and risk (known odds), which is crucial because of the ubiquity of probability throughout life and in games.[51] As an example, consider a choice between walking or driving and two possible states of the world, rain or dry weather.[52] A ranking of outcomes could be strictly ordinal, without incorporating intensity of preferences. Thus, the agent could have the following preference ordering from most to least preferred: walking while dry, driving while raining, driving while dry, and walking while raining.

In most games of interest to political economists, a mere ordinal ranking of preferences is insufficient.[53] Instead, actors must know the intensity of their preferences. This example stipulates that the actor has a ranking of 10 utils for walking while dry, 6 utils for driving while wet, 1 util for driving while dry, and 0 utils for walking while wet. Although these utils are not comparable across individuals, they do express information about the intensity with which the agent in question prefers the four possible outcomes. The concept of expected utility, over and above strict utility, arises in considering both the preference for the state of the world and the probability of that state occurring. Let us assume a 50 percent chance of rain and a 50 percent chance of dry weather.[54] The actor's expected utility for walking is equal to {50 percent likelihood dry

[49] See Heath's discussion for a complementary treatment, *Following the Rules*, 2011, 15–23.
[50] Myerson, *Game Theory*, 1991, 12–20.
[51] Hargreaves Heap and Varoufakis, *Game Theory*, 2004, 13. Note that expected utility theory neither uses the ordinal concept of utility from neoclassical economics nor the one assuming interpersonal comparisons of utility from Jeremy Bentham's utilitarianism. See also Nicola Giocoli, "The True Hypothesis of Daniel Bernoulli: What Did the Marginalists Really Know," *History of Economic Ideas*, 1998 (2), 7–43.
[52] This example comes from Hargreaves Heap and Varoufakis, *Game Theory*, 2004, 11–12.
[53] Some simple games can be considered with ordinal rankings on the basis of John Nash's equilibrium concept of "mutual-best-reply." However, to technically define the Prisoner's Dilemma, to guarantee a solution, and to apply it to multiplayer and repeating games, one needs cardinal utilities, see Luce and Raiffa, *Games and Decisions*, 1957, 106–109; Kuhn "Multiple Players and Tragedy of the Commons," in "Prisoner's Dilemma," 2014.
[54] Objective probabilities, used by von Neumann and Morgenstern, *Theory of Games and Decisions* (1944) need not necessarily be assumed in game theory, but the mathematics is easier on this assumption. See, e.g., Luce and Raiffa, *Games and Decisions*, 1957, 24. An alternative is subjective expected utility theory in which actors' knowledge of the frequencies with which outcomes occur is derived from individual experience and must be updated in a feedback, learning experience. For discussion, see Giocoli, "Do Prudent Agents Play Lotteries," 2006, 105–107; L. J. Savage, *The Foundations of Statistics* (New York: John Wiley, 1954). See also Nicola Giocoli, "From Wald to Savage: *Homo Economicus* Becomes Bayesian Statistician," *Journal of the History of Behavioral Sciences* (2012) 49, 1–33.

weather multiplied by 10 utils + 50 percent likelihood of wet weather multiplied by 0 utils}, or a total of 5 utils. The actor's expected utility for driving is equal to {50 percent likelihood of dry weather multiplied by 6 utils + 50 percent likelihood of wet weather multiplied by 1 util}, or a total of 3.5 utils. Based on this expected utility calculation, the actor has a greater expected utility by walking. These utils do not reflect any inherent metric but do stipulate a range of satisfaction with intensities.

Additionally, expected utility theory can accommodate an actor's attitude toward risk. However, only so long as this additional concern obeys an orderly transformation from the original evaluation of utility over certain outcomes can this information be incorporated.[55] The axioms of expected utility theory depend on actors having transitive preferences over certain outcomes, and transitive preferences over lotteries of outcomes.[56] Caveats, however, apply. Significantly, expected utility theory can be applied more effectively when actors are making recurrent decisions over the same outcomes with known probabilities because, in the long run, consistent decision making will yield a positive result.[57] The axioms of expected utility theory demand consistency of choice among lotteries of outcomes so that the property of transitivity applies not only to preferences over outcomes but also to preferences over lotteries comprised of various chances of outcomes. Another caveat is that average people make choices that deviate from these axioms.[58]

Both von Neumann and Nash assume individualistic maximization in their approach to games. They did recognize that actors could cooperate in coalitions. However, they believed that when it came time for subgroups of players to divide spoils of a collaborative venture, these actors would resort to individualistic maximization. Many game theorists champion individualistic maximization because it has the additional virtue of comporting with the dictates of methodological individualism, according to which interdependent actions are analyzed by reference to individuals' independent choices and actions. Thus, they eschew cooperative game theory and celebrate John Nash's noncooperative approach.[59] This approach to understanding social interactions identifies

[55] This formal restriction is that expected utility functions vary from certain outcomes, in view of attitudes toward risk, by adding a constant and multiplying the original utility by a constant. This is the definition of an affine transformation: $y = f(x) = Ax + B$, von Neumann and Morgenstern, *Theory of Games and Economic Behavior*, 1953, 24–25; on von Neumann and Morgenstern's original goal in introducing affine transformations, see Giocoli, "Do Prudent Agents Play Lotteries," 2006, 104–105.

[56] Luce and Raiffa specify the axioms, *Games and Decisions*, 1957, 23–31.

[57] Luce and Raiffa, *Games and Decisions*, 1957, make this point, 21; many introductions to these concepts also point out that actors' intuitions about consistent choice do not necessarily coincide with the mathematical consistency required by expected utility theory; see Hargreaves Heap and Varoufakis, *Game Theory*, 2004, 8–18; Luce and Raiffa, *Games and Decisions*, 1957, 19–37.

[58] Allais paradox, discussed by Luce and Raiffa, *Games and Decisions*, 1957, 25; on the Ellsberg paradox, see Hargreaves Heap and Varoufakis, *Game Theory*, 2004, 25–26.

[59] See Ken Binmore's introduction to Nash's *Essays on Game Theory* (Brookfield, VT: E. Elgar, 1996), ix–xx. See also Michael Bacharach's exploration and defense of team reasoning in

individual behavior as the source of collective outcomes. Although team reasoning does not violate methodological individualism, some game theorists worry that this alternative approach to rationality proposes that individual actors may comprise a corporate agent without clearly specifying who gets what after the team obtains its objective: how are the spoils divided? While classical liberals permit division via normative agreement, game theorists typically propose that even after initial cooperation is complete, noncooperative competition must characterize actors' subsequent pursuit of individual gain.

Summarizing, for most rational choice games, actors' expected utility functions are three times removed from reflecting every consideration that could be of value to them. First, the expected utility functions only reflect *outcomes* and not the processes by which they are obtained. Second, the expected utility functions directly correlate to the *observable and measurable reward* characterizing the payoffs. Third, this reward is often held to be *transferable* across agents; the default is precisely countable cash value.[60] In addition to maximizing some inherently scarce and objective feature of the world, strategic rationality typically recommends individualistic maximization.[61] Thus, standard game theory adopts an approach consistent with consequentialism, realism, and narrow individualism.

Learning game theory promotes a mindset that translates these fundamental tenets into guidelines for making rational choices, either in parametric environments involving risk and uncertainty or in strategic environments with other rational agents. In his advanced introduction of expected utility theory and game theory, Roger B. Myerson observes,

A *prize* in our sense could be any commodity bundle or resource allocation. We are assuming that prizes in X [a set of possible prizes that the decision could potentially achieve] have been defined so that they are mutually exclusive and exhaust the possible consequences of the decision-makers decisions. Furthermore, *we assume that each prize*

Beyond Individual Choice, 2006, which reveals how this hypothesis of shared intention and group action is a subaltern position in game theory.
[60] For the concept of interpersonally transferable utility, see von Neumann and Morgenstern, *Theory of Games and Economic Decisions*, 1953, Appendix II, 603–632. This concept should not be confused with the prohibited concept of interpersonally comparable utility discussed in Amadae, *Rationalizing Capitalist Democracy*, 2003, 100–105, 115–116. The claim is that the substance or property yielding expected utility is transferable, not that the respective agents' experiences or satisfactions thereof can be compared. For example, all agents seek money; however, they are not assumed to each experience the reward of allotments of money in the same way. For the ready acceptance of transferable utility, usually introduced by relying on cash rewards to represent payoffs, see Ken Binmore, *Game Theory: A Very Brief Introduction* (New York: Oxford University Press), 2007.
[61] Von Neumann's original two-person zero-sum game theory was individualistic, although the decision rule he supplied was the minimax rule of securing the best-possible worst outcome by minimizing the opposition's potential gain, which leads to a stable equilibrium if both actors select this strategy. For discussion, see Nicola Giocoli, "Do Prudent Agents Play Lotteries," 2006. Giocoli argues that in game theory, actors are "hyper-individualistic" and "hyper-rational" because they act independently to achieve their goals without any reliance on others' choices, other than as a means to secure their own ends, "Nash Equilibrium," 2004.

in X represents a complete specification of all aspects the decision-maker cares about in the situation resulting from his decisions. Thus, the decision-maker should be able to assess a preference ordering over the set of lotteries, given any information that he might have about the state of the world.[62]

This statement that all considerations impinging on choice are contained in actors' expected utility functions over prizes requires that preference rankings incorporate all considerations relevant to their choices. Therefore the mathematical characteristics of these functions purchase comprehensive hold over individuals' judgment at the price of excluding important features of the world from possible evaluation. Not only does the model itself become indistinguishable from the reality it models, but this superposition of the model over the lived world gives rise to the uniquely neoliberal subject who internalizes the limiting guidelines of what can count in a rational judgment.

The conceptual mapping required to operationalize the Prisoner's Dilemma game eclipses the classical liberal worldview because it categorically ignores the means by which outcomes are realized. The expected utility functions used throughout game theory assume "that agents only invest outcomes with motivational significance."[63] Canonical rational actors are thus unable to act on principle, with commitment to agreements or promises made, or on the basis of fair play or side constraints.[64] Although these alternative rationales for action entail different causal outcomes than those sustained by strategic rationality, they become void of motivational content because they do not directly contribute to the measurable gain of decision makers. This encompassing attention on outcomes to the exclusion of processes undermines classical liberalism's dependence on procedural justice and individual's self-incurred responsibility to avoid harming others.[65]

An additional consequence of the exclusive association of utility with outcomes is that communication becomes a signaling game in which "the meaningfulness of the speech act [is] dependent upon the payoff structure of the game."[66] Actors' interests and values exist prior to social interaction.[67] Actors can only use language effectively – that is, avoid deception – when their interests are favorably and extensively aligned.[68] Thus, the game theoretic understanding

---

[62] Myerson, *Game Theory*, 1991, 7–8.

[63] Joseph Heath effectively discusses this topic in *Communicative Action and Rational Choice* (Cambridge, MA: MIT Press, 2001), 137; see 137–139; Heath seeks to extend orthodox game theory so that deontic constraints could be reflected in models, Heath, *Following the Rules*, 2011, 6.

[64] Heath, *Communicative Action*, 2001, 86–92.

[65] Amartya Sen, *Rationality and Freedom* (Cambridge, MA: Belknap Press, 2004), 175–181, 232–239.

[66] Heath, *Communicative Action*, 2001, 70.

[67] Hargreaves Heap and Varoufakis, *Game Theory*, 2004, 209.

[68] For discussion of David Lewis's analysis of communication based on the Prisoner's Dilemma model, see S. M. Amadae, "Normativity and Instrumentalism in David Lewis' *Convention*," *History of European Ideas* (2011) 37, 325–335; see also David Lewis, *Convention: A Philosophical Study* (Cambridge, MA: Harvard University Press, 1969), 173–195.

of linguistic exchange views communication as action derived from payoff structures that permit persistent equilibria to emerge.[69] Game theory entails an instrumentalist view of language that insists both that the meaning and value of acts precede intersubjectively shared intelligibility and that communication itself is a strategic game.[70]

The formalized concepts of expected utility and individualistic maximization structure the possible horizons of meaning available to the neoliberal citizen and consumer. The canonical strategic actor must obey these guidelines of rational choice or become the experimental subject for behavioral economists who aim to systematically catalogue people's observed deviations from pure rationality. The two options available are to abide by the Platonic ideal of rational choice or to succumb to irrational behaviors that behavioral scientists can correct through choice architecture.[71] This latter approach denies the central tenet of classical liberalism, which holds that actors voluntarily participate in institutions and rule-governed practices that they tacitly or expressly agree have procedural validity.[72] Thus, the bourgeois classical liberal world of Adam Smith is displaced by a new interpretation of the meaning of action and individuals' relationships to other actors.

In this next discussion, I examine how a leading classic textbook imparts Prisoner's Dilemma pedagogy and show how learning the inherent impossibility of resolving the dilemma relies on explicitly accepting the characteristic assumptions underlying strategic rationality. These assumptions require restricting value to the horizons of game theoretic expected utility theory and accepting individualistic maximization in competition with others. This limiting perspective makes many social interactions appear to have the Prisoner's Dilemma structure. Once actors either internalize the guidelines for choice consistent with functional strategic rationality or are exposed to institutions designed in accordance with this logic, they experience numerous types of interactions as Prisoner's Dilemmas.

---

[69] Heath, *Communicative Action* 2001, 59–72.

[70] The opposite view is held by Donald Davidson, "The Folly of Trying to Define Truth," *Journal of Philosophy* (1996) 93, 263–278, and Jürgen Habermas, *The Theory of Communicative Action*, vols. 1 and 2, trans. by Thomas McCarthy (Cambridge, MA: MIT Press, 1984 and 1987, respectively); for discussion, see Heath, *Communicative Action*, 2001, 16–25.

[71] Richard Thaler and Cass Sunstein, *Nudge: Improving Decisions about Health, Wealth, and Happiness* (London: Penguin, 2009). Economist and Nobel Laureate Herbert Simon has put forward a middle ground, *Models of Bounded Rationality: Economic Analysis and Public Policy* (Cambridge, MA: MIT Press, 1984); for discussion, see Hunter Crowther-Heyck, *Herbert A. Simon: The Bounds of Reason in Modern America* (Johns Hopkins University Press, 2005).

[72] For discussion, see, e.g., John Rawls, "Justice as Fairness: Political Not Metaphysical," *Philosophy and Public Affairs* (1985) 14:3, 223–251.

TABLE 4. *Prisoner's Dilemma Game "G"*

		B_1	B_2
G:	A_1	(0.9, 0.9)	(0, 1)
	A_2	(1, 0)	(0.1, 0.1)

In Game "G," A is the row player with the choice of action A_1 (cooperate) and A_2 (defect).
B is column player with the choice of action B_1 (cooperate) and B_2 (defect).
Payoffs are the pair of numbers, highest is best, first number accrues to agent A; second number
accrues to agent B.

PRISONER'S DILEMMA PEDAGOGY AND NEOLIBERAL SUBJECTIVITY

Howard Raiffa and Duncan Luce articulate Prisoner's Dilemma pedagogy in their immediately influential and authoritative *Games and Decisions* (1957). The matrix reward structure of the game they discuss is the one presented at the beginning of this chapter. The payoff matrix, in terms of jail time, considers only outcomes, and not any of the circumstances by which they may come about or how the prisoners subjectively evaluate the different outcomes. From the start, only the tangible rewards register in individuals' preferences over outcomes and their judgments over the right course of action.

In their presentation, Luce and Raiffa next introduce the more familiar game payoff matrix that directly uses numbers to reflect the rewards structuring a game as specified by expected utility theory. In this case, numbers without any units reflect each individual's subjective evaluation of the game's outcomes. Again, the fact that these expected utility functions are over end states and even lotteries of end states must be kept in mind.[73]

In Table 4's Prisoner's Dilemma game "G," agent A and agent B can each choose between strategy 1 and strategy 2. The outcomes are jointly determined and deliver the quantity of numeric utility in the payoff matrix to each player in the form: (Expected Utility$_A$, Expected Utility$_B$). This payoff matrix has the characteristic Prisoner's Dilemma structure: every actor hopes to unilaterally opt for the first choice ("defect"), thereby leaving the other agent who cooperates with the least preferred outcome. Every agent prefers mutual cooperation to mutual defection. Every actor least prefers to be the sole cooperator, or "sucker."

At this point in learning to play the Prisoner's Dilemma game, readers have been supplied with a narrative about two conspirators who are given choices of action by a district attorney. This prosecutor apparently hopes that each will indict the other, either because each actually prefers unilateral success or because they are actually in a situation called an "Assurance Game," or "Stag Hunt," in which neither is confident in how the other will choose to act,

[73] Luce and Raiffa, *Games and Decisions*, 1957, 95.

TABLE 5. *Prisoner's Dilemma Game "H"*

		B$_1$	B$_2$
H:	A$_1$	(5, 5)	(−4, 6)
	A$_2$	(6, −4)	(−3, −3)

Game "H" has the same payoff matrix conventions as Game "G" in Table 4.

although both prefer joint cooperation over their unilateral defection.[74] Although this latter possibility is crucial to Thomas Schelling's application of the PD game to nuclear security, Luce and Raiffa's introductory PD pedagogy is consistent with orthodox game theory in disregarding the possibility that actors' subjective rankings deviate from material rewards.[75] Thus, even though theorists sometimes stipulate that assurance-seeking actors find themselves in a Prisoner's Dilemma because of risk, still they do not offer a means to disambiguate a situation in which actors really prefer to cooperate, despite the salience of tangible rewards, from the characteristic PD in which every actor has the first choice of suckering others. This becomes increasingly apparent as the authors' explanation progresses.

It is standard throughout most game theory to provide numbers that reflect a concrete source of value and simultaneously represent mathematically precise and well-ordered expected utilities. Luce and Raiffa illustrate this useful way of representing the Prisoner's Dilemma game as shown in Table 5.[76] The authors succinctly state in their explication of the Prisoner's Dilemma that the game referred to as "H" results in the aforementioned game "G." They provide the following description of game H's payoff matrix:

This will be given the interpretation that an entry (−4, 6) means player 1 loses $4 and player 2 receives $6, and we shall suppose that each player wishes to maximize his monetary return. Note that if we take the utility of money to be linear with money and set the utility of $6 to be 1 and of −$4 to be 0, then the game G results from H.[77]

---

[74] The Assurance Dilemma (or Stag Hunt Dilemma) matrix is derived from an Assurance Game with payoffs stipulated numerically to reflect each individual's assessment of the likelihood that the other actor may harbor Prisoner's Dilemma instead of Assurance Game preferences (preferring unilateral defection to joint cooperation). The Assurance Dilemma payoff values are those from the underlying Assurance Game multiplied by the likelihood with which each player evaluates that the other will play either "cooperate" or "defect." Thomas Schelling introduces his matrix in *Strategy of Conflict* (Cambridge, MA: Harvard University Press, 1960) to make an argument for mutual assured destruction, which is discussed in Chapter 3 and Figure 7.

[75] Giocoli makes clear that this direct correlation between subjective preferences and tangible rewards was a move made by von Neumann to make it possible to establish and objective science of choice, "Do Prudent Agents Play Lotteries," 2006, 102–105.

[76] Luce and Raiffa, *Games and Decisions*, 1957, 95.

[77] Ibid.

Monetary value and expected utility are interchangeable here as game theory often assumes for analyzing various social interactions.[78] Remarkably, this remains the game theoretic protocol used to analyze numerous social interactions. Sometimes, a tangible resource such as water, time, food calories, fitness value, or energy can substitute for money.[79]

Luce and Raiffa are well aware of the restricted elements of judgment permitted to enter into the strategic rational actor's logic for action. Only outcomes matter, particularly those observable and measurable features of the decision problem of direct relevance to each agent, and agents maximize individually. Moreover, in the most mathematically useful game in applied modeling throughout game theory, the measurable reward directly reflects expected utility. Luce and Raiffa introduce the necessary limitations on the value judgments defining the Prisoner's Dilemma game: only outcomes specified in cash value enter into actors' judgment and they maximize individualistically without regard for how their acts impact others. However, future authors introduce the game without carefully delineating these crucial assumptions.

Writing one of the first textbooks on game theory, Luce and Raiffa know that these limitations on judgment may strike some readers as far-fetched. What now strikes many as familiar, even necessary, seemed patently abusive in the 1950s.[80] The authors observe,

> We are assuming explicitly in the following discussion that ... [the players' utility stated in terms of cash value] does reflect their preferences. If this seems too gross an abuse of the utility notion, consider players who are *only* interested in the maximization of their *own* expected monetary return, and let the numbers in the payoff matrix represent money returns.[81]

Luce and Raiffa articulate in exacting terms both the restrictions on rational judgment for the standard game theoretic strategic rational actor and the manner in which a mathematical model of rationality can only readily be applied to many contexts of interest when these simplifying assumptions are introduced. In other words, the abstract mathematicized *Homo strategicus* is only relevant to the actual study of society in cases where agents are presumed to value only personally relevant outcomes, which represent instrumentally prominent concrete rewards such as cash. Yet the standard treatment remains like

---

[78] This is apparent in Binmore's *Game Theory*, 2007; for Giocoli's discussion of von Neumman on this point, see "Do Prudent Agents Play Lotteries," 2006, 102.

[79] For example, see Thomas Schelling, "Hockey Helmets, Concealed Weapons, and Daylight Saving: A Study of Binary Choices with Externalities," *Journal of Conflict Resolution* (1973) 17:3, 381–428. Cristina Bicchieri, "Covenants without Swords: Group Identity, Norms, and Communication in Social Dilemmas," *Rationality and Society* (May 2002) 14:2, 192–228; and Binmore, *Natural Justice*, 2005, 65.

[80] See the *Economist*'s claim that all decisions can be monetized, "Economic Focus: Never the Twain Shall Meet," *Economist*, February 2, 2002.

[81] Luce and Raiffa, *Games and Decisions*, 1958, 98.

TABLE 6. *High-Stakes Prisoner's Dilemma Game*

		YOU	
		The other prisoner does *not* confess	The other prisoner *does* confess
ME	You *don't* confess	Freedom + $1,000	Slow death by torture
	You *do* confess	Freedom + $10,000	Quick but painful death

"Me" is the row player, and "You" is the column player; payoff matrix only reflects the outcome for "Me," depending on what "You" chooses to do.

Luce and Raiffa's, and these models are used to design policies and institutions for neoliberal citizens and consumers.

In the landmark book *Paradoxes of Rationality and Cooperation*, Richmond Campbell introduces the concept of a "High-Stakes Prisoner's Dilemma" to capture the inexorable logic of the game that he believes "could arise in many other circumstances."[82] In this perilous game, the author asks us to "suppose that the first two possibilities are freedom plus $10,000 and freedom plus $1,000 while the second two are quick, but painful, death and slow death by torture" (Table 6).[83] In this game, Campbell stipulates a tangible reward system and automatically assumes that it matches up with the actors' subjective appraisal of expected utility. However, confronted with these outcomes, readers may find it obvious that, to the contrary, one should rank them in a different order: (1) neither confesses and both receive freedom plus $1,000; (2) only I confess; (3) both confess, gaining quick but painful death; and (4) only you confess. And yet throughout orthodox game theory, the mutually observable physical rewards are assumed, without a second glance, to directly reflect actors' expected utilities. The result is that actors are presumed to be in a state of Prisoner's Dilemma, whereas, in fact, they may interpret their situation to have more in common with an Assurance Game in which mutual cooperation is preferred to one's sole defection.

One's possible intuition that mutual cooperation actually results in the superior outcome for both players makes it possible to believe that the logical impasse resulting in mutual and painful death is due to each actor's doubt that the other can be depended on to remain silent. It is easy to suppose that the significance of the Prisoner's *Dilemma* rests in the fact that each would obviously cooperate to achieve mutual freedom *if assured the other would similarly cooperate*, as would be the case in either an Assurance Game, in which both actors most prefer to cooperate, or an Assurance Dilemma, in

[82] Richmond Campbell, "Background for the Uninitiated," introduction to Campbell and Snowden, eds., *Paradoxes of Rationality and Cooperation*, 1986, 3–41, at 6.

[83] Game taken from Campbell, "Background for the Uninitiated," in Campbell and Snowden, eds., *Paradoxes of Rationality*, 3–41, at 6.

which actors are unsure whether the other's first choice is unilateral defection or mutual cooperation.[84] In this latter game form, although both most prefer the mutually cooperative outcome over unilateral defection, neither is sure of the other's preferences, and both are aware of this lack of confidence. But the Prisoner's Dilemma, played between two perfectly rational agents who cannot achieve a mutually agreeable and available outcome, plays out the way it does because, from the perspective of the rational actor who independently maximizes personal expected utility consistent with the game's payoff matrix, the decision categorically and in principle has no relationship to risk or uncertainty. The canonical game theoretic Prisoner's Dilemma arises precisely because if either player were certain of the other's cooperation, her first choice would still be to confess, since this grants her freedom and financial gain by exporting the costs for her defection onto the other player. The bright-line test of whether agents' actual subjective preference rankings places them in a PD is whether they both would choose to defect when 100 percent guaranteed of the other's cooperation. Campbell's High-Stakes Prisoner's Dilemma game underscores how most games simply assume that the tangible payoff structure characterizing the game also determines individuals' preference rankings.[85]

This bizarre High-Stakes PD reaffirms the Prisoner's Dilemma pedagogy outlined earlier: actors will frequently find themselves in situations with a tangible reward structure reflecting the PD game in which the only rational outcome is to defect.[86] I doubt that most practitioners of game theory would accept that game theory necessarily endorses the view that predatory gain is preferable to reciprocal respect for others' rights to bodily integrity and private property. However, the standard assumptions used to operationalize strategic rationality in many contexts do, indeed, routinely reinforce the strategy of profiting by displacing costs on others without any discussion of either moral accountability for actions or subjects' possible preference for mutual cooperation over

---

[84] For example, consider the Batman movie *Dark Knight*, produced by Christopher Nolan, 2008, in which one of the plot developments climaxes with the apparent setup of a Prisoner's Dilemma standoff with a High-Stakes material rewards payout. Two groups of hostages have been told that they can win their freedom by killing the members of the other group. However, the ultimate resolution shows that both sides actually held Assurance Game preferences.

[85] The accepted equivalence of Newcomb's Paradox and the PD demonstrates that the negative causal impact of the unilateral victory in a PD on the sucker is fully treated as an externality with no relevance to an individual's choice, see David Lewis, "Prisoners' Dilemma Is a Newcomb Problem," in Campbell and Snowden, eds., *Paradoxes of Rationality and Cooperation*, 1985, 251–255.

[86] Numerous authors conclude that PD's abound: Richard Tuck, *Free Riding* (Cambridge, MA: Harvard University Press, 2008), 22; Hargreaves Heap and Varoufakis, *Game Theory*, 2004, 175–178; Heath, *Following the Rules*, 2011, 39. Note, however, that in his investigation of games played in the Old Testament, game theorist Steven Brams only finds at most one to two PD games, *Biblical Games: Game Theory and the Hebrew Bible* (Cambridge, MA: MIT Press, 2002), index, 212.

unilateral defection.[87] The claim that concerns of due process or other-regarding consideration can, in principle, register in actors' preference rankings may be true in parametric decision theory. However, in the expected utility theory required by game theory, only outcomes that register gain for agents count.[88] Therefore, the structures for action characteristic of classical liberalism – fair play, self-adopted rule following, commitment, perfect duty, and side constraints – are inconsistent with strategic rationality because they function independently of the material rewards that accrue to respective actors.[89] Concern for others can be assimilated into preference rankings over outcomes and also should be considered in the standard presentation of the Prisoner's Dilemma game. Yet, again, this admission would deviate from the standard game theoretic reliance on commonly sought after scarce and measurable rewards directly accruing to each actor to define expected utilities.[90]

Experts in game theory realize these caveats. Nevertheless, a prevailing PD pedagogy has emerged, which can be imparted by teachers with less perfect and thorough knowledge of game theory and absorbed by students who will not go on to become experts themselves. This readily transmitted indoctrination presents the dilemma without specifying the simplifying assumptions that learners must tacitly endorse to perpetuate the worry that strategic rationality is non-negotiable and mutually destructive, and that Prisoner's Dilemma situations abound.[91] A superior pedagogy would clearly outline the limitations of strategic rationality, explicitly acknowledging that expected utility functions can only exhaustively incorporate actors' subjective concerns by rendering some superfluous to rational choice. Additionally, astute teachers of the Prisoner's Dilemma should clearly highlight the standard shortcut of assuming that salient

---

[87] The language "externality" for the cost displaced onto others for personal gain is developed within this context; for discussion see Tuck, *Free Riding*, 24–27; see also Schelling, "Hockey Helmets," 1973.

[88] On how "standard decision theory" assumes that "agents only invest outcomes with motivational significance," see Heath, *Communicative Action*, 2001, 137–139.

[89] Amartya Sen contrasts these procedural considerations from the concerns about outcomes reflected in expected utility functions; see Sen, *Rationality and Freedom*, 2004, 175–181; see also Heath, *Communicative Action*, 2001, 86–92.

[90] Heath urges an expansion of orthodox game theory to encompass these considerations, *Following the Rules*, 2011. Binmore emphasizes that game theory can denote altruistic preferences and does not require cash value, yet both tend to use either cash or tangible value in their exposition of games, and Binmore notes that the instrumental consistency demands that rational actors must "necessarily behave as though maximizing the expected value of *something*," which grounds the payoff of games to a fungible existential property of existence, *Natural Justice*, 2005, 64–65.

[91] This is the conclusion drawn from the single-play PD, in addition to iterated PD games with a known termination point, or indefinite play scenarios in which an end point could be surmised; with respect to indefinitely played PDs with no discernable end point, many equilibria permit some degree of cooperation, with the two caveats that there is no single clear equilibrium for players to gravitate toward, and the only safe strategy in which every stage of the game has a self-contained rational strategy (coincident with mutual defection) is the only purely safe strategy; Luce and Raiffa, *Games and Decisions*, 1958, 72–102; repeated PDs are discussed in Chapter 11, along with the Tit for Tat strategy.

tangible rewards directly underlie the payoffs defining the game. Therefore, orthodox game theory does not admit as rational the type of agency characterizing classical liberalism or neoclassical economics. Both of these characteristic agents voluntarily constrain their action to be consistent with an internalized Pareto condition to act to make at least one person better off and no one worse off. In the PD, actors most prefer the state in which the other is less well off than had the two agents not interacted at all.

In the dire High-Stakes game, the classical liberal would cooperate if assured the other would also, even if paired with a stranger. If someone not only prefers, but also triggers, a stranger's slow death to get $9,000, instead of $1,000 and shared freedom, then this person is violating the central no-harm principle that recommends that agents respect the sanctity of each other's physical integrity. The individual who accepts the terms of this high-stakes Prisoner's Dilemma game unwittingly acquiesces to neoliberal subjectivity. The ground rules governing neoliberal subjectivity entail accepting the utility of sending another actor to death by slow torture for a $9,000 gain without taking any responsibility for the role one's own decision plays in the other person's fate. The ready presentation of this high-stakes "game," inviting those exposed to participate in its logic of financial gain at the cost of another individual's extreme harm, aptly reflects the change in gestalt from the classical liberal to the neoliberal paradigm of markets and government.

The standard apparatus for teaching the Prisoner's Dilemma fails to disambiguate a PD game from an Assurance Game in which actors' appraisal of existential significance may not track interpersonally transferable expected gain in the way typically assumed. In the Assurance Game, actors prefer mutual cooperation yet may defect out of anxiety that the other actor may fail to cooperate.[92] From a revealed preference perspective, from which an actor's preferences are only known once observed during choice, the only way to detect whether the actor's preferences reflect an Assurance Game stance or Prisoner's Dilemma stance is to see whether that agent cooperates after the other individual has. The difficulty lies

---

[92] In an Assurance Game (or Stag Hunt), as related in Table 3, both actors prefer to cooperate rather than defect. Yet this situation is difficult to capture if the preference for a cooperative outcome yields less tangible instrumental gain than defecting. Actors may opt to defect in an Assurance Game because this action guarantees the best worst outcome. In an Assurance Dilemma game, neither actor knows whether the other views the situation as a Prisoner's Dilemma or an Assurance Game. Using expected utility theory, sufficient suspicion about the preferences of the other actor can transform the expected payoff into the characteristic Prisoner's Dilemma game, making the rational strategy to defect. However, this game theoretic logic obscures the fact that in both the Assurance Game and the Assurance Dilemma game, each actor cooperates once it is certain that the other will or has. Chapter 3, "Assurance," follows how Thomas Schelling argued that the nuclear security dilemma, in its worst-case form, should be treated as a Prisoner's Dilemma derived from an Assurance Dilemma in view of each actor's doubt about the other's intentions. This treatment, along with the ensuing conventionalized pedagogy of introducing the PD game, has led to the characteristic confusion that the logical impasse of mutual impoverishment involves some dilemma over choice.

in determining agents' actual ranking of outcomes over and beyond the tangible payoff matrix. An effective teacher of the High-Stakes PD game, or any PD game for that matter, needs both to clarify the standard game theoretic default of permitting tangible rewards to directly reflect inherently reductionist expected utility rankings and to reaffirm that the bright-line test for whether actors actually perceive themselves to be in a Prisoner's Dilemma game instead of potentially an Assurance Game situation is if they choose to defect after the other person has already cooperated.

Campbell recognizes the apparent bizarreness of the high-stakes Prisoner's Dilemma. He writes, "If rational, you should both ... choose a quick but painful death rather than go scot-free with $1000 a piece in your pockets." Thus, he suggests that the PD game offers a logical imperative, an "ought," that actors confronted with tangible rewards characterizing the PD payoff matrix should defect. Campbell goes on to observe, *At this point it may appear that the dilemma, however tantalizing as a logical puzzle, is too fantastic to have any practical relevance.*[93] He agrees that *any* rendering of the Prisoner's Dilemma is "odd enough" in itself.

Nevertheless, he presses on to convince readers of the relevance of the Prisoner's Dilemma by applying it to the superpower standoff: "Two super-powers sign a nuclear disarmament pact on the shared belief that failure to disarm will result sooner or later in a nuclear holocaust in which each side will be quickly and painfully destroyed, while mutual disarmament will avoid this dreaded outcome." Campbell acknowledges that this problem has the structure of a Prisoner's Dilemma game because "each would say that its having complete nuclear superiority is a better guarantee of peace on earth than mutual nuclear disarmament, and each would explain its breaking of the agreement as a purely defensive maneuver."[94] He presents the Prisoner's Dilemma payoff matrix from the perspective of the United States as shown in Table 7.[95]

Campbell explains that "each side regards this vulnerability [of nuclear inferiority] as a fate worse than mutual destruction, while it regards a position of complete nuclear superiority as ideal." This example illustrates that in international relations, the classical liberal's intuitive preference for bilateral agreements and symmetric deterrence yields to a neoliberal's unapologetic predilection for unilateral success, asymmetric deterrence, and nuclear hegemony. For the neo-liberal approach to relationships, security is not a positive-sum good predicated on all parties striving to make choices that avoid incurring harm on other actors. Rather, strategic rationality assumes that the decision makers must displace the costs for their security and prosperity on other actors when they act to maximize their gain of scarce fungible goods in competition with others.

---

[93] Campbell, in Campbell and Snowden, eds., *Paradoxes of Rationality*, 1985, 6. Emphasis added.
[94] All the quotes in this are from Campbell, "Background for the Uninitiated," in Campbell and Snowden, eds., *Paradoxes of Rationality and Cooperation*, 1985, 6–7.
[95] This game reproduced in Table 7 is taken from ibid., 7.

TABLE 7. *Nuclear Security Dilemma Modeled as Prisoner's Dilemma*

		USSR	
		The other *adheres* to the agreement	The other *violates* the agreement
US	You *adhere*	No mutual destruction	A fate worse than mutual destruction
	You *violate*	The ideal upshot	Mutual destruction

US is row player; USSR is column player; payoffs are strictly considered from the perspective of what outcome the US receives and which it most prefers.

Finally, in Campbell's eyes, the nuclear security dilemma provides the rationalization for why the Prisoner's Dilemma game and, by extension, game theory are useful tools for understanding social relations. The Cold War relevance of the Prisoner's Dilemma, which results in counseling an offensive and aggressive stance justified by self-defense, makes this unconventional view that rationalizes dominating others seem not only pedestrian but also mandatory. Whereas assuming the toughest case and concentrating on actors' estimation of material gain may have provided initial impetus to delineating game theory, as subsequent chapters demonstrate, the advent of nuclear weapons does not necessarily provide a compelling reason to rethink the security of individual agents vis-à-vis one another in markets or states. Every practitioner of game theory should be clear how the reliance on individual maximization and the introduction of risk, worst-case planning, and the demands for commensurable and interpersonally transferable value to simplify calculations ultimately mire agents in a prison of strategic reason.

NONCOOPERATIVE GAME THEORY AND THE SHIFT TO COERCIVE BARGAINING

The Prisoner's Dilemma narrative, in conjunction with its name, conjures up images of stressful decision making, especially because it represents a miscarriage of a classical liberal exchange in which both actors seek a consensual, mutually beneficial trade. This section explains how the means of tracking value in much operationalized game theory normalizes the view that in routine market transactions, each actor not only prefers to but would indeed choose to sucker others if able to do so without consequences. This is the inevitable result of depending on expected utility functions that can only assess fungible payoffs independently of the means by which they are realized. This section also discusses how neoliberal theory attempts to mimic classic liberal exchange by two means. First, game theorists encompass the cooperative act of exchange within noncooperative game theory and thereby stipulate that the outcome of a bargain must derive from actors' ability to threaten negative

TABLE 8. *Neoliberal Car Sale: Exchange Modeled as Prisoner's Dilemma*

	Other	
Self	*Send* the car	*Keep* the car
Send the car	Mutual exchange	Other gets cash & car
Keep the car	I get cash & car	Cash and car harmed in skirmish

In this payoff matrix, the presumption is that each actor most prefers to obtain both the cash and the car; second best, both prefer to exchange; third best, both prefer mutual defection from exchange; each actor least prefers to be suckered by having neither the cash nor the car.

repercussions.[96] Second, game theorists envision that the bargainers may perpetually encounter each other over and over with the exact same decision problem and thus have the wherewithal to punish the other actor for failing to cooperate by defecting in their next encounter.[97]

Consider the neoliberal car sale depicted in the payoff matrix in Table 8.[98] In this now widespread Prisoner's Dilemma model of exchange, both oneself and the other most prefer to get the car and the cash and leave the other with nothing. Whereas the liberal actor pursues amicable exchange, the neoliberal actor most prefers to cheat the other.[99] In the Prisoner's Dilemma application, every agent is presumed to seek sole gain for herself, thereby implicitly hoping to leave all other actors with their worst outcome, because only outcomes, distinguished by instrumentally salient permutations of the phenomena, register in expected utility functions according to this model of exchange and bargaining.[100] Therefore, the *only* motive for carrying through on the terms of a contractual agreement is the threat of punitive sanctions, and this understanding stretches from international relations through international political economy to the social contract and routine bargaining. In their penetrating analysis of contemporary economic science, Daniel Hausman and Michael

[96] Ken Binmore has an excellent discussion of this in his introduction to Nash, *Essays on Game Theory*, 1997, ix–xx.

[97] Game theory texts move quickly from the Prisoner's Dilemma model of exchange to the repeated game (Hargreaves Heap and Varoufakis, *Game Theory*, 2004, 191–194; however, this theoretic move misses that in the classical liberal market, exchanges were often between individuals who did not know each other and would likely not encounter each other again; again, the primary difference is that in the neoliberal model of exchange, each most prefers to cheat the other and in the classical liberal exchange, each chooses to cooperate given the other's alike cooperation.

[98] This example is drawn from Campbell, "Background for the Uninitiated," in Campbell and Sowden, eds., *Paradoxes of Rationality*, 1985, 9; Russell Hardin uses the same example in "The Utilitarian Logic of Liberalism," *Ethics* (Oct. 1986) 97:1, 47–74, at 52. See also Axelrod, Conflict of Interest, 1970, 58–60.

[99] Adam Smith's concept of fair play, *Theory of Moral Sentiments*, section II.ii.2.2; see also Robert Nozick on side constraints, *Anarchy, State and Utopia* (Oxford: Basil Blackwell, 1974), 28–33.

[100] Roger Myerson draws attention to the requirement of most operationalized game theory to track instrumentally "salient permutations" of the world; see his *Game Theory*, 1991, 25.

McPherson note that from this perspective, "the only thing wrong with cheating is the risk of getting caught"; furthermore, "competitive pressures do not permit firms [and other actors] the luxury of moral scruples."[101]

This section discusses how game theorists developed this mutually compromising view of market exchange early on to analyze arms control. In 1967, future Nobel Prize winners Robert J. Aumann, John C. Harsanyi, and Reinhard Selten published, with three other authors, the report *Gradual Reduction of Arms*, under the auspices of the US Arms Control and Disarmament Agency.[102] From this point onward, the classical view of the free market, which is necessarily bounded by the respect for persons, property, and contracts, was increasingly displaced by the view that bargains are facilitated by agents' power to threaten harm on others to secure better terms and subsequently enforce them.[103] While in traditional liberalism, normative agreements are self-guiding and create patterns of constructive interdependence when actors are assured others will cooperate, in postmodern neoliberalism, regularized patterns of interaction are the by-product of individual preference satisfaction and may well harm individuals and squander resources. Moreover, whereas classical liberalism entails the achievement of prudential judgment and the wherewithal to attain third-person impartial assessment of the conduct of others and ultimately of one's own conduct, the neoliberal paradigm views strategic rationality as biologically programmed into agents as a condition of their survival and replication.[104]

Reinhard Selten's contribution, coauthored with Reinhard Tietz, is analytically distinct from the other articles. These authors model a "Class of Simple Deterrence Games," assuming that nuclear war represents an "irreversible game" with "one type of atomic bomb."[105] The conclusions of Selten and Tietz's study are intuitively plausible: nations with good will toward one another are less likely to attack one another and that fewer to no atomic weapons leads to more stability than increasing stockpiles past one to several weapons. Yet their modeling exercise does not dovetail with the other papers

[101] Hausman and McPherson, *Economic Analysis*, 2006, 72–73; contemporary reputation mechanisms for establishing transparency are neoliberal in the sense that they do not discover a disposition or character for integrity but only serve to demonstrate a past trend that may indicate a forward trend.

[102] I am indebted to Jerry Green and Helen Gavel for access to this report, written by Robert J. Aumann, John C. Harsanyi, John P. Mayberry, Michael Maschler, Herbert E. Scarf, Reinhard Selten, and Richard Stearns, *Models of Gradual Reduction of Arms*, Arms Control and Disarmament Agency, ACDA/ST-116, September 1, 1967.

[103] Luce and Raiffa, *Games and Decisions*, 1958, 97–101.

[104] John Nash articulated this view in his PhD thesis; for discussion, see Giacoli, "Nash Equilibrium," 2004. See also Binmore, who draws on Nash, *Natural Justice*, 2005, 23–27, 42, 73–75, 101–112.

[105] "Security Equilibrium for a Modified Version of Scarf's Deterrence Model," in *Models of Gradual Reduction of Arms*, Arms Control and Disarmament Agency, ACDA/ST-116, September 1, 1967, 501–551. For contrast, see John C. Harsanyi, "A Generalized Nash Solution for Two-Person Cooperative Games with Incomplete Information," also in this report, 71–286.

in the volume. In exploring the implications of bargaining, those other papers found it too mathematically cumbersome to incorporate how actors might demonstrate concern for how other actors feel about outcomes. The Selten-Tietz paper demonstrates a broader view of game theory, but it was the exploration of strategic bargaining for nuclear strategy taken up by the other authors that generated the neoliberal approach to political economy. The demands of mathematical tractability and the fact that game theory is an instrumental account of rationality that necessarily and directly associates expected gain with configurations of ontologically existing phenomena have encouraged theorists to standardize the noncooperative Prisoner's Dilemma model of contracts, bargaining, and exchange.[106]

John Mayberry's introduction to the report, "The Notion of 'Threat' and Its Relation to Bargaining Theories," sets forth the aggressive view of bargaining. Appropriate for international relations, actors cannot exit a state of nature. This means that actors gain advantage through posing credible threats to one another and that no bargain is safe unless exposed to the constant pressure of sanctions endogenously supplied by the participants themselves to address compliance failures. It is worth analyzing Mayberry's introduction to the Nash Bargaining Solution encompassed by noncooperative game theory in detail because it is paradigmatic of neoliberal market discipline. Mayberry's paper confirms that neoliberal political economy is predicated on the strategic rationality of game theory, first vindicated within the context of avoiding nuclear war by preparing to wage it. Mayberry notes that strategic rationality is indispensable for nuclear strategy and is likewise essential to decision making in other domains such as "the arms race (in non-nuclear weapons especially); the Viet-Nam conflict; price competition in capital-intensive industries" and for analyzing how castaways on an island may bargain over the food necessary to stay alive.[107] Bargaining problems, from simple exchange to military contestation, take the Prisoner's Dilemma structure and should be solved according to the logic of strategic rationality.

A graphic illustration of a bargain that equates players' utilities with the potential outcomes in a bargain is presented in Figure 2.[108] The bargaining space is defined by outcomes reflected by the respective units of gain each actor expects. Technically speaking, a cooperative bargaining game, by which game theorists specify that actors could reach a mutually preferred outcome by

[106] Von Neumann and Morgenstern articulate the view that value inheres in the world as an ontological property that obeys the laws of physics, *Theory of Games and Economic Behavior*, 1944, 1953, 2004, all editions, 16–24; for discussion, see Nicola Giocoli, "In the Sign of the Axiomatic Method," in Richard Arena, Sheila Dow, and Mathias Klaes, eds., *Open Economics: Economics in Relation to Other Disciplines* (London: Routledge, 2009), 129–149.
[107] Mayberry, in Aumann et al., *Models of Gradual Reduction of Arms*, 1967, 35, see also 29–32.
[108] Mayberry's "Notion of 'Threat,'" 1967, has thirteen figures demonstrating this concept; this is a common depiction; see also Rapoport, *Fights, Games, Debates*, 4th ed., 1970, 189; Luce and Raiffa, *Games and Decisions*, 1958, 118; Figure 2 is derived from John F. Nash Jr., "The Bargaining Problem," reprinted in Nash, *Essays on Game Theory*, 1996, 155–162, at 161 and integrates the moving threat point concept, see Robert Axelrod, *Conflict of Interest* (Chicago: Markham, 1970), 53.

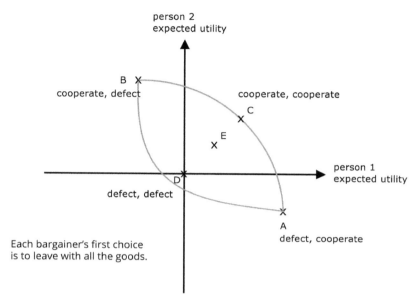

FIGURE 2. Nash Bargaining Solution as Prisoner's Dilemma Game

cooperating, rather than if they played noncooperatively achieving mutual defection, is encompassed by noncooperative game theory and typically exhibits the characteristic PD payoff matrix (including pure and impure variants in which actors may gain more by alternatively defecting than always cooperating). *The crucial point is that in all cases the cooperative bargaining settlement is a function of the disagreement point, which designates the outcome that ensues when all actors defect.*[109]

Points on the expected utility graph denote the expected gain from a specific outcome. Each pair of numbers representing a point specifies the expected utility received by person 1 on the horizontal axis and person 2 on the vertical axis. Given the normal linear relationship between expected utility and tangible goods obtainable through bargaining, each point could represent an expected value of money or some other fungible source of value. Each point in the bargaining space represents either a concrete payoff or a probabilistic lottery of two other outcomes that has equal numeric expected utility. For example, to technically define the bargaining space, a specific point E could represent a 50 percent probability of receiving the payoff for unilateral defection plus a 50 percent probability of receiving the payoff for being suckered (with an evaluation equivalent to ½ {Defect, Cooperate} + ½ {Cooperate, Defect} enumerated for each player).[110] Throughout zero-sum and

---

[109] Myerson refers to this derivation of cooperative outcomes from noncooperative game theory "Nash's program," *Game Theory*, 1991, at 371; on the "Nash Program" see also Binmore's introduction to Nash, *Essays on Game Theory*, 1996, xiv.

[110] For an effective discussion, see Kuhn, "Cardinal Payoffs," in "Prisoner's Dilemma," entry, 2014.

non-zero-sum game theory, solutions can often only be guaranteed to exist if the game is viewed through the lens of indefinite play and players are permitted to play each strategy in a fixed proportion to all the other strategies over time.[111]

In Figure 2, the point labeled A {Defect, Cooperate} designates player 1 defecting and player 2 cooperating; hence, player 1 gets all the possible value giving player 2 negative utility. The point labeled B {Cooperate, Defect} designates player 1 cooperating and player 2 defecting, with player 2 getting all possible value. The point labeled C designates the point at which both individuals cooperate {Cooperate, Cooperate}. The curve that traces through points B, C, and A represents the "Pareto frontier," which designates the points in the expected utility space from which one cannot improve any single actor's expected gain without diminishing that of the other actor. The inverse curve demarcating the joint set of the actors' least preferred outcomes lies in the bottom southwest quadrant of the figure.

The point designated D {Defect, Defect} represents mutual defection, which is the solution of the noncooperative game. In a bargaining game, the mutual defection point may be derived from the total set of possible outcomes as a mixed strategy solution to the noncooperative game.[112] This is a two-step process involving first establishing the disagreement point, and building up from that to the agreement point of the bargain. This disagreement point {Defect, Defect} can be determined in three manners and thus is not uniquely specified. It can be derived from each individual securing his or her best-worst case, or maximin, outcome; it could represent a focal point of mutual salience; or it can result from each individual choosing an action to cause the opposition the greatest damage. Because the cooperative agreement point is directly deduced from the disagreement point, and the worse the noncooperative outcome is for any actor, that actor will be content to settle for less in the final cooperative settlement. Therefore it is typically in each individual's interest to threaten the greatest harm on the other to achieve the superior cooperative outcome for oneself.[113] Whether one selects to threaten the other with the worst loss or protect oneself with the best worst-case outcome depends on which achieves the best outcome overall for the decision maker, and which strategies are credible insofar as they do not place oneself in an unrealistically vulnerable position.[114]

In the Nash bargaining solution, the identification of point D {defect, defect} is crucial because, along with the axioms defining Nash's approach, this point

[111] Luce and Raiffa, *Games and Decisions*, 1958, 106–109; Myerson follows Luce and Raiffa in first treating the repeating PD as the best example of repeating games in general, and immediately discussing bargaining as a cooperative game derived from the foundations of noncooperative game theory, *Game Theory*, 1991, 379–390.

[112] Luce and Raiffa, *Games and Decisions*, 1958, 106–109. On the three means to establish the default disagreement point see Myerson, *Game Theory*, 1991, 386. Note that Axelrod's original analysis permits the default disagreement outcome to be determined by an outside party, demonstrating the power of that agent over the outcome, *Conflict of Interest*, 1970, 52–53.

[113] Rapoport has the clearest discussion of this, *Fights, Games, Debates*, 1970, 186–192.

[114] Luce and Raiffa, *Games and Decisions*, 1958, 106–109; on rational threats see Myerson, *Game Theory*, 1991, 385–389.

determines the single-point solution of the bargain, which in Figure 2 coincides with mutual cooperation. An arbitrary point, designated E in the diagram, may not actually represent a concrete outcome, but instead a lottery ticket with an expected value based on, for example, a specific likelihood that one will be the sole defector and a specific likelihood that one will be the sole cooperator (a mixture of outcomes A {Cooperate, Defect} and B {Defect, Cooperate}). A Nash bargaining game requires that the field of points be filled in, and often the only way to make this possible is to invent lotteries among possible outcomes that can reflect the expected value for all conceivable points.[115] This consideration, standard throughout game theory to guarantee game solutions, restricts the features of the decision environment that can register subjective utility not only to outcomes independent of the means by which they are achieved but also to fungible properties, which accommodate attitudes toward risk.

The bargain takes on the characteristics of a Prisoner's Dilemma because, simply from the considerations of expected tangible gain, each actor is best off exiting the bargain with all the goods, leaving the other with none.[116] In this case, if one individual decides to cooperate, the other player will sucker that individual by defecting. If both fail to cooperate, then mutual defection occurs. The failure to cooperate is reflected as the "zero point," or default point signifying mutual defection in the Prisoner's Dilemma. This default can be manipulated if either or both actors can drive down the other's payoff in the case of noncooperative mutual defection, with the lower bound number for each agent reflected by the sucker's payoff of sole cooperation against the other's defection. If secured by an enforcement mechanism to forestall individuals' inherent tendency to renege, settlement will be somewhere on the northeast frontier of the diagram and, according to Nash, is strictly delimited by the default point, or what will come to pass if no agreement is reached.[117] Game theorist Roger Myerson explains how "the payoff to player 1 ... increase[s] as the disagreement payoff to player 2 decreases," and that therefore, "a possibility of hurting player 2 in the event of a disagreement may actually help player 1" should an agreement be

---

[115] There are three levels of specificity available to define a bargaining game, in which individuals tend to evaluate the worth of others' gain, that could suggest interpersonal comparison of utility, and may permit side payments, see Luce and Raiffa, *Games and Decisions*, 1957, 120, and Myerson, *Game Theory*, 1991, 381–85.

[116] Myerson, *Game Theory*, 1991, 370–371.

[117] The Nash bargaining solution is a highly technical mathematical result similar to Kenneth Arrow's impossibility theorem. Nash is able to obtain a solution to the bargaining game because he assumes a status quo point that disregards from consideration outcomes that are less appealing to either player than the status quo point. Much of the work to solve a Nash bargaining problem goes into establishing what outcome represents the status quo point. Leveraging threats requires an understanding of the relative costs of threatening someone to oneself and to the other agent. For understanding, the Luce and Raiffa discussion requires reading both chap. 5 on "Two-Person Non-Zero-Sum Non-Cooperative Games," 88–113, and chap. 6 on "Two-Person Cooperative Games," 114–154; note that solving a Prisoner's Dilemma–style bargaining game with asymmetric rewards requires a randomized strategy, 115. Roger Myerson's later treatment is more concise because he combines the Nash bargaining solution and noncooperative game theory in line with Mayberry's example, *Game Theory*, 1991, chap. 8, 371–394.

reached.[118] Therefore, the hope of a mutually cooperative outcome in game theory "may give players an incentive to behave more antagonistically before the agreement is determined." John Nash formalized this "chilling effect."[119]

Beyond Nash's original bargaining problem, which he illustrated between the fictitious figures Bill and Jack over objects, and Mayberry's exploration of bargaining over terms of arms reduction, having actors bargain over monetary value became standard.[120] In what became known as the ultimatum game, there is a total sum of cash value to be distributed to players if the two can agree on how to share it. One person chooses a distribution, and the other has the power to accept or reject it. If there is agreement, the money is shared; if not, neither receives any reward.[121] Game theorists notice that it is rational for the second individual to even accept just $1 of a total $100 because that single dollar is still worth more than nothing. However, rather than settle for a perceptible though small gain, each actor has the capability to threaten that the other will get nothing by presenting a willingness to defect unless personal stakes are sufficiently high. The ability and credibility to threaten other actors is thus crucial to how much one can gain oneself. In terms of understanding the development of a uniquely neoliberal approach to interactions, markets, and governance, the key point is that in the move to apply abstract formal game theory to the lived world, all that can register in actors' rational preference rankings over outcomes and lotteries thereof is their mathematically consistent appraisal of tangible outcomes irrespective of the processes and intentions that brought them about: this neoliberal market "is therefore the final step in a process that first leaches out the moral content of a culture and then erodes the autonomy of its citizens by shaping their personal preferences."[122] This is obviously true in the case of a bargain structured by relative ratios of fungible goods accruing to each individual. Value is significant and measurable independent of social relationships and context; actors' preferences reflect structural environmentally given rewards and are not autonomously determined.[123]

Game theory extrapolates from lived experience to provide a tidy mathematical analysis of conflict.[124] However, it is less clear how formal strategic rationality oversteps the boundary from being a thought experiment to becoming a categorical imperative directly linked to survival in more mundane contexts other than a military contest.[125] The early Cold War embrace of

[118] This and the following three quotes are from Myerson, *Game Theory*, 1991, at 386.

[119] "Two-Person Cooperative Games," *Econometrica* (1953) 21, 128–140.

[120] "The Bargaining Problem," *Econometrica* (1950) 18, 155–162.

[121] Note that it is possible to define variations of this game. For example, the players could play repeated rounds with a discount factor reducing the total sum to be distributed in each round. Hargreaves Heap and Varoufakis, *Game Theory*, 2004, 196–204.

[122] Binmore, *Natural Justice*, 2005, at 184.

[123] Hargreaves Heap and Varoufakis, *Game Theory*, 2004, 209.

[124] For an early treatment, see Robert Axelrod, *Conflict of Interest: A Theory of Divergent Goals with Applications to Politics* (Chicago: Markham, 1970).

[125] Binmore discusses a "sharing food" example, *Natural Justice*, 2005, 119–121, 160–161.

hard-nosed strategic rationality to confront the Soviet Union was subsequently applied across multiple domains of decision making and choice by the late twentieth century. War, which strategists may view as a state of potentially unbounded conflict, invites the belief that command of resources is necessary for survival and propagation. In this case, the power granted by such resources stems from natural properties, governed by the laws of physics, as opposed to specific patterns of interactivity dependent on intricate norm-governed social arrangements or specific intersubjectively perpetuated interpretations of worth and significance. I further elaborate on this theme in the Chapter 4, "Deterrence," by showing how game theory's dependence on interpersonally transferable utility renders it compatible with the international relations of realism, neorealism, and neoliberalism.

Mayberry culminates his analysis by using game theory to draw a sharp distinction between a classical liberal and neoliberal approach to bargaining. Whereas the former is normative and obeys the no-harm principle, which could be interpreted in view of individuals' maximin strategies in which each seeks to secure his best possible worst outcome, the latter deploys coercive threats to gain the advantage over opposition even though each may be worse off than otherwise should both players defect. Mayberry clearly specifies how the Nash bargaining solution can be incorporated into noncooperative game theory to support the nuclear strategy of preparing to fight and win a nuclear war. He reasons:

Nash's concept of threat and solution can reconcile and illuminate for me the inconsistent extreme views of those ultra-pacifists who say, "War is unreasonable, and we are reasonable; therefore let us not prepare for war, nor consider it as an option" and those extreme hawks who say, "If we do not prepare for war, we shall be forced to surrender, and it is ridiculous to *prepare* for war unless we *intend* to fight."[126]

Mayberry's central idea, derived from Nash's bargaining solution, is that the outcome of strategic arms control, or any settlement among protagonists, will be a function of their willingness to leverage credible threats to achieve an outcome in their favor.[127] Upholding the normative no-harm principle, which effectively represents "protect[ing] oneself against the worst the opponent can do," is a weaker strategy than the one that becomes the crux of neoliberal bargaining: "to ensure that the opponent is injured as much as possible even if his main effort is to defend himself."[128] Mayberry makes clear he views bargaining as part of relentlessly competitive non-zero-sum game theory that, within the context of arms control, prescribes leveraging coercive threats to achieve national security. He pithily states the central tenet of the nuclear utilization targeting strategy: "it is ridiculous to *prepare* for war unless we *intend* to fight."

Mayberry's analysis helps introduce Part I of *Prisoners of Reason* because he shows how game theory rationalizes the case for preparing to fight a nuclear

---

[126] Mayberry, "The Notion of 'Threat,'" 1967, 46.
[127] Luce and Raiffa provide an effective discussion of this material, *Games and Decisions*, 1957, 106–109.
[128] Mayberry, "The Notion of 'Threat,'" 1967, 43.

war by arguing for the analytic necessity of making credible threats to the Soviet Union to improve US bargaining power.[129] Cooperative games, in which outcomes are the function of agreements, are encompassed by noncooperative game theory for three reasons that are consistent with the game theoretic neoliberal orientation. First, no strategic rational actor voluntarily abides by agreements made. Second, leveraging threats secures the most favorable settlement for oneself. Third, others will only uphold terms of a bargain with the constant pressure of credible threats for noncompliance.

CONCLUSION

In the vast intellectual landscape of game theory, the Prisoner's Dilemma game has become accepted as a discovery of a core puzzle at the heart of all manner of cooperative ventures: Adam Smith's invisible hand, joining trade unions, participating in public vaccinations, standing at football games, and even marriage.[130] Game theory scholars frequently present the PD as though it were as simple and straightforward as tic-tac-toe. As long as two actors' subjective preference rankings conform to the characteristic PD payoff structure, then the rational choice for each individual is to fail to cooperate, leaving both with their second-worst preference. However, many layers of analytic complexity are involved in setting up the Prisoner's Dilemma game, which, if not rendered explicit, become part of the pedagogic baggage relied on to transmit the acuteness and inevitability of the Prisoner's Dilemma trap. The result then is that teaching the PD game as a particularly useful exemplar of noncooperative game theory and using it to model situations throughout markets, governance, and international relations, and to generate blueprints for institutions and practices, will shape the social world in accordance with the tacit and limited assumptions required to operationalize the model.[131]

Thus, in teaching the Prisoner's Dilemma, I offer the following four recommendations.[132] These explanatory strategies make explicit the otherwise implicit assumptions packaged into standard practicable game theory. Once these suppositions, which are necessary to becoming trapped in the Prisoner's Dilemma impasse in the first place, are rendered explicit, initiates may subject

---

[129] In this informative RAND report, Jack L. Snyder analyzes the impact of the US nuclear strategy of flexible response in view of its likely reception by the USSR, *The Soviet Strategic Culture: Implications for Limited Nuclear Options* (RAND Corporation, September, 1977).

[130] These examples are taken from Hargreaves Heap and Varoufakis, *Game Theory*, 2004, 175–180.

[131] In the essay on the PD, Mary Morgan concludes that the use of the PD model has become sufficiently routinized that analysts treat the model as though it were existence itself, "The Curious Case of the Prisoner's Dilemma: Model Situation? Exemplary Narrative?" in *Science without Laws* (Duke University Press, 2007), 157–188.

[132] The Prisoner's Dilemma, of course, is a pertinent and central game in noncooperative game theory; see, e.g., Luce and Raiffa, *Games and Decisions*, 1957, 88–113.

them to inspection and can opt into or out of them at will. Hence, individuals may select to become neoliberal citizens and consumers voluntarily, instead of unknowingly succumbing to or being pressured into the mentality of *Homo strategicus*.

First, educators should clarify that orthodox game theoretic payoffs only reflect outcomes and exclude the means by which outcomes are realized. Thus, it is consistent with utilitarian philosophy but distinct from Jeremy Bentham's original approach because it denies interpersonal comparability of utility or the rationality of joint instead of individualistic maximization. By itself, this restriction on the evaluation of worth negates ethical and normative characteristics of action correlating to classical liberals' first principle of mutual respect and reciprocal no-harm, whether in the form of Adam Smith's negative virtue, Immanuel Kant's perfect duty, Robert Nozick's side constraints, or John Rawls's fair play.[133] It is then self-evident that if strategic rationality is limited to its current form, neoliberal institutions built on its premise will necessarily break with the modern approach to markets and justice. The urgency associated with the nuclear security dilemma, inviting worst-case analysis and emphasizing the raw power of resources existing prior to communication, provided the precedent for discarding classical liberalism first in international relations and subsequently in the social contract, markets, and democracy.[134] Game theorists' newfound prestige and the momentum propelling strategic rationality forward shielded the nascent paradigm of neoliberal political economy from scrutiny.[135]

Second, most operationalized game theory also relies on an expected utility metric that represents not only an ontologically salient feature of the decision environment, but also a scarce, commonly sought after, interpersonally

---

[133] This is a point that Amartya Sen has emphasized as a heterodox amendment to rational choice theory; see *Rationality as Freedom*, 175–181, 232–239; for another approach, see Joseph Heath, *Following the Rules*, 2011; and Margaret Gilbert, *Joint Commitment: How We Make the Social World* (Oxford: Oxford University Press, 2013).

[134] Hargreaves Heap and Varoufakis make a point of noting that Roger Myerson has argued that "game theory makes pessimistic assumptions regarding the nature of rationality because its role is to study the sort of social institutions that might work well … even when peopled by instrumentally rational egoists," *Game Theory*, 2004, 184 (uncited), and Myerson makes a point of referring to the nuclear deterrence on the first page of his *Game Theory*, 1991. However game theory is prescriptive, and individuals exposed to it will learn its rules of conduct, see, e.g., Professor Peter Nonacs, UCLA, "Why I Let My Students Cheat on Their Exam: Teaching Game Theory Is Good. Making People Live It Is Even Better," www.zocalopublicsquare.org/2013/04/15/why-i-let-my-students-cheat-on-the-final/ideas/nexus/, accessed January 6, 2015.

[135] John Rawls was one of the most adept theorists to span classical liberalism and the new rational choice liberalism, and it soon became clear that a commitment to both approaches was difficult, if not impossible, to entertain. See his "Philosophy and Public Affairs," 1985, 14:3, 223–251, and commentary in S. M. Amadae, *Rationalizing Capitalist Democracy* (Chicago: University of Chicago Press, 2003). For other such acknowledgments, see David Gauthier, *Morals by Agreement* (Oxford: Clarendon, 1987); Martin Hollis, *Trust within Reason* (Cambridge: Cambridge University Press, 1998).

transferable feature such as nutritional calories, energy, or cash value. This interpersonally transferable utility is surmised to have value prior to establishing intersubjective agreement on the social significance of the decision context. Thus, game theory, promoted as an exhaustive science of choice, ends up eviscerating from intelligible meaning all but affine transformations of some intersubjectively evident ontological property subject to the laws of physics. This ignores the creation of positive-sum value that can arise from complex, norm-governed patterns of social interdependence, not to mention well-being associated with potentially unlimited sources of value such as hope, healing, reconciliation and understanding.

Third, most game theorists assume that individuals must maximize payoffs individually, in strategic competition with others. Given the excision of moral scruples and accountability, because of the superfluity of processes to the judgment of rewards under the rules of standard game theory, a population of strategic rational actors resembles dueling Maxwellian Demons, both striving to accumulate as much finite and vanishing utility as possible on their side of a partition.[136] These neoliberal subjects cannot realize common goals of achieving a globally vibrant and sustainable world conducive to all individuals attaining the basic goods represented by Abraham Maslow's hierarchy of needs, and generating inclusive cultural wealth.[137] A classical liberal or post-neoliberal agent may accept that some decision problems are indeed defined strictly by scarce, ontologically prior resources. However, classical liberal or post-neoliberal actors may elect to collaborate against natural scarcity and maximize resources as a group, rather than against one another in an incessant, mutually undermining contest.[138]

Fourth, not only does traditional political economy depend on the no-harm principle, but it also recognizes a role for the imperfect duties of charity and beneficence.[139] For classical liberals, acceptance of the moral obligation of the better-off to ensure that the less well-off are not pushed to the brink of ruin is a touchstone of personal independence and autonomy. In a neoliberal world order, in which strategic rationality has the pedigree of reason, actors presume the prerogative to cannibalize others' life expectancies and qualities of life as an

[136] In axiomatizing quantum thermodynamics before axiomatizing strategic rationality, John von Neumann used the same symbolic designation for both the expected energy of quantum particles and for individuals expected utility, "EU"; see his *Mathematical Foundations of Quantum Mechanics* (Princeton, NJ: Princeton University Press, 1996).

[137] Abraham H. Maslow, "A Theory of Human Motivation," *Psychological Review* (1943) 50:4, 370–396; George Simmel, *The Philosophy of Money* (London: Routledge Classics, 2011), 291.

[138] Bacharach, *Beyond Individual Choice*, 2006.

[139] Adam Smith, *Theory of Moral Sentiments* (Indianapolis: Liberty Fund, 1982), VI.conl.2; for discussion, see Amadae, *Rationalizing Capitalist Democracy*, 2003, 215–216; see also Peter Singer, "Famine, Affluence, and Morality," *Philosophy and Public Affairs* (1972) 1:1, 229–243, to see how the classical liberal approach to justice and political economy lingers into the neoliberal era.

external cost to their own success.[140] Additionally, neoliberal strategic rational actors will only conduct charity as recommended by Richard Dawkins's selfish gene theory: to secure their immortality through conspicuous and memorable acts of generosity.[141]

Neoliberal subjectivity arises from the intricate pedagogy of game theory that comes to the fore in the Prisoner's Dilemma game and is interchangeable with contemporary paradigmatic instrumental rationality.[142] Rational choice is promoted as an exhaustive science of decision making, but only by smuggling in a characteristic confusion suggesting that everything of value to agents can be reflected in their appraisals of existential worth even though this is patently not the case in life viewed as a "fixed game."[143] Without a critical and scrupulous pedagogy that carefully identifies as optional the assumptions necessary to operationalize strategic rationality, a new neoliberal understanding of capitalism will dominate the worldview of the student of game theory and inhabitant of neoliberal institutions. This reductionist perspective on agency first proved itself useful for projecting the power of national sovereignty through wielding deterrent threats of destroying other nations using nuclear weapons. Here are the barebones central elements of this worldview:

1.   It entails coercive bargaining: threatening harm on others whether they cooperate or not, instead of bargaining consistent with the no-harm principle.

2.   It entails the inadvertent commodification of all value, and by considering that all goods of value are ultimately scarce fungible resources, it thereby negates the possibility of positive-sum and unlimited experiential goods including security, social capital, and friendship.

3.   It entails the view that only sanctions keep people in line with agreements they voluntarily make or laws they view as reasonable.

4.   It implies that cheating and free riding, if one can get away with them, are rational.

5.   It implies that information and language are purely signaling devices deployed to realize preferences over world states with value independent of social relations.

6.   It implies, finally, that agents must comply with this neoliberal view because the price for resisting is either bankruptcy or the failure to survive.

---

[140] Hausman and McPherson begin their book *Economic Analysis, Moral Philosophy and Public Policy*, 2006, by showing how Lawrence Summers's articulation of the principles underlying contemporary economic theory permit, even necessitate, negatively impacting the qualities and quantities of life of less well-off individuals, at 12–13; this is also true for the Kaldor-Hicks compensation principle widely used in contemporary law and economics, discussed in Chapter 8, "Consent."

[141] Richard Dawkins, *The God Delusion* (New York: Houghton Mifflin Harcourt, 2011), 241–267.

[142] See, e.g., Hargreaves Heap and Varoufakis, *Game Theory*, 2004, 8–12.

[143] Game theorists tend to refer to all life contexts as independent games or one supergame, e.g., see Binmore, *Natural Justice*, 2005, 184, who also considers that "the game of life is the infinitely repeated Prisoners' Dilemma," at 96.

PART I

WAR

Introduction

It is not usually supposed that intellectuals possess significant political or social power. However, the circumstances linking the atomic bomb and the American state have given specific academics and intellectuals more power than is usually recognized. The bomb was an intellectual project. Its very feasibility was suggested to Roosevelt by Einstein as a counter to a possible Nazi bomb. Subsequently, a scientist became head of the Manhattan Project and Oppenheimer and others then attained enormous power through the Atomic Energy Commission. But what is even more unusual is that social scientists acquired significant power within the US state ... Many of these were European *émigrés* whose experience of totalitarian regimes in Europe gave them strong reasons to support the ideology of the US style liberalism. The new global responsibilities of the US and the security dilemmas created by the bomb greatly increased the intellectual work needed in the area of policy advice and guidance ...

There was a need for a mode of discourse which removed such anxiety from nuclear speculation. Anxieties about nuclear security are not just a matter for the general population; they also pervade political elites ... The nuclear state in the United States did secure a consensus on nuclear security, and a key element in the articulation of the nuclear security consensus has been the discourse of the civilian intellectuals who fashioned a cohesive ideology of nuclear rationality ... Hegemony was attained by a liberal/technocratic discourse creating a fusion of "reasonableness" with scientism and instrumental rationality.

Philip K. Lawrence, 1996[1]

Neoliberal researchers strive to place the institutions of free markets and democratic governance on a firm foundation based on rigorous analysis of human behavior. However, these theorists, who employ game theory to defend a social order reminiscent of classical liberalism, make two concessions that cannot be reconciled with the classical liberal worldview. First, game theorists propose that standard strategic rationality captures the only available logic of action for

---

[1] Philip K. Lawrence, "Strategy, Hegemony and Ideology: The Role of Intellectuals," *Political Studies* XLIV (1996), 44–59, at 44, and 45–47.

explaining behavior.[2] And second, over the past half-century, they have argued that the security dilemma, social contract, arms race, and collective action have the structure of a Prisoner's Dilemma (PD) game. It may appear that it would be possible to maintain the first assumption but reject the second, thereby possibly rescuing classical liberalism's mutually beneficial free market system regulated by appropriate governance. However, the PD game represents the central paradox within game theory: that strategic self-interest demonstrably does not result in mutual benefit, as Adam Smith had proposed.[3] Thus, accepting strategic rationality as one's sole understanding of coherent choice and action will necessarily make the Prisoner's Dilemma the focal social challenge to resolve.[4]

Strategic rationality condemns the rationality of promising and voluntarily abiding by consensual agreements and recommends instead that subjects influence each other's actions by applying incentives. Actors who perceive themselves to be limited to strategic rationality and caught in a Prisoner's Dilemma payoff scheme pursue their interests by displacing costs onto others. In the attempt to avoid the mutually destructive outcome, deterring the other from defecting will depend on the issuance of credible threats rather than the assurance of cooperative intention. Reassuring the other of one's intent to cooperate fails because it contradicts one's preference for dominance, which in the PD is reflected as seeking foremost to defect while the other cooperates. Instead of contributing to a mutually constituted normative order, such as is evident in the modern traditions of capitalist contracting and international maritime law, rational actors must take self-help in an anarchic world to an extreme, pursuing their self-interest at the expense of others.[5] When cooperation does emerge,

[2] This is consistent with John von Neumann and Oskar Morgenstern's initial statement in *Theory of Games and Economic Behavior* (Princeton, NJ: Princeton University Press, 1944), 31.

[3] This challenge is the topic throughout Richmond Campbell and Lanning Sowden, *Paradoxes of Rationality and Cooperation* (Vancouver: University of British Columbia Press, 1985).

[4] Game theory texts tend to progress from two-person zero-sum to two-person non-zero-sum games, which immediately lead to the PD and repeating PD game, and then bargaining, evident in the tables of contents in Howard Luce and Duncan Raiffa, *Theory of Games and Decisions* (New York: John Wiley, 1958), xiv–xv; and Roger Myerson, *Game Theory* (Cambridge, MA: Harvard University Press, 1991), viii–ix; Shaun Hargreaves Heap and Yanis Varoufakis, *Game Theory*, 2nd ed. (London: Routledge, 2004), move from bargaining to the PD and repeating PD, vi–vii.

[5] Joseph S. Nye Jr. and Robert O. Keohane argue that neoliberal institutionalism can provide explanations helping to sustain a normative order. See Robert Axelrod and Robert O. Keohane, "Achieving Cooperation under Anarchy: Strategies and Institutions," in *Neorealism and Neoliberalism*, ed. by David A. Baldwin (New York: Columbia University Press, 1993), 85–115; and Joseph S. Nye Jr., *Soft Power* (New York: Public Affairs, 2005). Nye's concept of "soft power" remains ambiguous, suggesting reason-giving norms and action in accordance with commitment, but it remains unclear whether in the final analysis soft power is put forward as an alternative to strategic rationality or a more subtle manifestation of its expression of purposive agency. *Prisoners of Reason* argues that whereas classical liberals accept Immanuel Kant's asocial sociability, those who accept noncooperative game theory as their sole coherent standard for rational choice are bound to inadvertently embrace pure "asociability." On Kant's asocial sociability, see Michael W. Doyle, "Kant, Liberal Legacies, and Foreign Affairs," *Philosophy and Public Affairs* (1983) 12:3, 205–235 at 228.

according to this analysis, it can only be the result of punitive institutions or individual policing of others' conduct through endogenous Tit for Tat punishment if the Prisoner's Dilemma scenario is indefinitely and perfectly repeating.

Chapter 3, "Assurance," and Chapter 4, "Deterrence," trace the nuclear policy debate between two strategic plans for maintaining US military security. Mutual assured destruction (MAD) attempted to keep the United States in a posture of reciprocal deterrence reminiscent of classical liberalism. Breaking with this tradition, nuclear utilization targeting selection (NUTS) placed the US deterrent footing on the pursuit of escalation dominance and strategic bargaining. Thomas Schelling pioneered modeling the security dilemma, which he proposed was populated by hopeful cooperators, as a Prisoner's Dilemma game. This seemed to be a good choice to win over military hard-liners to his position of limiting nuclear warheads to undetectable submarines that would withstand a first strike and thereby maintain a secure deterrent second-strike posture with minimum deterrent power. Chapter 3 concludes that Schelling's neoliberal defense of MAD should have won the nuclear security debate. Assured destruction seemed to mimic the classical liberal assurance of the intention to cooperate after the other party had already done so, by threatening to retaliate if the other attacks.

Chapter 4 recounts how Schelling's defense of MAD lost the nuclear security debate during the 1970s, both as an argument waged in the form of rational deterrence theory and as a policy during Jimmy Carter's presidency. Prior to Carter's administration, as Secretary of Defense under President Richard Nixon, James R. Schlesinger had already compromised MAD's counter-city targeting scheme pursing flexible response and treating nuclear warheads as conventional weapons without prohibitions for first use. Carter additionally adopted the NUTS pursuit of escalation dominance and the capability to prevail in protracted nuclear war by integrating the offensive MX missile into the US war plan. Thus, instead of offering the assurance of recognizing the insurmountable futility of fighting thermonuclear war, Carter sought to supersede the existential condition of mutual assured destruction through developing the capability and credibility of issuing nuclear threats.

As a set of purely intellectual arguments, MAD lost the nuclear security debate to NUTS because strategic rationality cannot change course from its initial assumptions: individualistic competition despite others; single criterion valuation of outcomes, typically in terms of a scarce tangible resource; and a consequentialist framework that negates ethical conduct, the logic of appropriateness, or commitment. Furthermore, Schlesinger, a seasoned US Department of Defense official and long-term proponent of flexible response, played a formative role on Carter's national security team. Schlesinger advocated US hegemony among allies and escalation dominance over opponents. Thus, MAD was outflanked logically and pragmatically.

Part I offers the following insights. First, rational deterrence theory is the same as rational decision theory, which was initially developed as the science of military decision making. Although this coincidence may appear to be mere happenstance, in fact noncooperative game theory represents a relentlessly combative decision technology. It proved its usefulness in providing a means to project US nuclearized military force in a rational manner with the added benefit of blurring the distinction between human actors and artificial intelligence that could execute the sovereign's commandments throughout the mayhem and confusion of atomic war. Second, *Prisoners of Reason* argues that the development of neoliberal governance and markets is inseparable from game theory's rise to become the established methodology in social science. This late 1970s development followed two decades of entanglement of US strategic policy with rational deterrence theory. In the 1960s, Thomas Schelling offered what was essentially a neoliberal defense of MAD and simultaneously established the precedent of using the Prisoner's Dilemma game to model the security dilemma underlying international relations anarchy and the social contract. In the 1970s, rational deterrence theory failed to marshal an argument for MAD because strategic rationality did not permit offering assurances of cooperation as a means to solve the nuclear security dilemma modeled with the PD. Instead, the only logical solution was to credibly signal the awesome intent and capability of fighting nuclear war, and to maintain escalation dominance over every level of military engagement.

President Jimmy Carter's endorsement of the nuclear war-fighting posture demarcates both the preeminence of strategic rationality and the US embrace of military ascendance as the primary means to maintain security. Carter's presidency is characterized by an illiberal capitulation from his early commitment to win the Cold War by enticing the East to recognize the superiority of free markets, democratic governance, and Western culture, and maintain a secure second strike nuclear capability consistent with assured destruction. We can best understand the widespread acceptance of strategic rationality, and its default fixation on the Prisoner's Dilemma imbroglio, as the noble attempt to maintain classical liberal values in the late-modern era. Proponents of game theory, from Thomas Schelling and Robert Axelrod to Ken Binmore, strove to secure the foundations of a familiar classical liberal social order on what they perceived to be the more secure building materials of strategic rationality.

However, as Part II argues, the embrace of strategic rationality as the exemplary subjectivity of citizens and consumers, and the superior means to model not only security but also markets, governance, and collective action, yields a neoliberal order. This world order, like NUTS' ascendancy over MAD, is one in which no actor can ever exit the state of nature. The NUTS' protocol of escalation dominance and coercive bargaining consistent with noncooperative game theory appears to be required for survival and success and therefore displaces alternative logics from action from actors' range of possibilities.

3

Assurance

We may now combine our analysis of PDs [Prisoner's Dilemmas] and commitment devices in discussion of the application that first made game theory famous outside of the academic community. The nuclear stand-off between the superpowers during the Cold War was exhaustively studied by the first generation of game theorists, many of whom worked for the US military ... Both the USA and the USSR maintained the following policy. If one side launched a first strike, the other threatened to answer with a devastating counter-strike. This pair of reciprocal strategies, which by the late 1960s would effectively have meant blowing up the world, was known as 'Mutually Assured Destruction', or 'MAD'. Game theorists objected that MAD was mad, because it set up a Prisoner's Dilemma as a result of the fact that the reciprocal threats were incredible.

Don Ross, 2014[1]

Some writings on international cooperation have applied game theory – particularly the Prisoner's Dilemma game – to security issues to identify the conditions under which cooperation is likely to emerge ... [S]ecurity dilemmas ... are often modeled as single play Prisoner's Dilemma.

Joseph S. Nye Jr. and Sean M. Lynne-Jones, 1988[2]

In the late 1940s and 1950s, defense intellectuals developed game theory to anchor military strategy in scientific analysis. The last chapter discussed how game theorists developed strategic rationality as a comprehensive science of decision making, which entailed plausibly identifying "a set of rules for each participant which tell him to behave in every situation which may conceivably arise," and which was immediately developed for application to military and nuclear strategy.[3] Analysts

---

[1] Don Ross, "Game Theory," *Stanford Encyclopedia of Philosophy*, 2014.
[2] Joseph S. Nye Jr. and Sean M. Lynne-Jones, "International Security Studies," *International Security* (1988) 12:4, 5–27, at 19.
[3] John von Neumann and Oskar Morgenstern, *Theory of Games and Economic Behavior* (Princeton, NJ: Princeton University Press, 1944, 1947, 1953), 31.

perceived that military strategy could not be contained by zero-sum game theory and John von Neumann's minimax concept of securing the least bad outcome for oneself. In the Prisoner's Dilemma model of bargaining, John Mayberry discovered that threatening the adversary with the worst outcome that one could credibly muster was the superior approach.

In 1960, Thomas Schelling used a Prisoner's Dilemma model to defend the nuclear strategy of mutual assured destruction (MAD). The Prisoner's Dilemma model of the nuclear security impasse implies that one actor's self-defense necessarily compromises the other's security because each has the first preference of achieving supremacy through preferring the outcome least acceptable to the opposition. Schelling's hope was that in addressing the worst-case scenario for achieving security, MAD would provide a blueprint for achieving peaceful coexistence in a nuclear age. Schelling's allegiance to strategic rationality and his attendant acceptance that the pursuit of mutual security was best represented by a Prisoner's Dilemma game made his approach to international security and economics consistent with the neoliberal school of international relations theory as well as with neoliberal theory as generally understood. In addressing the worst-case scenario and limiting himself to game theory, Schelling sought to provide a strong foundation for rescuing a classical liberal world order rooted in reciprocal security and mutual prosperity.

However, the PD model of security departs from classical liberalism in suggesting that the pursuit of security is an antagonistic exercise pitting competitors against each other. Schelling's approach to security set a precedent in international relations as well as in political economy and civil society.[4] The Prisoner's Dilemma model seemed germane to reflect both the ultimate unknowability of another actor's intentions and actors' instrumental imperative to compete for scarce resources required to realize goals.[5] However, Schelling's double concession – first to the overarching relevance of strategic rationality, and second to the PD model of the security dilemma – made it impossible for him to salvage classical liberalism. Thus, as the pages ahead discuss, MAD ultimately capitulated to pro-nuclear-use theory, and a strategy of offering the assurance of cooperative intention yielded to one of deterring predators by adopting their stance. Thus, during the 1970s, deterrence would replace classical liberal assurance, incentives would replace reasons and legitimacy, and coercive threats would replace shared normative expectations.

Although consequential and intensely riveting, the hard-fought nuclear security debate between MAD proponents and nuclear utilization targeting selection

---

[4] For Schelling's treatment of Jean Jacques Rousseau's Stag Hunt in terms of a Prisoner's Dilemma model, see his "Strategy, Tactics, and non-Zero-Sum Theory," in *Theory of Games: Techniques and Applications*, ed. by A. Mensch (New York: Twentieth Century Fund, 1966), 469–480 at 475.

[5] Recall that economic science has traditionally analyzed the production and distribution of scarce goods that enable purposive actors to realize goals. See, e.g., Lionel Robbins, *Nature and Significance of Economic Science* (London: MacMillan, 1932), 12–20, 23–27, and 45–47.

(NUTS) enthusiasts is barely known to anyone aside from nuclear security experts. Even though MAD guided US nuclear strategy through the end of Robert S. McNamara's tenure as Secretary of Defense, MAD increasingly lost the nuclear security debate to NUTS from the late 1960s until the Carter administration's 1980 adoption of countervailing strategy formalized in Presidential Directive 59. Whereas MAD relied on bilateral deterrence, NUTS promoted asymmetric deterrence, escalation dominance, coercive bargaining, and hegemony. NUTS is predicated on sustaining the intent and capability to fight and win nuclear war among superpowers.[6] Rather than promoting the assurance of reciprocal no-harm, NUTS maintains asymmetric leverage of the power to harm at all levels of conflict as the basis for achieving US security independently from that of its superpower rivals. The intellectual ascendancy of NUTS over MAD logically followed from Schelling's double admission of the exhaustive reach of game theory and its implicated Prisoner's Dilemma model of the security dilemma.

In hindsight, the loss of MAD to NUTS in intellectual debate seems puzzling. As international relations theorist Robert Jervis insists, MAD is a stubborn, practical fact, whereas NUTS relies on the fantasy that it is possible to win an all-out nuclear conflict.[7] Many nuclear strategists concede that stockpiling more nuclear warheads than those already sufficient to destroy civilization to demonstrate the intent and ability to prevail in nuclear war is ultimately counterproductive.[8] Even theorists holding the view that international security is inherently antagonistic agree on the self-defeating implications of NUTS.[9] Not only will this offensive strategy contribute to proliferation, especially if regimes crumble and weapons are unaccounted for, but it also fails to acknowledge the potential threats that errors of judgment and technical accidents pose to all actors.[10] Nevertheless, despite the practical impossibility of being victorious in a nuclear conflict among superpowers, the aggressive NUTS strategy managed to win over even the liberal-minded Carter administration.

This triumph of NUTS over the classical liberal MAD invites thorough exploration.[11] Not only do actors fail to secure the status quo, but they further

---

[6] This strategy and its gradual ascendance over MAD is the topic of the next chapter, "Deterrence."

[7] Robert Jervis, *The Illogic of Nuclear Strategy* (Cornell University Press, 1984).

[8] Lawrence Freedman, *The Evolution of Nuclear Strategy* (New York: St. Martin's Press, 1981); and Charles Glaser, "Why Do Strategists Disagree about the Requirements of Strategic Nuclear Deterrence?" in *Nuclear Arguments*, ed. Lynn Eden and Steven E. Miller (Cornell University Press, 1989), 109–171.

[9] Repeatedly, theorists are sympathetic toward MAD and yet find that the logical weight of argument couched in terms of rational deterrence theory yields the ground of argument to NUTS over MAD, see, e.g., Glaser, "Why Do Strategists Disagree," 1989.

[10] See, e.g., Eric Schlosser, *Command and Control: Nuclear Weapons, the Damascus Accident, and the Illusion of Safety* (London: Penguin, 2014).

[11] One of the earlier investigators of this question, who also happens to reach a similar conclusion to that offered here, is Douglas P. Lackey, "The American Debate on Nuclear Weapons Policy: A

devote precious resources to pursing an arms race by engaging in the ceaseless and costly development of armaments. This chapter explores the implications of Schelling's dual concession to the PD game and to strategic rationality and, more broadly, seeks to understand why this framework was adopted in the first place.

The chapter is divided into five sections and a conclusion. The first two, "A Brief History of Game Theory" and "Rational Deterrence and Game Theory's Ascendance," recount the development and early reception of game theory. Examining the original applications of orthodox game theory, which subsequently has been accepted as humanity's most authoritative statement of rationality, helps us understand the current role rational choice theory plays throughout the social sciences and professional programs. Game theory was devised to provide a comprehensive theory of all multi-person decision making involving competition over scarce resources, independent of empirical examples. Rational deterrence theory is equivalent to rational decision theory or rational choice theory and can be used as a descriptive model, a normative theory of rationality, and a prescriptive policy tool.

The next two sections introduce Herman Kahn's and Thomas Schelling's strategic postures of NUTS and MAD, respectively. These two nuclear strategies, only two of four possible positions (the others being nuclear disarmament and Dwight D. Eisenhower's massive retaliation), predominated over nuclear policy debates throughout the 1960s and 1970s. These two sections present the MAD vs. NUTS security debate in light of their original expositors, who were both luminaries in the field of nuclear deterrence.[12] The fifth section discusses US nuclear defense policy from the beginning to the end of Robert S. McNamara's tenure as Secretary of Defense. This section makes evident that McNamara developed the policy of mutual assured destruction to maintain the US's deterrence posture in view of the Soviet's achievement of destructive parity.

Review of the Literature, 1945–1985," *Analyse and Kritik* (1987) 9, 7–46; John Mueller raises the question of why the constant obsession with security at a cost that far outpaces returns; see *Atomic Obsession: Nuclear Alarmism from Hiroshima to Al-Qaeda* (New York: Oxford University Press, 2009). Cold War revisionists understand "American policy not as a response to an expansionist Soviet Union but as growing out of internally generated forces, with Soviet policy as a reaction to American intelligence" (104). This revisionist position leaves room to grant to Mikhail Gorbachev a similar acknowledgment on the part of the USSR, accepting that "the belligerent Soviet security policies had been responsible for the Cold War and that an unambiguously defensive posture could increase Soviet security" (fn 8, p. 122). Furthermore, along these lines of communicating the recognition of the hopelessness of waging nuclear war, President Ronald Reagan "was ready to sign an agreement to abolish nuclear weapons" (p. 115). All three quotes are from Robert Jervis, "Security Studies: Ideas, Policy, and Politics," in *The Evolution of Political Knowledge: Democracy, Autonomy, and Conflict in Comparative and International Politics*, ed. Edward D. Mansfield and Richard Sisson (Columbus, OH: Ohio State University Press, 2004), 100–126.

[12] Patrick Morgan, "New Directions in Deterrence Theory," in *Nuclear Weapons and the Future of Humanity*, ed. Avner Cohen and Steven Lee (Totowa, NJ: Rowman and Allanheld, 1986).

The "Conclusion" acknowledges Schelling's legacy in theoretical analyses of the nuclear security dilemma and in discussions on security dilemma scenarios throughout international relations and civil politics. Schelling's approach is driven by the analytic structure of game theory and his objective of countering the worst-case scenario. In his application of the Prisoner's Dilemma model to both nuclear security and social contract theory, Schelling unhesitatingly accepts that hopeful cooperators must adopt the first preference of taking advantage of all other actors. He ignores the option already available in classical liberalism of offering assurance of cooperation backed by the threat of sanctions. Both MAD's focus on the mutual assurance of destruction and NUTS's asymmetric threats of deterrence, escalation dominance, and coercive bargaining depart from classical liberalism.

BRIEF HISTORY OF GAME THEORY

In the mid-twentieth century, Western civilization faced a triple crisis. Enlightenment philosophy was implicated in the rise of National Socialism and the failure to quarantine its potently destructive elements.[13] Western institutions of democracy and capitalism were brought to their knees by economic depression, fascism, totalitarianism, and communism.[14] For four grueling years, the Allied powers promoting what was left of Enlightenment liberal values faced a military juggernaut with global dimensions. Enlightenment reason confronted enemies from without and nightmares from within. World War II closed with America's cataclysmic detonation of the West's greatest scientific invention on ostensibly innocent Japanese civilians.[15]

Game theory and its corporeal creators were at the center of the unfolding drama, and none more so than the theory's brainy powerhouse John von Neumann.[16] A Hungarian Jew by birth and a mathematician by profession, von Neumann generated an axiomatized treatment of quantum mechanics in the 1930s. He reacted to anti-Semitic policies in the 1940s and simultaneously devised the mathematics of game theory. He was as aware as any of the potential emptiness of political promises and the possible ease with which self-evident rights of personhood and property could be rescinded.[17] Von Neumann was central to the Manhattan Project and served on the select committee that

---

[13] Max Horkheimer and Theodore W. Adorno, *Dialectic of Enlightenment*, trans. John Cumming (New York: Continuum Press, 1969). For discussion of the relationship between modernity and German National Socialism, see Jeffrey Herf, *Reactionary Modernism: Technology, Culture and Politics in Weimar and the Third Reich* (Cambridge: Cambridge University Press, 1986).

[14] This theme is central to S. M. Amadae, *Rationalizing Capitalist Democracy* (Chicago: University of Chicago Press, 2003).

[15] See John Gaddis for a comprehensive history of the Cold War followed hard on the heels of World War II; *The Cold War: A New History* (London: Penguin, 2006).

[16] Robert Leonard, *Von Neumann, Morgenstern, and the Creation of Game Theory: From Chess to Social Science* (Cambridge: Cambridge University Press, 2010).

[17] See Leonard, "Mathematics and Social Order," in ibid., 185–223.

determined the target cities of Hiroshima and Nagasaki.[18] In 1944, von Neumann coauthored *Theory of Games and Economic Behavior* at the Princeton University Institute for Advanced Studies with the Austrian economist Oskar Morgenstern.

Game theory was the product of the most exalted mathematical minds of pre-WWII Europe. It grew amid the intense ferment over the philosophy of mathematics and its relationship to social theory.[19] At the same time that the physics revolutions of relativity and quantum mechanics arrested determinism and reveries about Newtonian space-time, mathematicians wrestled with the source of certainty anchoring their discipline. Although Bertrand Russell and David Hilbert sought foundations for mathematical knowledge, it was Kurt Gödel's theorems on completeness and consistency that ended mathematicians' quest for certitude.[20] The search for positive knowledge and certain foundations proved a Holy Grail, even in the realm of pure math.[21] Game theory grew out of Morgenstern's dearly held opinion that the social sciences, especially economics, suffered from an abysmal deficit of rigor because of their lack of mathematical formalism and their oversight of strategic interdependence.[22]

Rather than being just a body of mathematical expertise, game theory captured the mentality of competitive dueling exercised in parlor games and military combat. It was, in fact, Europe's chess-playing culture, in which champions displayed their intellectual supremacy not just over losers but also over the general public that provided the basis for von Neumann's first academic paper on game theory. Masters played multiple concurrent games, blindfolded, against sighted opponents. The early nineteenth-century chess master and mathematician Emanuel Lasker viewed chess as exemplary of struggle and the science of strategy. The "perfect strategist" is free from emotion and will "calculate in advance ... the optimal path ... of the upcoming struggle."[23]

Von Neumann was also a chess player. His treatment is general, being applicable to all definable "games." He likens his analysis to all manner of circumstances because "there is hardly a situation in daily life into which this

---

[18] Discussed in ibid., see also Richard Rhodes, *Making of the Atomic Bomb* (New York: Touchstone, 1986).

[19] See Leonard, *von Neumann*, 2010, especially his "Ethics and the Excluded Middle" chapter, 110–139.

[20] Kurt Gödel, "Über formal unentscheidbare Sätze der Principia Mathematica und verwandter Systeme, I," *Monatshefte für Mathematik und Physik* (1931), 38, 173–198.

[21] For a detailed discussion of the relationship between the hope for and abandonment of formal and causal determinism in Germany prior to World War II, see Paul Foreman, "Weimar Culture, Causality, and Quantum Theory, 1918–1927: Adaptation of German Physicists and Mathematicians to a Hostile Intellectual Environment," *Historical Studies of the Physical Sciences* (1971), 3, 1–115. Apparently, von Neumann was not overly troubled by the collapse of the Hilbert program to axiomatize the foundations of mathematics.

[22] Robert J. Leonard, "Creating a Context for Game Theory," in *Toward a History of Game Theory*, ed. by E. Roy Weintraub (Duke University Press, 1993), 29–76.

[23] Leonard, quoting Lasker, *von Neumann*, 2010, at 14.

problem does not enter."[24] He explains, "It is easy to picture the forces strug-
gling with each other in such a two-person game. The value of g(x,y) [person 1's
(S_1's) winnings] is being tugged at from two sides, by S_1 who wants to maximize
it, and by S_2 who wants to minimize it. S_1 controls the variable x, S_2 controls
variable y."[25] In this zero-sum context, von Neumann demonstrates that each
player can select a strategy that will secure a minimum security threshold below
which the other player cannot force him, an early expression of his "minimax"
concept.

Whereas economists for almost three decades shunned game theory because
of its recondite mathematics and zero-sum focus, the theory found fertile
ground at the RAND Corporation, the military think tank where John von
Neumann was a frequent consultant in the late 1940s.[26] Game theory was
applied to strategic problems as early as World War II. Von Neumann himself
supplied the reasoning of his minimax theory to Philip Morse's analysis of
submarine detection in the English Channel.[27] Abraham Wald applied von
Neumann's minimax theorem to statistical decision theory to develop a generalized approach to games with a continua of strategies.[28] Merrill Flood, one of the
originators of the Prisoner's Dilemma game along with Melvin Dresher, applied
game theory to airborne attacks, authoring "Aerial Bombing Tactics: General
Considerations" in 1944.[29] Game theory's initial application to warfare shows
how well suited it is to analysis of strategic competition.

To apply game theory effectively, it was necessary to devise a metric of value
to stand in as inter-agentively valid on par with money in an exchange economy.
Flood's investigation of this problem in his 1944 report anticipates the work
done to appraise military worth at RAND during the early Cold War to pursue
the science of strategy.[30] At an early RAND meeting in 1948, von Neumann was
present at a discussion of fighter aircraft dueling to which his minimax theorem
was pertinent.[31] Leonard notes, "Insofar as RAND was created to perpetuate

---

[24] Quoted from Leonard, *von Neumann*, 2010, at 62, from John von Neumann, "Zur Theorie der Gesellschaftsspiele," *Mathematische Annalen* (1928) 100, 295–320, trans by S. Bargmann as "On the Theory of Games of Strategy," in A. W. Tucker and R. D. Luce, eds., *Contributions to Game Theory*, vol. 4 (Princeton, NJ: Princeton University Press, 1958), 13–42.

[25] Von Neumann, "Zur Theorie der Gesellschaftsspiele," 1928, at 21.

[26] Leonard, "Creating a Context for Game Theory," 1993.

[27] Graphically discussed by Leonard, *von Neumann*, 2010, 270–275.

[28] Abraham Wald used von Neumann's 1928 "Zur Theorie der Gesellschaftsspiele" in his two papers "Statistical Functions Which Minimize the Maximum Risk," *Annals of Mathematics* (1945) 46, 265–280; and "Generalization of a Theorem by v. Neumann Concerning Zero Sum Two Person Games," *Annals of Mathematics* (1945) 46:2, 281–286, discussed by Leonard, *von Neumann*, 2010, 279–280.

[29] See Leonard, *von Neumann*, 2010, 281.

[30] See ibid., 282–283, 303.

[31] Ibid., 293–294, fn 2. Leonard notes that "at RAND, beginning in 1948, month-long summer gatherings were held for discussion of game theory, utility theory, linear programming, decision theory, and related subjects," and this resulted in H. F. Bohnenblust et al., "Mathematical Theory of Zero-Sum Two-Person Games with a Finite Number or a Continuum of Strategies"

operations research work of the kind applied during World War II, game theory was part of its *raison d'etre*."[32]

Von Neumann's intellectual presence towered over RAND's researchers.[33] Game theory was applied to aerial attacks and nuclear strikes. Ed Paxson addressed "Games of Tactics." Richard Bellman analyzed "A Bomber-Fighter Duel," "Application of Theory Games to Identification of Friend and Foe," and "Games Involving Bluffing."[34] The list of such papers stands as testimony to the ready application of game theory to warfare: "Local Defense of Targets of Equal Value," "Optimal Tactics in a Multistrike Air Campaign," "Optimal Timing in Missile Launching: A Game Theoretic Analysis," and "A Game Theory Analysis of Tactical Air War." Melvin Dresher wrote the text *Games of Strategy: Theory and Applications*, which had appeared as a classified manual for Air Force staff in the decade prior to its 1963 publication.[35] Leonard observes that "at the early stages, game-theoretic models were thought likely to be useful in solving tactical military problems to be encountered in a war with the USSR."[36]

RATIONAL DETERRENCE AND GAME THEORY'S ASCENDANCE

The exceptional nature of nuclear warfare calls for a nonempirical science prior to any experience of nations sparring with weapons of mass destruction 1000 times more powerful than the US atomic bombs Fat Man and Little Boy that annihilated Hiroshima and Nagasaki. It was thus widely perceived to require the expertise of blackboard researchers specializing in purely esoteric theory. The unknown proportions of thermonuclear war opened a window for conquering uncertainty with pure intellect.[37] In the words of RAND's Herman Kahn, "There is no one with experience in the conduct of thermonuclear war," hence the need to rely on theoretical studies.[38]

(Santa Monica, CA: RAND Corporation, 1948), September 3, 1948, discussed in Leonard, *von Neumann*, 2010, 295.

[32] Leonard, *von Neumann*, 2010, at 298.

[33] Ibid., 300–301.

[34] Ibid., 310.

[35] Ibid.; Melvin Dresher, *Games of Strategy* (Upper Saddle River, NJ: Prentice Hall, 1963).

[36] Leonard, *von Neumann*, 2010, 298.

[37] Civilian analysts with no prior military experience pushed aside seasoned military officers in leadership roles; see Amadae, *Rationalizing Capitalist Democracy*, 2003, 27–85. For an engaging and illuminating analysis of the abstract yet unethical character of rational decision and deterrence theory, see Philip Green, *Deadly Logic: The Theory of Nuclear Deterrence* (Columbus: Ohio State University Press, 1966), especially 118–128 and 213–254.

[38] Herman Kahn, *On Thermonuclear War* (Princeton, NJ: Princeton University Press, 1960), 162. Robert Jervis recognizes this point, observing, "All the fundamental ideas about nuclear strategy came from civilians, and their dominance probably helps explain the break with much previous thinking. Nuclear weapons also seemed so stark as to lend themselves to abstract analysis, and the match with the recently developed game theory was very attractive, although critics argued

The story of Robert S. McNamara's ascent to the office of the US Secretary of Defense under John F. Kennedy's celebrated administration is told and retold with unremitting enthusiasm.[39] Like the civilian strategists whom McNamara appointed to work under him, the new Secretary of Defense could boast of analytic expertise, but no experience in the realm of military affairs. Hedley Bull's 1968 *World Politics* article, "Strategic Studies and Its Critics," notes the "abstract and speculative character" of "the present ... strategy thinking." He states the obvious: "There has not yet been a nuclear war, and the possibility that there will be one has not yet existed long enough for it to have become clear how the structure of international life will be affected."[40]

The direct implication of this new reality was that "the civilian experts have made great inroads" into the formerly closed military strategy war room:

They have overwhelmed the military in the quality and quantity of their contributions to the literature of the subject; no one would now think of turning to the writings of retired officers rather than to the standard academic treatments of deterrence, limited war, or arms control, for illumination of the problems of the nuclear age.[41]

Bull enunciates the displacement of military staff at war colleges by these civilian defense intellectuals and notes that "most prominently in the United States, the civilian strategists have entered the citadels of power and have prevailed over military advisers on major issues of policy."[42] Fortunately, the empirical experience with nuclear warfare did not increase. In 1977, political scientist Jack L. Snyder observed that this lack of experience continually led to an emphasis on deductive game theoretic analysis. The burden of proof thus fell on any strategist proposing an alternative to the "generic rational man" codified by game theory because "it is well known that the established strategy of deterrence in the United States is explicitly based upon the rational theory of decision."[43] Alluding to Thomas Schelling's corpus and the approach it

that this influence was malign." "Security Studies: Ideas, Policy, and Politics," in *The Evolution of Political Knowledge: Democracy, Autonomy, and Conflict in Comparative and International Politics*, ed. Edward D. Mansfield and Richard Sisson (Ohio State University Press, 2004), 100–126, at 109.

[39] David Halberstam, *The Best and the Brightest* (New York: Ballantine Books, 1993); Fred Kaplan, *Wizards of Armageddon* (Stanford University Press, 1983); Gregg F. Herken, *Counsels of War* (New York: Knopf, 1985); Deborah Shapley, *Promise and Power: The Life and Times of Robert McNamara* (Boston: Little, Brown 1993); Stephen W. Twing, *Myths, Models and US Foreign Policy: The Cultural Shaping of Three Cold Warriors* (London: Lynne Rienner, 1998), 145–186; Amadae, *Rationalizing Capitalist Democracy*, 2003, 27–82.

[40] Hedley Bull, "Strategic Studies and Its Critics," *World Politics* (1968), 20:4, 593–603, at 594–595.

[41] Ibid., 594.

[42] Ibid.

[43] Jack L. Snyder, "The Soviet Strategic Culture: Implications for Limited Nuclear Operations," RAND Report R-2154-AF, September 1977, especially 4. Second quote is John Steinbruner, "Beyond Rational Deterrence: The Struggle for a New Concept," *World Politics* (1976) 28:2 223–245 at 225.

inspired, Snyder makes clear that "the current U.S. selective-options policy is an intellectual offspring of the distinctively Western notions about deterrence, coercive diplomacy, and war."[44] Thus, in the1960s and early 1970s, the rational actor model was initially embraced in the domain of national security.

Observers acknowledged that military strategy as it was developed first at RAND and subsequently within the Pentagon evolved as a hybrid of decision and game theory, systems analysis, operations research, and computerized command structures.[45] Bernard Brodie, Herman Kahn, Morton Kaplan, Glenn Synder, Daniel Ellsberg, Walter Kaufman, Albert Wohlstetter, Henry Kissinger, Oskar Morgenstern, Malcolm Hoag, Thomas Schelling, and James R. Schlesinger were prominent defense analysts. Of these theorists, Schelling, Morgenstern, Kahn, Kaplan, Synder, Hoag, and Ellsberg drew on game theory. Discussing their work, Frank Zagare observes, "In time, the theoretical edifice they created came to be seen as the Rosetta stone of nuclear theory," and remained so into the 1990s.[46] Robert Jervis notes that this "second wave" of deterrence theory "soon became conventional wisdom."[47] Thomas Schelling would be awarded the 2005 Nobel Prize for economics for his work "enhancing our understanding of conflict and cooperation through

---

[44] Ibid., 18.

[45] See Robert Leonard, "Creating a Context for Game Theory," 1992, 29–76; Philip Mirowski, *Machine Dreams* (Cambridge: Cambridge University Press, 2002), 184–190. See also William Thomas, *Rational Action: The Sciences of Policy in Britain and America, 1940–1960* (Cambridge, MA: MIT Press, 2015). Robert Jervis observes subjects of inquiry labeled security studies were elevated in "standing, legitimacy, and [had] a claim on national resources and priorities," "Security Studies," 2004, 106; in response to this essay, Thomas Schelling adds that "Jervis states that 'seeing security studies as being created at the end of World War II is very much a simplification, but one with some validity.' I would have said, 'not very much of a simplification,'" "Academics, Decision Makers, and Security Policy during the Cold War: A Comment on Jervis," in *The Evolution of Political Knowledge: Democracy, Autonomy, and Conflict in Comparative and International Politics*, ed. Edward D. Mansfield and Richard Sisson (Columbus: Ohio State University Press, 2004), 137–139, at 137.

[46] Frank C. Zagare, "Classic Deterrence Theory: A Critical Assessment," *International Interactions: Empirical and Theoretical Research in International Relations* (1996) 21:4, 365–387, at 366, 376. Schelling flatly rejects the independent significance of game theory in the development of security studies, in his "Academics, Decision Makers, and Security Policy during the Cold War: A Comment on Jervis," in *The Evolution of Political Knowledge: Democracy, Autonomy, and Conflict in Comparative and International Politics*, 2004, 137–139, at 138–139. However, in correspondence to me, Robert Jervis notes that "this is quite contrary to what he [Schelling] told me in several conversations." Jervis adds that the core ideas of the two nuclear security positions of assured destruction and pro-nuclear use were articulated by Bernard Brodie, *The Absolute Weapon* (New York: Harcourt Brace, 1946) and William Liscum Borden, *There Will Be No Time* (New York: Macmillan, 1946).

[47] Robert Jervis, "Deterrence Theory Revisited," *World Politics* (1979) 31:2, 89–324, at 289; see also Morgan, "New Directions in Deterrence Theory," 1986.

game-theory analysis."[48] By 1963, Schelling's research directly helped secure the nuclear policy rationale for "assured destruction," later derided as "mutual assured destruction," or MAD, by its detractors.[49] Assured destruction typified the McNamara doctrine that shaped US nuclear strategy from 1963 until 1973, when RAND alumnus James R. Schlesinger became Secretary of Defense under President Richard M. Nixon.

Following Schelling's 1960 treatment in "Reciprocal Fear of Surprise Attack," defense analysts routinely viewed the nuclear arms race as a Prisoner's Dilemma.[50] Reviewing E. S. Quade's *Analysis for Military Decisions*, F. M. Scherer worried about the over-reliance on the Prisoner's Dilemma model that uncritically accepts the proposition that the Soviets hope to annihilate America, and vice versa.[51] Scherer laments that

On the whole, however, there is too little concern for the fact that strategic deterrence, decisions to develop new weapons technologies, and arms races are typically Prisoner's Dilemma games, in which minimax is a notoriously bad strategy. The deeply pessimistic *Weltanschauung* of this work is characterized most vividly by [Albert] Wohlstetter's assertion that the Soviets' "fondest desires" include launching a successful surprise nuclear attach against the United States. No one can prove such pessimism is wholly unwarranted, but the expected value of the world's survival game almost certainly suffers if we continue pursuing a minimax approach to the problems of deterrence, arms control, and disarmament. All might benefit if military systems analysts spent less time worrying how we can make the best of the worst our rivals may do, and more time inventing ways of cooperating to avoid new self-defeating steps in the perfection and spread of mass-destruction weapons.[52]

In this prescient quote, Scherer astutely sets forth the main argument of *Prisoners of Reason*: the embrace of the game theoretic mentality creates a self-fulfilling prophecy, as worst-case planning builds the institutional infrastructure for a dystopian world. In his opinion, game theory itself seems to drive a mutually suboptimal outcome and leads strategists to view the Soviet Union as "the enemy."[53]

KAHN'S DEFENSE OF NUCLEAR USE THEORY (NUTS)

When analyzing Herman Kahn's pro-nuclear-use stance, one must begin with an understanding of the state of the American military capability by the

---

[48] Quote from original press release, available online Official Nobel Award website: Nobelprize.org.

[49] For discussion of this, see Gregg F. Herken, *Counsels of War* (New York: Knopf, 1985), 342.

[50] Essay is in Thomas Schelling, *Strategy of Conflict* (Cambridge, MA: Harvard University Press, 1960); for a retrospective account of the development of nuclear strategy, and the importance of the Prisoner's Dilemma game, see Freedman, *Evolution of Nuclear Strategy*, 1981. Note, too, Robert Jervis's discussion of the Prisoner's Dilemma game in "Cooperation Under the Security Dilemma," *World Politics* (1978), 30:2, 167–214.

[51] Review of *Analysis for Military Decisions* by E. S. Quade, *American Economic Review*, 55:5, Part 1, December 1965, 1191–1192, at 1192.

[52] Ibid.

[53] Ibid. Regarding Wohlstetter's unfounded pessimism about a potential strategic imbalance favoring the Soviets, see Amadae, *Rationalizing Capitalist Democracy*, 2003, 44–57.

mid-1950s. Not only had the Americans successfully detonated the world's first thermonuclear warhead in 1952 and exploded a fully functional hydrogen bomb in 1954, but under Project Vista orchestrated during the Korean War, the United States developed the atomic cannon, atomic mortar, and atomic mine.[54] During the WWII campaign against Germany, the United States had rejected the British military policy of the aerial bombing of cities to produce massive civilian casualties. Yet in the Pacific theater, the United States had adopted a city bombing policy even prior to the detonation of the atomic bombs over Hiroshima and Nagasaki, which produced 120,000 and 80,000 casualties, respectively. Early US Cold War policy likewise planned to target cities and their inhabitants with hydrogen bombs. This strategic stance was consistent with General Curtis LeMay's leadership over the Strategic Air Command and underwrote President Eisenhower's policy of massive retaliation to respond "massively, at times and places of our choosing" to any provocation deemed worthy of substantial reaction.[55] Notably, during these early days of the Cold War, when the United States enjoyed unparalleled military supremacy, some of President Eisenhower's high-level national security advisors recommended initiating preemptive nuclear war against the USSR to prevent the Soviets from developing thermonuclear weapons of their own. Schelling's protracted analysis of the reciprocal fear of surprise attack and Kahn's proposal to maintain escalation dominance were thus responsive to the state of 1950s military science, technology, and strategic policy.

Herman Kahn (1922–1983) was a nuclear physicist who worked at the University of California Lawrence Livermore Laboratory and helped create the hydrogen bomb. His 600-page *On Thermonuclear War* (1960), prepared from his countless lectures, provides excruciating analysis of the logistical exercise of fighting and winning a nuclear war. Kahn aspired to achieve the gravitas and legendary quality of Prussian military strategist Carl von Clausewitz's classic *On War*.[56] He was an early employee of the RAND Corporation, joining the military think tank in 1947. He thrived on RAND's exploitation of gaps: the bomber gap, the missile gap, and the vulnerability gap. Fred Kaplan's description of the grandiloquent physicist's ponderous tome is unsurpassed for its fidelity: "It was a massive, sweeping, disorganized volume, presented as if a giant vacuum cleaner had swept through the corridors of RAND, sucking up every idea, concept, metaphor and calculation that anyone in the strategic community had conjured

---

[54] See Douglas P. Lackey's extraordinary report on "The American Debate on Nuclear Weapons Policy," 1987; these topics are discussed on 14–15.

[55] Quoting US Secretary of State John Foster Dulles in 1954; Lackey, "American Debate on Nuclear Weapons Policy," 1987, 13.

[56] For a telling commentary on Herman Kahn, see "Fat Man; Herman Kahn and the Nuclear Age," a book review of Sharon Ghamari-Tabrizi's *The Worlds of Herman Kahn* (Cambridge, MA: Harvard University Press, 2005), *The New Yorker*, June 27, 2005; Kahn, *On Thermonuclear War*, 1960; Carl von Clausewitz, F. N. Maude, and Anatol Rapoport, *On War* (London: Penguin, 1968).

up over the previous decade."[57] Still, the text became a best seller and incited the imaginations of both fellow strategists and the public. Kahn's position would become known as nuclear utilization target selection (NUTS, sometimes spelled out as Nuclear Use Theorie[S] in the 1980s). Kahn's approach represented one side of the nuclear strategy debate, articulated in Robert McNamara's 1962 "no-cities" "flexible response" policy, then subsequently in Lyndon Johnson's Foster panel, Richard Nixon's Schlesinger doctrine codified in NSDM-242, Gerald Ford's Team B Report, Jimmy Carter's Presidential Directives 59, and Ronald Reagan's Direct Guidance.[58] Brashly promulgating rationalist strategy, *On Thermonuclear War* represents what would become the essence of the NUTS position: the attempt to achieve military superiority, even at the risk of instigating a hard-fought arms race punctuated with the incentive for preemptive war and the continual worry of an unstoppable escalation into an endless arms buildup and spasm warfare.[59]

Kahn's *On Thermonuclear War* is consistent with Albert Wohlstetter's 1959 "Delicate Balance of Terror" and the 1957 Gaither Report in urging the United States to shore up any vulnerabilities in its Strategic Air Command and develop a coherent plan to prevail in any conceivable military showdown with the Russians.[60] Building his case for analytic deterrence theory, Kahn reminds readers that because "there is no one with experience in the conduct of thermonuclear war ... we must depend on hypotheses, i.e., paper studies."[61] He announced the "technical break through" at the RAND Corporation in the science of warfare, of which game theory was one component. In *On*

---

[57] Fred Kaplan, *Wizards of Armageddon* (Stanford University Press, 1983), at 227.

[58] With respect to Kennedy's initial adoption of a pro-nuclear-use stance, see Kahn, *On Thermonuclear War*, 1960, 174–175; for commentary, see Freedman, *Evolution of Nuclear Strategy*, 1981, 235, especially fn 10.

[59] For a succinct discussion of Kahn's position, see Freedman, *Evolution of Nuclear Strategy*, 1981, 208–219; see also Kahn's contribution to Klaus Eugene Knorr and Thornton Read, eds., *Limited Strategy War* (New York: Praeger for Princeton University, 1962); see also Freedman, *Evolution of Nuclear Strategy*, 1981, 161, on the linkage of the Gaither report to Albert Wohlstetter and the pro-nuclear-use philosophy in opposition to the Killian Report's philosophy of nuclear stalemate. Freedman describes Kahn's strategic insights in the following manner: "It was possible that controlled, discriminating patterns of behaviour would continue even in a conflict being resolved in the presence, and with the occasional use, of nuclear weapons. Where 'irrationality' was in evidence it would be for some 'rational' reason, because advantage could be gained from irrational conduct or the expectation of irrational conduct ... Nor did he [Kahn] believe that the various moves up the escalation ladder would be undertaken out of a narrow risk-calculus. There would be a 'competition in risk-taking' (the original term was first used by Schelling) and the need for 'nerve, skill, and courage.' The essential point was that even up to the very end – the spasm – the crisis leaders maintained some semblance of being the masters instead of the victims of events" (217).

[60] Security Resources Panel of the Scientific Advisory Committee of the Office of Defense Mobilization, *Deterrence and Survival in the Nuclear Age* (*The Gaither Report*), November 1957; Wohlstetter's article appeared in *Foreign Affairs*, XXXVII:2 (January 1959). See also Amadae, *Rationalizing Capitalist Democracy*, 2003, 27–82.

[61] Kahn, *On Thermonuclear War*, 1960, 162.

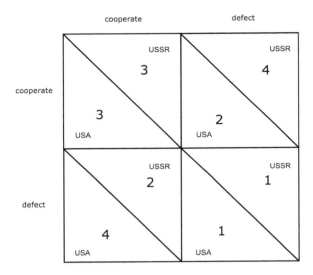

cooperate

defect

cooperate

defect

USSR
3

USSR
4

3
USA

2
USA

USSR
2

USSR
1

4
USA

1
USA

FIGURE 3. Nuclear Chicken Game Matrix

Thermonuclear War, Kahn set the precedent of referring to nuclear deterrence, or brinkmanship, as a "Chicken game," drawing on the writings of the pre-eminent philosopher and renowned pacifist Bertrand Russell.[62] In the nuclear Chicken game, the US, in striving for its first choice of supremacy (DC), will settle for mutual cooperation (CC) or even surrender (CD), rather than accept mutual annihilation (DD) (Figures 3 and 4).

This nuclear Chicken game plays an important role in the ongoing debate between proponents of NUTS and advocates of MAD. Both groups view nuclear deterrence as a form of coercive bargaining best reflected by this game.[63]

[62] Bertrand Russell, *Common Sense and Nuclear* Warfare (New York: Simon & Schuster, 1959). For discussion, see Lackey, "American Debate," 1987, 19–20. Russell was not only a pacifist and a famous philosopher, but his commentary on the morality of nuclear weapons also had significant impact. Russell makes the case that nuclear war cannot serve any aim of purposive agency and is immoral besides; consequently, it must be a common goal to end the likelihood that nuclear warfare could occur. Russell's position is in opposition to that which would prevail over the US military.

[63] How to resolve the question of strategic bargaining so as not to be the loser in a perpetual contest of wills was central to the debate between MAD and NUTS. On Chicken games and bargaining, see Jervis, "Deterrence Theory Revisited," 1979, 291–92; Glaser, "Why Do Strategists Disagree," 1989, at 145. For commentary, see also Patrick James and Frank Harvey, "The Most Dangerous Game: Superpower Rivalry in International Crises, 1948–1985," *Journal of Politics* (1992), 54:1, 25–53. Note that Herbert Gintis argues that the Chicken game explains the origin of property rights, *Bounds of Reason: Game Theory and the Unification of the Behavioral Sciences* (Princeton NJ: Princeton University Press, 2009).

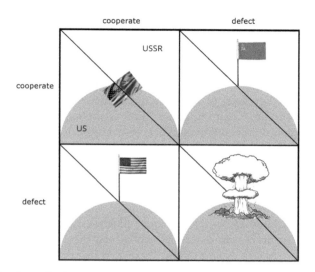

FIGURE 4. Nuclear Chicken Game Diagram

Once the mutual threat of assured destruction is achieved, the initial concession to the Prisoner's Dilemma payoff matrix of each seeking unilateral defection and strategic domination remains: each side prefers to sucker the other, even though, of course, both seek to avoid mutual destruction. Thus, the key difference is that in the Chicken game, both agents would choose their own unilateral cooperation paired with the other's unilateral defection rather than face mutual annihilation.

Kahn outlines and advocates for three types of deterrence: (1) passive, or the ability to marshal a second strike against the aggressor; (2) active, or first-strike capability; and (3) "tit-for-tat."[64] The third, which is also referred to as a graduated, or flexible, response, represents the centerpiece of the NUTS policy. Its central idea is to command the ability to respond, not in kind, but with a strategy of "escalation dominance" to trump any aggressive move made by the opposition. Achieving (3) made it difficult to avoid (2), or at a minimum being perceived to seek first-strike capability. By contrast, as will become apparent ahead, Schelling's position on deterrence recognizes the need to be able to signal the threat of counteraction but focuses on the ability to marshal a destructive second strike that does not mimic the first strike.

In 1965, Kahn published his sequel, *On Escalation: Metaphors and Scenarios.*[65] In it, he defines escalation as "an increase in the level of conflict in [an] international situation."[66] He provides an analysis of the step-by-step

[64] Kahn, *On Thermonuclear War*, 1960, at 282.
[65] Herman Kahn, *On Escalation: Metaphors and Scenarios* (New York: Praeger, 1965).
[66] See Freedman, *Evolution of Nuclear Strategy*, 1981, 210; Kahn also wrote *Thinking about the Unthinkable* (New York: Horizon Press, 1962).

qualitative growth of force up the "escalation ladder." His ladder has 44 rungs; nuclear exchange is initiated at the 15th rung, and he prescribes 30 ways to wield nuclear bombs.[67] The book's articulation of the logic of "escalation dominance" anticipated the late twentieth-century US military mandate of "full spectrum dominance."[68] This strategy recommends pressing one's military advantage on any level of the escalation ladder on which one enjoys an asymmetric advantage. Kahn explains, "This is the capacity, all other things being equal, to enable the side possessing [escalation dominance] ... to enjoy marked advantages in a given region of the escalation ladder."[69] In Freedman's words, "Thus success through escalation dominance depended on a favorable asymmetry of capabilities."[70] Ideally, one would enjoy such asymmetric advantage on each rung of the escalation ladder or at least be able to control events such that the conflict is concluded on a rung over which one dominates. Kahn addresses the possibility of "eruption" into full-scale thermonuclear war, which seems to be an inevitable consequence of climbing the escalation ladder.[71] He refers to the final rung on the ladder as "spasm war," indicating a total loss of control and thought. The sexual connotation was not foreign to Kahn who lectured high-level SAC officers on "wargasm."[72] NUTS breaks with the classic liberal security posture of reciprocal noninterference by resting security on the achievement of dominance; its deterrence can only be asymmetrical.[73]

SCHELLING'S DEFENSE OF MUTUAL ASSURED DESTRUCTION (MAD)

President Eisenhower's nuclear strategy of massive retaliation signified that the Strategic Air Command was already prepared to devastate Russia to a "smoking, radiating ruin in about two hours."[74] By 1955, this strategy began to receive scrutiny in view of its potential immorality and inefficacy. At this time, as a RAND analyst and US Department of Defense consultant, Kahn had access to military information not generally available to the American public, and his *On Thermonuclear War* frightened the lay public into recognizing the dangers

---

[67] See Keith Krause, "Rationality and Deterrence in Theory and Practice," in *Contemporary Security and Strategy*, ed. Craig Snyder (New York: Routledge, 1997), 120–148, at 125; Freedman, *Evolution of Nuclear Strategy*, 1981, 216–217.

[68] *Joint Vision 2020*, US Department of Defense, May 30, 2000; referenced by USDOD at http://www.defense.gov/news/newsarticle.aspx?id=45289, accessed July 4, 2015.

[69] See Freedman, *Evolution of Nuclear Strategy*, 1981, 218.

[70] Ibid.

[71] Freedman worries about this entailment; ibid., 382.

[72] See Carol Cohn, "Sex and Death in the Rational World of Defense Intellectuals," *Signs* (Summer, 1987), 12:4, 687–718; Freedman, *Evolution of Nuclear Strategy*, 1981, 222; see also Kaplan, *Wizards of Armageddon*, 1983, 222–223.

[73] See Patrick Morgan, *Deterrence Now* (Cambridge: Cambridge University Press, 2003), at 51; see also Charles Glaser, "The Security Dilemma Revisited," *World Politics* (1997), 10:1, 171–209, 193.

[74] Quoted from Lackey, "American Debate," 1987, 15.

Assurance

and realities of nuclear confrontation.[75] Pointedly, US superior military might had done nothing to prevent or stop the Soviets from invading Hungary in 1956. All that a massive retaliating strike was bound to secure was the Soviet's nuclear devastation of Western Europe, and even potentially the United States itself.

Eisenhower's aversion to budget deficits led him to reject building up conventional forces, and he left office with his Single Integrated Operational Plan (SIOP), a war plan of simultaneous massive nuclear retaliation against all Soviet (and Chinese) cities, in place. This strategic reality, coupled with the Soviet's newly acquired thermonuclear capability, set the stage for a two-decade policy debate between advocates of mutual deterrence and proponents of flexible response, counterforce, and escalation dominance. Hence, the Cold War challenge, based on achieving a bipolar strategic deadlock in which neither side could reasonably afford to wage nuclear war, stood in place.

Unlike Kahn, Schelling played an active role in McNamara's Department of Defense.[76] The Kennedy administration had come to office arguing that Eisenhower was weak on defense. McNamara's first move, which he brought to public attention with his famous public speech in Ann Arbor, Michigan, in June 1962, was to introduce the new SIOP 63, set to take effect the following year, predicated on maintaining "Flexible Response."[77] McNamara was attempting to respond to the perceived weaknesses of Eisenhower's massive retaliation: the immorality of targeting innocent civilians, and the lack of credibility of only being able to achieve efficacy at the price of suicide. Flexible Response worked by responding to a less-than-all-out attack against the United States with a less-than-all-out-attack against the Soviet Union and targeting harder to hit military installations and forces rather than simply attacking vulnerable and large cities.[78] Although this policy was ostensibly more moral for targeting weapons systems instead of people, counterforce weapons could be used in a first-strike assault. Therefore, this strategy was inherently destabilizing because it put the Soviets in the position of worrying about preemptive attack and how to secure their own second-strike capability. Nor was it more certain that this new form of flexible deterrent threat of

---

[75] Discussed in ibid., 18. The most interesting point of Lackey's investigation is that on the surface, between Kahn's *On Thermonuclear War* and Ronald Reagan's presidency, "the notion of preparing to fight and 'win' nuclear wars did not enter the mainstream of American political thinking." However, it is obvious in following the bureaucratic movement and strategic writings of James R. Schlesinger through the Nixon and Carter administrations that behind the scenes, securing Kahn's goal of escalation dominance was a key focus of this RAND alumnus and nuclear strategist, discussed in Chapter 4.

[76] Schelling's role in McNamara's Department of Defense is discussed in Amadae, *Rationalizing Capitalist Democracy*, 2003, 27–82; see also E. M. Sent, "Some Like It Cold: Thomas Schelling as a Cold Warrior," *Journal of Economic Methodology* (2007) 14:4, 455–471.

[77] Discussed by Lackey, "American Debate," 1987, 22.

[78] Lackey makes clear that McNamara's policy still left room to target cities, but only after American cities had been subject to attack; "American Debate," 21–23.

retaliation was any less suicidal that Eisenhower's because there was no guarantee that escalation could be capped.[79]

Secretary of Defense Robert S. McNamara's 1963 reconstitution of US nuclear strategy along Flexible Response lines and his subsequent backpedaling to assured destruction made it possible to recognize Thomas Schelling's contributions to what would become the then-prevailing conceptual framework for organizing military capability and posture.[80] Schelling's essay "Diplomacy of Violence," in which he directly responds to McNamara's 1962 Ann Arbor address, as well as his essays "The Reciprocal Fear of Surprise Attack" and "Surprise Attack and Disarmament" set forth his view that the nuclear standoff resembles a Prisoner's Dilemma situation and his resolution of this otherwise mutually disastrous scenario with the deterrent policy of assured destruction.[81]

According to Schelling, two unprecedented features of nuclear war forever altered the nature of warfare. First, concurring with Bernard Brodie's 1950s estimation, he held that the primary characteristic of nuclear war is that it makes coercive force more effective as bargaining leverage to deter or compel specific behavior; nuclear weapons are not useful for achieving the military defeat of one's enemy. Second, he argued that nuclear weapons make it possible to exert a phenomenal threat of human destruction even without the power to stand as the military victor of a conquest.[82] Fully on board with strategic rationality, Schelling insisted that the ability to inflict pain was a significant attribute of bargaining. If the other is persuaded by coercive threats of punishment, then the application of physical harm need not be imposed. The brute force of physical domination is thus replaced by carefully calibrated threats measured against the other's expected gain from prospective choices of action: "The threat of pain tries to structure someone's motives, while brute force tries to overcome his strength." Schelling concludes, "Unhappily, the power to hurt is often communicated by some performance of it . . . It is the expectation of *more* violence that gets the wanted behavior, if the power to hurt can get it at all."[83]

Notwithstanding Schelling's seeming pessimism that the grand power of nuclear weapons signaled a military stalemate because neither side could gain by using them, he was sanguine that these weapons could be deployed to one's advantage through coercive bargaining:

The power to hurt is nothing new in warfare, but for the United States modern technology has drastically enhanced the strategic importance of pure, unconstructive, unacquisitive pain and damage, whether used against us or in our own defense. This in turn

[79] These points are discussed ibid.
[80] See Freedman, *Evolution of Nuclear Strategy*, 1981, 231; note Schelling's Nobel Prize lecture that stresses his contributions to nuclear strategy and security; available online, Nobelprize.org.
[81] First essay in *Arms and Influence* (New Haven: Yale University Press, 1966); second two in *Strategy of Conflict*, 1960.
[82] Schelling, *Arms and Influence*, 1966, 21, 22.
[83] Ibid., 3; for discussion, see Sent, "Some Like It Cold," 2007.

enhances the importance of war and threats of war as techniques of influence, not of destruction; of coercion and deterrence, not of conquest and defense; of bargaining and intimidation ... War no longer looks like just a contest of strength. War and the brink of war are more a contest of nerve and risk-taking than of pain and endurance.

Schelling, like Kahn, accepts that with deterrence in place, superpowers will face each other in a Chicken game contest of nerves. He observes that "military strategy can no longer be thought of ... as the science of military victory." War is better regarded as "the art of coercion, intimidation and deterrence." He therefore concludes, "military strategy... has become the diplomacy of violence."[84] Schelling's diplomacy of violence embedded in a nuclear stalemate held out the hope of retaining military control and international stature by wielding nuclear threats. Schelling advocated the use of a demonstration explosion of a nuclear device during the Berlin crisis, "over an isolated place in Russia like the island of Novaya Zemlya," as an attempt to show that the United States would retaliate if the Soviets did not back down.[85] Therefore, although the strategist did not endorse escalation dominance because it is inherently self-defeating, still he did support a flexible use of nuclear weapons in what he hoped would be limited war to maintain deterrent brinkmanship.[86]

"The Reciprocal Fear of Surprise Attack" and "Surprise Attack and Disarmament" initiated the convention of analyzing the fear of nuclear preemption in terms of a Prisoner's Dilemma.[87] Schelling begins by applying a Stag Hunt, or Assurance Game, to a standoff between a homeowner and a burglar, each of whom wants a peaceable resolution without escalation to a shooting incident.[88] In game theory, a Stag Hunt has two possible equilibrium outcomes in which neither actor can alone improve on his payoff by selecting a different course of action: mutual cooperation and mutual defection. Figures 5 and 6 present a matrix and graphic depiction of the Stag Hunt game. Using von Neumann's reasoning, some argue that choosing to defect is always superior in a Stag Hunt because this choice at least achieves the maximin outcome of a hare for certain rather than the slim pickings that would be available after attempting to cooperate but being stood up by the other actor.

Schelling introduces the idea that each is uncertain about the preferences of the other, whether he seeks a peaceable solution or prefers to shoot the other, and similarly worries that the other may incorrectly guess his own intent. Introducing an 80 percent likelihood that one believes the other is a predator and multiplying the cardinal utility matrix of the Stag Hunt by the anticipated likelihood of each act, Schelling demonstrates that the Stag Hunt transforms

---

[84] Schelling, *Arms and Influence*, 1966, 34.

[85] Note Schelling's organization of such war-gaming simulations at Camp David in the early 1960s (Herken, *Counsels of War*, 1985, 158).

[86] For astute discussion, see Green, *Deadly Logic*, 1966, 129–152.

[87] The first essay appears in Schelling, *Arms and Influence*, 1966, chap. 1; the second two appear in Schelling, *Strategy and Conflict*, 1960, chaps. 9 and 10.

[88] Schelling, "Reciprocal Fear of Surprise Attack," in *Strategy and Conflict*, 1960.

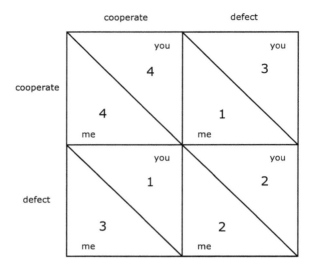

FIGURE 5. Stag Hunt Game Matrix

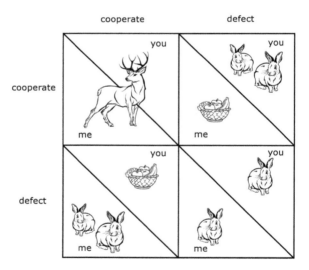

FIGURE 6. Stag Hunt Game Diagram

into a Prisoner's Dilemma, in which the row actor is better off defecting than cooperating no matter what the other side does (DC>CC and DD>CD) (Figure 7). Schelling's transformation of an Assurance Game into a Prisoner's Dilemma as a function of probability is a crucial moment in the analytic development of game theory, nuclear security, and the future use of the PD game to model actors' hopes to achieve security. It set a precedent for suggesting

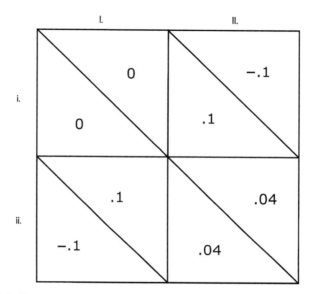

FIGURE 7. Schelling's Prisoner's Dilemma Payoff Matrix Derived from Stag Hunt

that one's sufficient doubt about the other's intention warrants adopting the preferences of an aggressor. Doves must become hawks in self-defense.[89]

The reciprocal fear of surprise attack emerges in an environment characterized by a high degree of suspicion and large preemptive incentives. It is a situation captured by a one-shot PD, in which both sides must defect regardless of what the other does. Because the only route to security lies in offensive action, even pure security seekers must assume the worst of others and act like aggressors. The assumed impossibility of signaling one's own benign intentions combined with the staggering costs of guessing wrong about the intentions and future actions of the other explains how unpredictability can lead to aggressive behavior and preemptive wars that no one wants.[90]

States operating in an uncertain environment of large first-strike advantages confront the same problem faced by two gunslingers in a small town lacking a capable sheriff. Both gunslingers may prefer a bargain whereby each leaves the other alone, but neither can credibly commit not to shoot the other in the back.[91]

---

[89] Chapter 2 discusses how introducing formal expected utility to capture the degrees of doubt about the other agent's intent is necessary to transform a Stag Hunt payoff matrix into a PD. Matrix in figure 7 is taken from Schelling's Strategy of Conflict, 1960, at 213; note at this time there is no convention for establishing figures of games, and he has mutual defection in the top left corner and mutual cooperation in the bottom right corner.

[90] See Jervis, "Cooperation Under the Security Dilemma," 1978.

[91] See James Fearon, "The Offense-Defense Balance and War since 1648," available at http://www.seminario2005.unal.edu.co/Trabajos/Fearon/The%20offense%20-%20defense%20balance%20and%20war%20since%201648.pdf, accessed July 4, 2015.

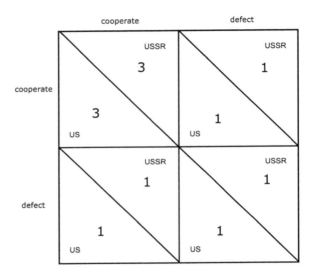

FIGURE 8. MAD Payoff Matrix

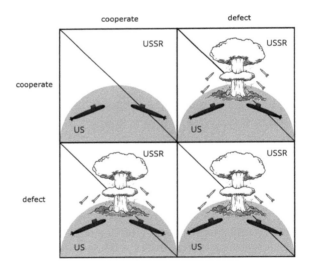

FIGURE 9. MAD Payoff Diagram

Assured destruction solves this problem by eliminating all preemptive incentives. The key is for both superpowers to possess a survivable second-strike force that can wreak unacceptable damage on the other. If, for instance, the two gunslingers now only have slow-acting bullets that take minutes to work, a shootout makes no sense for either side, for it has become tantamount to mutual suicide (Figures 8 and 9). In Figure 9, even if the US initially cooperates and receives the USSR's preemptive first strike, its second strike counterforce ensures the USSR's destruction.

Schelling addresses the considerations of the incentive to strike a preemptive blow in self-defense in "Surprise Attack and Disarmament." It is in this essay that Schelling outlines the philosophy underlying Assured Destruction: "The innovation in the surprise-attack approach ... has to do with what scheme is designed to protect and what armaments it takes for granted. An anti-surprise-attack scheme has as its purpose not just to make *attack* more difficult but to reduce or to eliminate the advantage of striking *first*."[92] In essence, if the advantage of striking first is entirely reduced by the opposition's reserve of devastating second-strike force power, then neither side will be inclined to attack first. This eventuality depends entirely on the United States and the USSR building weapons that cannot be mistaken for first-strike forces designed to destroy military targets. Therefore, "schemes to avert surprise attack have as their most immediate objective the safety of weapons rather than the safety of people," rejecting Herman Kahn's type II deterrence.[93] The other nation's weapons must remain sacrosanct so that its leaders feel secure in being able to marshal a devastating counterattack at the first sign of being assaulted. Submarine forces are ideal for this, because they are difficult to detect and destroy, and they can maintain a counterstrike capability targeted at vulnerable population centers.

Schelling speaks in terms of the logic of stalemate, for even if one side could mount an attack that could knock out the other in a devastating blow subverting their counter-strike ability, the opposition would only have the heightened incentive to strike first. In the language of the Prisoner's Dilemma game that Schelling argues captures the problem of the reciprocal fear of surprise attack, the capability of second-strike reprisal serves to alter the payoff for a unilateral attack to that of mutual devastation signifying the inherent impossibility of obtaining dominance. Schelling and Kahn differ in their appraisals of nuclear strategy: whereas the latter holds out the hope for achieving nuclear supremacy, the former espouses the belief that because the capability for mutual devastation has already been achieved, the best that can be hoped for is a stable "balance of terror."[94]

Persuaded by Schelling's logic, Robert McNamara capped the US nuclear arsenal at 400 megatons of explosive potential, regarding this as sufficient to achieve a devastating counter strike in case the Soviets launched a preemptive attack. Harkening back to Eisenhower's administration, McNamara used a 1960 military study called WSEG-50, which advocated for a policy of "finite deterrence" through city destruction.[95] He officially announced his reformulated strategy to President Lyndon B. Johnson two weeks after John F. Kennedy's

---

[92] Thomas Schelling, "Surprise Attack and Disarmament," in *Strategy and Conflict* (Cambridge, MA: Harvard University Press, 1960), 231.

[93] Ibid., 233.

[94] Ibid., 239.

[95] See Kaplan, *Wizards of Armageddon*, 1983, 316.

assassination, calling it "assured destruction."[96] It rested on a calculation of the destructive force of megatons of explosive power arrayed at the Soviets in terms of "industry destroyed" and "population destroyed."[97] McNamara's crucial insight was that after a certain amount of money has been allocated to building the US stockpile of nuclear warheads for land-based ICBMs (intercontinental ballistic missiles) and strategic bombing, in addition to deployment on undetectable submarines, any more allocation of funds could not significantly increase the destructive force already unleashed because there would be few if any remaining targets, including cities, industry, or military sites.

McNamara's move was twofold. First, he set the rhetoric of US defense policy in terms of deterrent city busting versus counterforce capability, there-fore diminishing the readiness with which the United States might be perceived as clandestinely seeking first-strike power.[98] Second, he maneuvered to deter-mine precisely how much destructive power was sufficient to deter the Soviets from attacking the United States, and thus exactly how large the US missile arsenal needed to be. His civilian defense rationalists put the number at 400 megatons, which guaranteed that after a Soviet first-strike on US territory, there would be sufficient counter-strike potential to kill 30 percent of the Soviet population and half of its industrial base.[99] Even though this policy resembled President Eisenhower's strategy of massive retaliation, to the liberal wing of the defense rationalists, it seemed reasonable because they found it "very hard to believe that any country would deliberately accept the *certainty* of severe retaliatory damage in preference to the *uncertain* prospect of being the victim of a first strike."[100] Whereas McNamara equipped each arm of the US military triad of air, land, and sea with 400 megatons of destructive capacity to build redundancy into the US defenses, Schelling argued for a minimum deterrent

---

[96] Herken, *Counsels of War*, 1985, 170.

[97] Figure in Kaplan, *Wizards of Armageddon*, 1983, 318; note figure is based on economic analysis of diminishing marginal utility and a computer model, not on an actual study of what would "really" deter Soviet aggression.

[98] For discussion see Lackey, "American Debate," 1987, 27.

[99] Kaplan, *Wizards of Armageddon*, 1983, 317; according to Patrick Morgan, this 400 megatons of destructive power was redundantly held by all three branches of the US military triad of air, land, and sea. McNamara thereby articulated a decisive logic for limiting US nuclear warheads to 1054 ICBMs and 656 SLBMs, where the force would basically remain until 1980. However, even after having announced a definite "counter-city" nuclear stance, McNamara still retained in place his 1963 Single Integrated Operational Plan (SIOP) for nuclear war that articulated the strategy of "Damage-Limiting," or counterforce. Significantly, now any deployment of war-heads against military targets would fall within the 400 megatons of explosive force range, instead of providing a rationale for additional weapons. With its 400-megaton arsenal, the United States would ensure the Soviets' destruction were they to advance a first strike. The basic insight was to cap the US arsenal at the level of assured destruction capability, but to preserve the latitude to deploy some of these weapons against military, instead of civilian, targets.

[100] Glen Snyder *Deterrence and Defense* (Princeton, NJ: Princeton University Press, 1961), 108, quoted by Freedman, *Evolution of Nuclear Strategy*, 1981, 164.

restricted to submarine missiles that are virtually impossible to detect and destroy but have less accurate targeting capability and thus cannot be mistaken for preemptive weapons.[101]

Schelling's ideas were severely tested in the Vietnam War. The war itself, organized in terms of escalating bouts of punitive damage on the country's civilian population, provided an empirical case study of Schelling's idea of diplomatic violence through the threat of pain.[102] This war, interpreted by US officials as a crucial stand against communism and Soviet imperialism, became a test case for maintaining American credibility. Schelling had written that in Chicken-style diplomacy, maintaining one's credibility is key, for once it is challenged, it will forever be possible to be out-maneuvered. All the US nuclear forces, on 24-hour deterrent alert, were apparently useless in controlling the outcome of the Vietnam War.

EARLY US RESPONSES TO STRATEGIC NUCLEAR PARITY

As the 1960s dragged to their conclusion, and Vietnam stood as an indictment of US military ineptitude on the world stage, the US conservative policy wing formed around the figures of Kahn, Wohlstetter, and Paul Nitze and became energetically revitalized. The liberals, who had wielded power for almost a decade, had presided over military fiasco and the embarrassment of the United States. However, no strategist could alter the fact that both superpowers had the destructive power to eliminate the other as a viable nation-state. Thus, viewed from a wider lens, the Cold War was unfolding according to this simple logic of Assured Destruction, assuming that the superpowers had achieved parity as a consequence of the hydrogen bomb.

From 1964 to the late 1970s, the Soviet Union added significantly to its offensive nuclear forces, changing the Soviet-American strategic nuclear balance

---

[101] Schelling's discussion of the reliance on submarine missiles is in *Strategy of Conflict*, 1960, 238–239.

[102] Schelling wrote that he was certain that the Gulf of Tonkin bombing was justified and warranted, although he admitted that by the late 1960s, he was as yet unclear that the escalating bombing raids on Vietnam were "working." Schelling's position is criticized by Theodore Draper in Herken, *Counsels of War*, 1985 223; see also, Herken, *Counsels of War*, 1985, 209; later Schelling stands against the bombing. These attacks seemed to be tailor made from his strategic thought. There is historical evidence that he was at least indirectly implicated in the bombing raid's invention through his close colleague John McNaughton, who implemented Schelling's strategy in operation Rolling Thunder, which commenced March 2, 1965. For discussion, see Kaplan, *Wizards of Armageddon*, 1983, 332–336; see also. Sent's "Some Like It Cold," 2007. Schelling was clear that controlling a population could be achieved through threatening it with "pain," and yet the steadily intensifying bombing raids on the Vietnamese people, if anything, seemed to make them more recalcitrant to agreeing to American terms. Despite the consistent evidence that civilian bombing only hardens a people against their enemy, Schelling's strategic theory suggested the opposite: that rational behavior will modify its course if violence stands in its way and can be avoided by a different path of action.

from one of American superiority to one of parity, or even Soviet superiority.[103]
In concrete terms, the Soviet Union moved from a position in which it could
inflict 30 million fatalities on the United States in a second strike to one
where it could inflict roughly 100–140 million; by comparison, the fatalities
that the United States could inflict on the Soviet Union remained constant at
100–120 million.[104] However, consistent with MAD logic of mutual deterrence, the United States did not contest the Soviet buildup. Instead, the two
states first signed the Strategic Arms Limitation Treaty (SALT I) in 1972,
freezing the number of ballistic missile launchers, and went on to sign the
SALT II treaty in June 1979, placing a common ceiling of 2400 on each side's
total number of delivery vehicles (ICBM launchers, SLBM [submarine-
launched ballistic missile] launchers, and long-range bombers).[105]

Given its options, the United States had little choice but to accept nuclear
parity. To maintain its 1964 ratio of superiority over the Soviets, the United States
would have needed 12,000 delivery vehicles and more than 50,000 warheads
(when the two sides formally agreed to the 2,400-vehicle limit, the Soviets
actually had 2,504 deployed delivery vehicles and 6,000 warheads, whereas the
United States had 2,058 vehicles and 9,200 warheads).[106] And even if the United
States were willing to pay the costs required to build such a force, it could not
have maintained its strategic nuclear superiority over the Soviet Union in terms of
military consequence – that is, the actual amount of death and destruction a
nuclear war would produce on any given set of targets. As Secretary of Defense
McNamara underscored with his "curve" in 1965, the act of dropping nuclear
weapons on an enemy is subject to diminishing returns: there are only so many
meaningful urban-industrial targets in a country. Thus, the destruction of 33 percent of the population and 66 percent of the industry of the Soviet Union became
the figures identified with assured destruction simply because the allocation of
additional forces to hit further target sets yielded only the death and destruction
of sheep, cows, trees, grass, and insects and comparatively few Soviet citizens and
factories.[107]

Nevertheless, some argued that a substantial buildup of US offensive nuclear
forces would have salvaged a measure of American superiority by buttressing its

---

[103] Joseph S. Nye Jr., *Nuclear Ethics* (New York: Free Press, 1986) makes it clear that MAD does
not require parity because at some point each side has sufficient weapons to destroy its
opponent; no more weapons are necessary.

[104] Warner S. Schilling, "US Strategic Nuclear Concepts in the 1970s: The Search for Sufficiently
Equivalent Countervailing Parity," *International Security* (1981) 6:2, 51.

[105] This is not to minimize the complications wrought by McNamara's decision to pursue multiple
independently targetable reentry vehicles (MIRVs) and the Soviet's development of an anti-
ballistic missile (ABM) system to protect Moscow; for discussion, see Lackey, "American
Debate," 1987, 26–29. Notwithstanding these developments, Lackey concludes that SALT I
demonstrably ended the nuclear arms race former Prisoner's Dilemma structure due to the
mutually realized impossibility of winning this race, 30.

[106] Schilling, "US Strategic Nuclear Concepts in the 1970s," 1981, 52.

[107] Jerome H. Kahan, *Security in the Nuclear Age: Developing US Strategic Arms Policy*
(Washington DC: Brookings Institute, 1975), chap. 2.

damage-limiting, counterforce capabilities. This argument has little merit, however. Even if the United States had employed 50,000 warheads (maintaining its 1964 ratio) and launched all of them in a first strike against Soviet ICBMs and bomber bases, Soviet SLBMs would still survive, as there was no certain way to destroy them. Moreover, the Soviet Union would surely offset such a massive preemptive strike against its land forces by means of a "launch on warning" strategy. The futility of a first strike of any kind – no matter how massive or unexpected – is, after all, the essential logic of MAD and should, therefore, surprise no one.

Arguably, assured destruction should have obviated both Schelling's fear of reciprocal attack and the need to engage in an arms race. The United States and the USSR entered into arms control talks in 1969 and concluded SALT I in 1972, leading Douglas Lackey to observe that this served as "proof that by 1972 the *nuclear* arms race had ceased to be a Prisoner's Dilemma."[108] Lackey explains:

In a genuine Prisoner's Dilemma, when two parties are co-operating, the benefits of unilateral defection (to the defector) are greater than the benefits to each of mutual cooperation, while the penalty of being the second to defect is worse than the penalty for mutual noncooperation. In 1972, the advantages of defecting from cooperation (building more offensive missile launchers, building an ABM [anti-ballistic missile] when the opponent has one) were in fact less than the penalties to both of both built more ABMs and more offensive weapons.[109]

Thus, the actual configuration of force options at the close of the 1960s removed the positive payoff of being the lone defector from an arms control treaty, leaving the lone arms racer to spend more on nuclear armaments than their value could offer in effective destructive power. The Prisoner's Dilemma payoff matrix ceased to apply because it seemed impossible to reasonably evaluate that either the potential for surprise attack or the possibility for winning an arms race was feasible.

Strategists agree that an offensive approach to nuclear security will inevitably be self-defeating.[110] In a discussion of the international relations security dilemma, defensive realist Shiping Tang observes:

Both [John] Mearsheimer (2001) and [Dale] Copeland (2000) explicitly acknowledge not only the existence of the security dilemma but also the impact of nuclear weapons on state behavior. For Mearsheimer, nuclear weapons make global hegemony seem impossible. Likewise, Copeland emphasized that nuclear weapons mean that preventative war by a reigning hegemon is no longer a viable option, and a far more plausible scenario will be that a reigning hegemon initiates crises in order to forestall the growth of a rising

---

[108] Lackey, "American Debate," 1987, 30.
[109] Ibid.
[110] This topic is addressed at length in Chapter 4, but for a summary conclusion to this effect, see Lackey, "American Debate," 1987, 31–37.

power ... Nuclear weapons are generally understood as the ultimate weapons that have brought defense [vs. offense] dominance to international politics (Jervis 1989).[111]

Both theorists who deny that mutual security can be a realistic goal among nation states under any circumstances (Mearsheimer and Copeland) and those who hold the opposite view (Schelling, Jervis, Tang) agree that nuclear weapons cannot be constructively deployed, even by a hegemon.[112] This, however, did not stop NUTS from winning not only the academic debate over nuclear policy but also the actual policy debate.

CONCLUSION

Schelling's legacy should have been winning the nuclear security debate in favor of MAD, thereby limiting the US arsenal of nuclear weapons to second-strike counterforce missiles on submarines. Instead, even with the competent although less well-known Jervis perpetuating Schelling's case in the 1970s, MAD lost the policy debate to NUTS during President Carter's *détente*-minded administration.[113] Schelling's greatest impact thus lies in how he set the precedent for mobilizing game theory and the Prisoner's Dilemma to represent the international relations security dilemma of achieving peaceful coexistence of nations in a state of anarchy.

Glenn Snyder's powerful essay "'Prisoner's Dilemma' and 'Chicken' Models in International Politics" (1971) popularized the idea that the Prisoner's Dilemma appropriately models the fear of preemptive nuclear attack, arms racing, and other problems associated with uncertainty throughout international politics. Linking Schelling's landmark "Reciprocal Fear of Surprise Attack" to Robert McNamara's 1967 assessment of American nuclear strategy, Snyder forcefully argued that the Prisoner's Dilemma exemplified the high-stakes Cold War stand-off. Quoting McNamara at length, Snyder conveys that uncertainty over the other's intentions is sufficient to spark an arms race:

In 1961, when I became Secretary of Defense, the Soviet Union possessed a very small operational arsenal of intercontinental missiles ... [and very little technological and industrial capacity to augment their military capability.]

Now, we had no evidence that the Soviets did in fact plan to fully use that capability. But as I have pointed out, a strategic planner must be "conservative" in his calculations; that is, he must prepare for the worst plausible case ...

Since we could not be certain of Soviet intentions – since we could not be sure that they would not undertake a massive buildup – we had to insure against such an

[111] Shiping Tang, *A Theory of Security Strategy for Our Time: Defensive Realism* (New York: Palgrave Macmillan, 2010), 17; John Mearsheimer, *Tragedy of Great Power Politics* (New York: W. W. Norton, 2001); Dale Copeland, "The Constructivist Challenge to Structural Realism: A Review Essay." *International Security* (2000) 25:2, 187–212; Robert Jervis, "Rational Deterrence: Theory and Evidence," *World Politics* (1989) 41: 2, 183–207.

[112] Charles Glaser, who also seems sympathetic to MAD, outlines why the asymmetric deterrence policy of NUTS wins the academic debate, "Why Do Strategists Disagree," 1989; see also Nye, *Nuclear Ethics*, 1986, 4.

[113] This is the topic of the next chapter, "Deterrence."

eventuality by undertaking ourselves a major buildup of the Minuteman and Polaris forces ...

Clearly, the Soviet buildup is in part a reaction to our own buildup since the beginning of this decade. Soviet strategic planners undoubtedly reasoned that if our buildup were to continue at its accelerated pace, we might conceivably reach, in time, a credible first-strike capability against the Soviet Union.

This was not in fact our intention. Our intention was to assure that they – with their theoretical capacity to reach such a first-strike capability – would not in fact outdistance us.

But they could not read our intentions with any greater accuracy than we could read theirs. And thus the result has been that we have both built up our forces to a point that far exceeds a credible second-strike capability against the forces we each started with ...

It is futile for each of us to spend $4 billion, $40 billion, or $400 billion – and at the end of all the spending, at the end of all the deployment, and at the end of all the effort, to be relatively at the same point of balance on the security scale that we are now.[114]

Snyder concludes that this regrettable situation is best captured by the Prisoner's Dilemma model: "Thus, the Secretary of Defense, with remarkable clarity, and in a tone which can only be described as wistful frustration, expounded the essence of the prisoner's dilemma in the nuclear age,"[115] a state of frustration driven by both sides' "belief that one's own country is peaceful ... [and that] one's own arms are only defensive reactions to the other's threat."[116] Mysteriously, despite the fact of MAD and the inability to win the nuclear arms race, still strategists viewed the Cold War nuclear security dilemma as a virulent Prisoner's Dilemma.

Secretary McNamara's interpretation of the dynamics fueling the Cold War were inseparable from Schelling's early game-theoretic application of the Prisoner's Dilemma to nuclear war and his solution (to this otherwise mutually suboptimal situation) of deterrence in the form of a second-strike, counter-city capability. Following Schelling's analysis, McNamara understood that nuclear weapons forever altered warfare by guaranteeing that the destroyed nation could itself launch a devastating counterattack on the would-be victor. Also following Schelling, Snyder asserts that the security dilemma – which, by definition, must be populated only by pure security-seeking agents, who, none-theless, are haunted by the specter that the other is seeking to get the upper hand – is best captured by a Prisoner's Dilemma: "It provides a more complete portrayal of the consequences of anarchic system structure because it allows for the possibility of illusory conflict engendered by mutual suspicion and fear *and* for the possibility of actual incompatibilities of interest and aggressive intent not motivated by security considerations."[117] And because it is impossible "ever to be sure of the other party's intentions" and guessing wrong can have terrible consequences, uncertainty without any actual predatory intent is sufficient,

---

[114] Snyder, "'Prisoner's Dilemma,'" 1971, 73.
[115] Ibid.
[116] Ibid., 78.
[117] Ibid., 82.

according to PD-security dilemma logic, to mimic real incompatibility of interests.[118]

Thomas Schelling's and Glenn Snyder's use of the Prisoner's Dilemma model shows how game theory, also known as rational deterrence theory, coevolved with the US deployment of nuclear forces to maintain a nuclearized form of sovereignty. Strategists developed rational decision theory to maintain US security by providing deterrence with a formal basis that theorists subsequently used to revisit human relations throughout society much more broadly. This chapter has provided evidence to sustain the following points:

1. After World War II through to the present, nuclear warfare has been a nonempirical subject ideal for applying purely analytic models of strategic rationality.
2. These strategists, and those who followed, deferred to game theory as the standard for rational action and established the norm of treating the high-stakes nuclear security dilemma and arms race in terms of a PD game.
3. Consistent with neoliberal theory, Schelling set out to salvage what he could of classical liberalism and sought to defend the bilateral position of mutual assured destruction. Yet, instead, as the next chapter explains, his initial concession to strategic rationality, and its entailed acceptance of the PD model, made it logically impossible to vindicate MAD against the more aggressive war fighting strategy of NUTS.
4. This game theoretic framing of the nuclear security dilemma, although logically impeccable, has the implication of giving sole credence to strategic rationality to imbue action with coherence. Furthermore, it cedes reciprocal respect and the assurance of cooperation based on upholding the no-harm principle in favor of viewing the other as a relentless strategic combatant who must be countered with coercive bargaining and asymmetric deterrence.

Part II, "Government," shows how social scientists extended this nuclearized exercise of national sovereignty into the interstices of civil society when they accepted the overarching reach of strategic rationality and lifted its core logical puzzle of the Prisoner's Dilemma from the domain of the nuclear security dilemma and applied it to the social contract. Therefore, strategists' tendency to "adopt ... the stance that there is an *enemy* or an opponent, and that whenever there is a choice between maximizing our values and his, we maximize ours even if that involves injuring him" was normalized throughout markets and government when social scientists and policy analysts adopted rational decision theory as the comprehensive understanding of intelligible behavior.[119]

---

[118] Ibid., 74.
[119] Quote is from Philip Green, *Deadly Logic*, 1966, 213.

4

Deterrence

Rational deterrence is a highly influential social science theory. Not only has it dominated postwar academic thinking on strategic affairs, but it has provided the intellectual framework of Western military policy in the same period as well. The theory's success drives largely from its clearheaded logic, which is as persuasive as it is elegant.

The power of rational deterrence theory is conceptual, not mathematical. It derives from the underlying logical cohesion and consistency with a set of simple first principles, not from the particular language in which it is expressed. In consequence, the model has been astonishingly fecund, both for theory and policy.

No other theoretical perspective has had nearly the impact on American foreign policy ... Far from being an abstract, deductivistic theory developed in a policy vacuum, rational deterrence theory has repeatedly taken inspiration from the most pressing policy questions of the day, from decision of bomber-basing in the 1950s to SDI [Strategic Defense Initiative] in the 1980s. It has set the terms of the debate, and has often influenced the outcome.

Christopher Achen and Duncan Snidal, 1989[1]

So far we have seen that strategic rationality, which endorses the logic of consequences, accepts an underlying philosophical realism about value in the form of interpersonally transferable utility, and rejects joint maximization, seemed tailor-made to address the as yet counterfactual hecatomb of waging

---

[1] Christopher Achen and Duncan Snidal, "Rational Deterrence Theory: Comparative Case Studies," *World Politics* (1989) 41:2, quotes at 143, 153, 164. Achen and Snidal argue that rational deterrence theory supports escalation equivalence. Robert Jervis makes the same point, associating "deterrence by denial" with the views of the proponents of nuclear utilization targeting strategy who include Albert Wohlstetter, Colin Gray, and Herman Kahn," in "Security Studies: Ideas, Policy, and Politics," in *The Evolution of Political Knowledge: Democracy, Autonomy, and Conflict in Comparative and International Politics*, ed. Edward D. Mansfield and Richard Sisson (Ohio State University Press, 2004), 100–126, at 115.

nuclear war.[2] Thomas Schelling had sought to defend mutual assured destruction (MAD), reminiscent of reciprocal security under classical liberalism, by modeling a high-stakes nuclear security standoff with the recalcitrant Prisoner's Dilemma game. Given the existential reality of assured destruction in a nuclear war among superpowers, and the shared goal of avoiding Armageddon, by all counts MAD should have won the theoretical security debate and prevailed over US nuclear strategy.

Yet, with hindsight, observers may now be tempted to conclude both that the nuclear security debate vindicated nuclear utilization targeting selection (NUTS) theoretically and helped the United States win the Cold War in practice. Readers may thus wonder, "Why revisit the nuclear security debate, especially given the successful denouement of the superpower standoff?" The choice of adopting orthodox game theory as the exhaustive statement of coherent action necessarily pronounces escalation dominance logically superior to reciprocal deterrence, despite the fact that maintaining nuclear ascendance over another superpower is impossible. Opting for a policy of disproportional deterrence instead of mutual assurance marks a clear rupture with the classical liberal resolution of a security dilemma. These implications have gone far beyond security itself into the interstices of civil society and the social contract.

This chapter follows how the least likely of US presidents to exercise national sovereignty through wielding nuclear threats, Jimmy Carter, took the biggest step by making NUTS official US strategic policy in his Presidential Direction 59 in 1980.[3] Against the grain of his initial commitment to deescalate the Cold War arms race at least maintaining minimum deterrence consistent with assured destruction and possibly even through progressive disarmament, Carter left office having signed into effect the US preparedness to wage and prevail in prolonged nuclear conflict. Carter's presidency culminated the consequential and yet widely unknown MAD vs. NUTS debate. In brief, Thomas Schelling had used game theory to defend the posture of mutual assured destruction by first assuming that both the United States and the USSR sought peaceful coexistence rather than strategic dominance. However, given the high stakes of nuclear confrontation and the uncertainties regarding the other's intentions, Schelling concluded that the nuclear security dilemma is most accurately modeled by the Prisoner's Dilemma (PD) game instead of an assurance game. He presented a solution of minimal deterrence through each side maintaining secure second-strike capability to mount a devastating counterattack. He established the PD logic that normalized that security seekers most prefer to sucker others because in the Prisoner's Dilemma, every actor will defect even if the other actor cooperates. Furthermore, he initiated the familiar PD pedagogy suggesting that self-defense warrants the pursuit of ascendance and coercive

---

[2] Neal J. Roese and James M. Olson, *What Might Have Been: The Social Psychology of Counterfactual Thinking* (Mahwah, NJ: Erlbaum, 1995).

[3] PD 59 is available as RAC Project number NLC-132-23-8-29 and NLC-12-37-4-8-6 at the Jimmy Carter Presidential Library (JCPL).

bargaining. However, as this chapter argues, Schelling's defense of MAD necessarily failed because of the logical structure of strategic rationality. From within the paradigm of rational deterrence, the only means of resolving the paradoxical Prisoner's Dilemma of mutual mishap was to move away from assuring peace for peace and war for war to a posture of deterrence through demonstrating the intention and capability to prevail in military conflict at all levels including even prolonged nuclear war.

This chapter focuses on the nuclear security dilemma that offered the initial proving ground for game theory. Theorists viewed the Prisoner's Dilemma game as analytically equivalent to the paradox of nuclear deterrence: in both cases, the intractable paradox resides in promising an action that, at the time of its enactment, violates instrumental rationality because at that moment of causal intervention, the action has no power to realize the protagonist's preferences. In the case of nuclear deterrence, mutual assured destruction relies on a promise to destroy the other nation once deterrence has already failed, and no purpose could be served other than to murder countless innocent civilians. In the case of the Prisoner's Dilemma, given that both actors most prefer an outcome of unilateral defection, even in view of any commitment to carry through on an agreement made, once the other agent cooperates, the protagonist has no reason to likewise cooperate.

This chapter makes clear the parallel structure of MAD and the PD arises if actors concede that strategic rationality, which upholds consequentialist logic, expected utility theory, and individualistic maximization, necessarily governs all coherent choice. We can understand nuclear strategists' concern to address the toughest case of national security, and hence their tendency toward accepting a *realpolitik* approach to international relations. This chapter analyzes the logical basis for ignoring the factual reality of mutual assured destruction in favor of pursuing security through adopting a nuclear war fighting posture. Thoroughly understanding how strategic rationality inevitably sustains the counterfactual NUTS approach to deterrence through demonstrating the capability and intention to wage nuclear war is important in itself.[4] Moreover, this exercise further helps us confront the implications of extending the domain of strategic rationality beyond nuclear politics into social contract theory. The result has been social scientists' inadvertent embrace of strategic combat as the basis for organization at all levels of interaction throughout the interstices of civil society, markets, and governance. Rather than exit a state of nature, strategic rationality views all social and civilizational order to be built up from acts of individual choice to secure fungible gain irrespective of its repercussions for other actors. This envisioned social order, consistent with orthodox game theory, reflects nuclearized sovereignty. The commitment problem

---

[4] For illuminating discussion of how a war fought with 100 or 1000 nuclear weapons would likely be indistinguishable, and how one nuclear weapon alone could alter the significance direction and purpose of all the rest, see Daniel Volmar, PhD dissertation, "The Power of the Atom: US Nuclear Command, Control, and Communications, 1945–1965" Harvard University, forthcoming 2016.

defying the credibility of exercising an immoral deterrent threat of mass destruction came to challenge the coherence of moral promises and agreements once theorists accepted the all-encompassing reach of strategic rationality. Hence, whereas game theory relies on the logic of consequences, single criterion valuation, and individualistic maximization to be operationalized, extending this method to all types of relationships and interactions entails stripping them of any type of significance or coherence not susceptible to expected utility theory and individualistic choice despite others.

This chapter first discusses the 1970s US nuclear strategy and James R. Schlesinger's long-term role in securing escalation dominance and flexible response. The next section follows President Carter's conversion from initially pursuing disarmament to finally leaving office having fully embraced his Presidential Direction 59 (PD 59), which placed the United States on a nuclear war–fighting stance.[5] The third section examines Carter's security dilemma, which uniquely issued from his scrupulous moral countenance: no one could believe him possible of presiding over the massive nuclear retaliation on which the deterrent posture of MAD relied. The fourth section discusses the counsel available to Carter in the late 1970s from moral philosophers, who began analyzing the nuclear security dilemma through the lens of rational decision theory. This section clarifies how only NUTS could satisfactorily resolve the nuclear security dilemma once it was modeled by orthodox game theory. The fifth section explores what amounts to a tacit theoretical alliance between the offensive realist school of international relations theory and standard game theory. The concluding section relies on philosophical exposition in the early 1980s to show how the way out of the nuclear security dilemma modeled using the Prisoner's Dilemma is to clearly perceive that adopting strategic rationality as the final statement of purposive agency rules out alternative modes of action. These types of action include the logic of appropriateness and legitimacy, in addition to incommensurable domains of value and solidarity. Where deterrence relies on issuing negative sanctions, classical liberal assurance builds on mutual recognition, self-ratified norms, and voluntary compliance with agreements made.

NUTS AND THE TRIUMPH OF PRISONER'S DILEMMA LOGIC

Despite both the US and the Soviet development of sufficiently powerful nuclear weapons that could withstand a first strike by the late 1960s, and the ensuing agreement to SALT I, which resolved any trace of a Prisoner's Dilemma, military hard-liners continued to press for the replacement of assured destruction with

---

[5] See memo from Fritz Ermarth to Jasper Welch and Victor Utgoff, subject "Countervailing Strategy and the Targeting Policy," March 20, 1980, and attached report "Countervailing Strategy for General War," "3/80–4/80," Box 35, Brzezinski Collection, Jimmy Carter Presidential Library (JCPL), especially p. 2 of 4 of the latter outlining "the credible capability of the US to wage general war of any plausible scope or duration."

escalation dominance.[6] Fearing the incredibility of deterrence predicated on recognizing an inherent limit on the constructive purposes of thermonuclear weapons, US leaders were not satisfied to have nuclear or conventional military parity. Rather than accept the obsolescence of large-scale war in light of the reciprocal fear of uncontrollable escalation into mutual annihilation and the increasing superfluity of perpetrating violence at other levels of engagement, the United States sought to preserve the prerogative to engage in effective armed combat at all levels of conflict. By 1980, President Carter solidified the US policy as maintaining escalation control, or escalation dominance, not just in prolonged nuclear contestation but also amid all military rivalry.

Carter's adoption of the countervailing strategy can be traced back to National Security Defense Memorandum 242 (NSDM-242), which PD 59 mentions and supersedes.[7] This earlier document, signed by President Nixon, with James R. Schlesinger at the helm of his Department of Defense, on January 17, 1974, directed that further plans "for limited employment options which enable the United States to conduct selected nuclear operations" be developed and formally incorporated into the Single Integrated Operational Plan (SIOP). While the public debate over NSDM-242 focused on its "reemphasis" on counterforce targeting, the SIOP had, since 1962 and including the period when assured destruction became avowed US policy, already contained most of these targets. Strategists acknowledge that the novelty of the NSDM-242 lay in "the notion of targeting those Soviet assets that would be critical to Soviet postwar recovery and power."[8] This meant that even though the pro-nuclear-use strategy seemed to win the moral upper hand by removing civilian populations from nuclear targeting, the intention was, in fact, to place these population centers on hold for ultimate extermination if warranted to undermine the Soviets' prospects of recovery.

The strategic rationale for the decision at the core of NSDM-242 not to give priority to population centers as targets was the concept of "escalation control," defined as the maintenance of "our capability to effectively withhold

---

[6] Douglas P. Lackey, "The American Debate on Nuclear Weapons Policy: A Review of the Literature 1945–85," *Analyse and Kritik* 9 (1987), S. 7–46, 24–30; note that Lackey draws his conclusion even in view of MIRVs and ABMs.

[7] Both Gregg Herken and Fred Kaplan concur that Carter's acceptance of the countervailing strategy was prefigured by and wholly consistent with the Schlesinger doctrine of flexible response underlying NSDM-242, Herken, *Counsels of War* (New York: Alfred A. Knopf, 1985), 300; Kaplan, *Wizards of Armageddon* (New York: Simon & Schuster, 1983), 384. The term "countervailing" seems to be Harold Brown's, although it is clear from the archival record that its stipulation of flexible response and escalation dominance was introduced by Zbigniew Brzezinski, whose team states that the term "comes from Harold's own pen"; memo from William E. Odom to Brzezinski, March 22, 1980, quote p. 4 of 5, see also p. 3, "5/80–1–81" (filed out of sequence), Box 35, Brzezinski Collection, JCPL. See the language of PD 59, NLC-12-37-4-8-6.

[8] Desmond Ball, "The Development of the SIOP, 1960–1983," in *Strategic Nuclear Targeting*, ed. Desmond Ball and Jeffrey Richelson (Cornell University Press, 1986), 73.

attacks from additional hostage targets vital to enemy leaders, thus limiting the level and scope of violence by threatening subsequent destruction."[9] Operational planning for this new guidance meant providing the National Command Authorities (NCA) the ability to execute options in a controlled and deliberate manner, to "hold some vital enemy targets hostage to subsequent destruction," and to control "the timing and pace of attack execution, in order to provide the enemy opportunities to consider his actions" so that "the best possible outcome" might be achieved for the United States and its allies.[10] For these purposes, NSDM-242 introduced the concepts of "withholds" and "non-targets." Centers of Soviet political leadership and control, for instance, would be withheld from destruction for the purpose of interwar deterrence and bargaining, whereas "population *per se*" had now been exempted absolutely from targeting – oddly enough, given that the definition of assured destruction rested on the ability to wipe out 33 percent of the Soviet Union's population in a second strike.[11] The capitulation of MAD to NUTS, therefore, depended on finding the stance of reciprocal deterrence incredible because massive retaliation would be immoral and thus pointless in the case of deterrent failure. However, retaliation could be moral in case it was intended as a constructive remedy on the path to US victory. Surely, strategists concluded, the United States could increase its chances of recovering more quickly than the Soviets.

Driven by the new reality of rough strategic equality, the United States changed its formal targeting criterion for Assured Destruction from the destruction of Soviet urban-industrial centers to the prevention of the Soviet Union's gaining any advantage from a nuclear exchange, that is, from recovering, economically or militarily, more rapidly than the United States after a nuclear war. As Secretary of Defense Harold Brown later expressed in 1981, according to the logic of this new strategic plan, American strategic nuclear forces were designed to cripple the military and political power of the Soviet state, not strictly its industry and people.[12] The US military planners refused to concede a reciprocal deterrent footing to the Soviets. They openly pursued a stance of ascendance consistent with both the Prisoner's Dilemma model of security and the arms race and the Chicken game representation of bargaining. Even though moving the United States to a footing that accepted entering into and winning a

[9] Senate Armed Forces Committee, Authorization for Appropriations for Fiscal Year, 1980, 1437 (cited by Ball and Richelson, eds., *Strategic Nuclear Targeting*, 1986).

[10] Ball, "The Development of the SIOP," 1986, 73.

[11] This change in targeting policy was first announced by Secretary of Defense Elliot Richardson, who testified in April 1973: "We do not in our strategic planning target civilian population *per se*." Quoted in Ball and Richelson, eds., *Strategic Nuclear Targeting*, 1986, 241. Of course, given the collocation of population and industry, any attack designed to cripple Soviet recovery would produce massive civilian casualties. Late Carter administration documents confirm that civilian populations remained targets; see, Memo, Harold Brown to The President, undated, "Nuclear Targeting Policy Review," "8/78–4/79," Box 35 (PD 59), Brzezinski Collection, JCPL.

[12] Warren C. Schilling, "US Strategic Concepts in the 1970s: The Search for Sufficiently Equivalent Countervailing Parity," *International Security* (1981), 6:2, 60, 65.

protracted nuclear conflict drawn out over "weeks or months" was a decisive extension of the flexible response originally implemented by Schlesinger, still its logic was entailed in NSDM-242.[13]

CARTER'S CONVERSION FROM MAD TO NUTS

Carter's presidency offers a pivotal case study of how the NUTS military stance came to win the intellectual and policy debate. Of all presidents, we would expect Carter to maintain a deterrent posture consistent with the classical liberal stance of peace for peace and war for war, defying the Prisoner's Dilemma model of the nuclear security dilemma. In the PD model of the security dilemma, each actor assures the other of his intention to pursue dominance even if guaranteed the other's peaceful cooperation. In his monograph *Carter's Conversion: The Hardening of US Defense Policy*, Brian Auten investigates Carter's transformation from being opposed to fighting the Cold War via military might to openly embracing this hard-line position.[14] Auten argues that Carter's defense team members came to appreciate the wisdom of NUTS and a combative defense policy because, over their time in office, they learned to grasp the strategic realities validating the offensive neorealist approach to international relations.[15] According to Auten, Carter's team came to comprehend the actual constellation of power dynamics and material facts comprising global security and shifted its defensive posture accordingly. Although Auten is correct that an offensive realist perspective came to dominate Carter's White House, the source of this transformation was not factual and logical truths but rather James R. Schlesinger's doctrine, as his flexible response, escalation control, approach is referred to.

As a seasoned chief executive officer with prior experience leading the US Department of Defense, Schlesinger had the ability get his perspective heard and implemented.[16] Schlesinger's approach has signature features consistent with strategic rationality. In 1967, he argued that the United States must assert

---

[13] For discussion see Herken, *Counsels of War*, 1985, 300; Carter administration documents confirm; quote is from memo William E. Odom to Zbigniew Brzezinski, March 22, 1980, "Draft PD on Nuclear Targeting," 1-5 at 4, "5/80–1/81," Box 35, Brzezinski Collection, JCPL.

[14] Brian Auten, *Carter's Conversion: The Hardening of US Defense Policy* (Columbia: University of Missouri Press, 2009).

[15] Most academic international relations "realism" is "neorealism" because it accepts that there are structures beyond individuals' control that are important to understand in analyzing global affairs. The two main schools of neorealism are "offensive" and "defensive." Throughout this chapter, I use "realism" as a shorthand designation for neorealism and modify the term by its offensive or defensive variant as required (see David A. Baldwin, ed., *Neorealism and Neoliberalism*, New York: Columbia University Press, 1993).

[16] Schlesinger articulated his position on RAND's brand-name systems analysis and planning-programming-budgeting (PPB) as a means of bureaucratic administration in "Uses and Abuses of Analysis," US Senate Memorandum, 90th Congress, 2nd Session, 1968, published in *Survival: Global Politics and Strategy* (1968) 10:10, 334–342; for discussion, see Amadae, *Rationalizing Capitalist Democracy*, 2003, 62–75.

hegemony over its allies.[17] Of course he had already played the pivotal role in implementing the flexible response nuclear posture as President Nixon's Secretary of Defense. He advocated the single criterion means of appraising value consistent with rational choice theory and was well aware of how it departed from the constrained maximization characterizing neoclassical economic theory.[18] And as a RAND analyst, he was thoroughly familiar with wargaming simulations that applied strategic rationality in the way anticipated to guide actual decision making in time of war. The simulated war game buttressing the Carter administration's rationale for moving decisively beyond MAD to NUTS, called the "Red Integrated Strategic Offensive Plan Version-5C," stated this claim outright: "The RISOP is built on an annual basis as a hypothetical Soviet counterpart to the SIOP ... The RISOP is not a lightly disguised version of the real thing. It is the result of an operational planning exercise in which we apply capabilities in ways in which we believe to be in their best interests."[19] The memorandum putting forward the implications of this simulation demonstrates the need for a new, land-based ICBM system, states that assured destruction is equivalent to a "1914 war plan," calls for war-fighting capability in case deterrence fails, and demands crisis bargaining capability.[20] In preparing to engage in nuclear warfare, the simulations provided the basis for the actual strategies that would be implemented.

Archival documents reveal Schlesinger to be a key figure in Carter's administration. Corroborating the view that Schlesinger's strategic perspicacity was only possibly eclipsed by his administrative acumen, President Carter had initially hoped to appoint Schlesinger to be his incoming Secretary of Defense. However, he soon realized that Schlesinger would not pass muster among his more liberal cabinet nominees and advisors. He thus chose to appoint Schlesinger to head the Department of Energy (DOE), which he created in August 1977. Like the Department of Defense, the DOE was responsible for managing atomic secrets and materials. Regarding the DOE's role, despite any congressional attempts to limit it, Schlesinger observes, "They are going to continue to produce nuclear fuel, and only the government can do that. They are going to produce a hell of a lot of nuclear weapons and do the research and development on nuclear weapons and the national labs are going to stay within the Department of Energy."[21]

[17] "European Security and the Nuclear Threat since 1945," RAND Report P3574, April 1967. Duncan Snidal discusses hegemonic stability theory based on a Prisoner's Dilemma analysis in "The Limits of Hegemonic Stability," *International Organization*, (1985) 39:4, 579–614.
[18] "Systems Analysis and the Political Process," RAND Report P3464, June 1967, 4 and 7.
[19] Report is called "The Red Integrated Strategic Offensive Plan Version-5C," the office is Studies, Analysis and Gaming Agency; in, "4/24/79," Box 35, "PD 59," Brzezinski's collection, JCPL.
[20] Memo from Vic Utgoff, William Odom, and Fritz Ermarth, to Zbigniew Brzezinski and David Aaron, April 24, 1979, "Targeting ... SAC," "4/24/79," Box 35, Brzezinski Collection, JCPL, page 1 of 2.
[21] "Interview with Dr. James R. Schlesinger," Carter Presidency Project, Miller Center of Public Affairs, July 19–20, 1984, 109; available at JCPL, and http://millercenter.org/president/carter/oralhistory/james-schlesinger, accessed June 2014.

Whereas Carter was attracted to Schlesinger because of his former tenure under President Nixon as Secretary of Defense and had a general idea of his strategic view of international affairs and military security, he was likely unaware that Schlesinger had been a virtual fellow traveler with the ultra-hawks of the Committee on Present Danger (CPD): Paul Nitze, Albert Wohlstetter, Richard Pipes, and Colin Gray.[22] This privately organized circle of defense analysts would burden Carter's attention throughout his term in office. In 1976, Schlesinger had ties to this pro-nuclear-use advocacy alliance, yet he determined that maintaining his distance from this organization gave him more independence as a government official and freedom from branding that could compromise his effectiveness by mere association.[23] Schlesinger's particular form of pro-nuclear strategy took the linguistic form of escalation control, instead of escalation dominance, although the two positions are indistinguishable once implemented.[24]

Thus, close inspection thus verifies that a prominent member of Carter's cabinet with extraordinary bureaucratic sagacity was a leading proponent of flexible response.[25] Schlesinger noted that "unlike most of the people in the Cabinet," he had a relationship with Carter characterized by a "degree of intimacy ... and rapport" and that the president "tended to regard [him] as a universal authority."[26] Indeed, on exiting Carter's administration, Schlesinger openly expressed both his incredulity at the "weak and parochial" nature of Carter's incoming White House staff and his assessment that he stood head and shoulders above everyone with respect to his own experience, knowledge, and Washington connections.[27]

[22] There have been at least two incarnations of this group, in the 1950s and 1970s; see Jerry S. Sanders, *Peddlers of Crisis: The Committee on Present Danger and the Politics of Containment* (South End Press, 1999).

[23] For Schlesinger's association with the CPD, see CPD mailing list dated October 14, 1976, "Master Copy of List of Possible Board Members," p. 6, folder "CPD: Board Names," Box 284, Collection, "Committee on the Present Danger," Hoover Institution Archives, Stanford University. There is a second document in this folder also bearing Schlesinger's name as a potential member.

[24] On this latter point, see Charles Glaser, "Why Do Strategists Disagree about the Requirements of Strategic Nuclear Doctrine," in Lynn Eden and Steven E. Miller, eds., *Nuclear Arguments: Understanding the Strategic Nuclear Arms and Arms Control Debates* (Cornell University Press, 1989). For discussion see Lackey, "American Debate," 1987, section titled "Schlesinger and the Rise of Counterforce," 31–35.

[25] Brezinski's papers contain the academic paper, "The Nuclear Warfighting Dimension of the Soviet Threat to Europe," by Joseph D. Douglass Jr., and Amoretta M. Hoeber, in the *Journal of Social and Political Studies* (1978), 3:2, which makes clear "Schlesinger's Strategy" is consistent with nuclear war fighting, p. 141, NLC-12-58-2-5, JCPL. Bzezinski's files also find the typed document, "P[residential] D[irective] Questions," no date, with exacting discussion of the demands of maintaining "escalation equivalence," and mention of NSDM-242, NLC-31–220-4-1-8, JCPL.

[26] "Interview with Dr. James R. Schlesinger," 1984, 39.

[27] Ibid., "September of 1976, I thought that Jimmy Carter had this immensely quick intelligence, and that he would quickly learn – reasonably quickly learn – what he needed to know for the job, because he seemed to have judgment and quickness of mind. That may have been my own self-flattery because he responded so well to the advice that I tendered. But in any event that was my

Additional archival evidence further suggests that Schlesinger's fingerprints are on the contents of PD 59. Carter's closest cabinet confidant and National Security Advisor Zbigniew Brzezinski relied on his military assistant and crisis coordinator, General William Odom, to consult with Schlesinger as a "source of support."[28] Brzezinski personally wrote to Carter stating, "The basic direction toward more flexibility was set by the Schlesinger effort in 1974 which led to NSDM-242."[29] And perhaps the most telling archival evidence is that not only did Brzezinski's staff denigrate Secretary of Defense Brown, but that it was the National Security Advisor who drafted PD 59, and not Secretary Brown as is typically assumed because of the auspices of his office.[30] The internal Carter administration documents reveal that behind Carter's back, Brzezinski "dragged Brown along on this PD [59]."[31] Brzezinski's team referred to

view early on, and it did not change in the … let's say, for the first six or seven months that I knew him. After a while it became clear to me, regrettably, that the lack of experience that I had initially undervalued just was very important, and could not be rapidly repaired even in the Presidency," p. 9.

[28] E.g., William Odom, March 17, 1978, NLC-12–53-5–12-2, JCPL; see mention about "kibitzed with Jim on PRM-38"; Fritz Ermarth's report to Brzezinski, October 5, 1978, NLC-17–51-1–7-9, JCPL, p. 2 of 4, note this memorandum covers a lot of key areas of nuclear policy; another memorandum from David Aaron to Brzezinski, July 25, 1978, reports "Hunter consulted with Vic and Jim on PRM-38," p. 1 of 1, NLC-10–13-6–9-1, JCPL. Schlesinger, "Interview with Dr. James R. Schlesinger," 1984, 58.

[29] Brzezinski's memorandum to President Carter urging him to endorse PD 59 makes clear that this directive is directly continuous with NSDM-242, July 24, 1980, subject "Nuclear Targeting Policy," "5/80–1/81," Box 35, Brzezinski Collection, JCPL.

[30] This attribution acknowledges that it seems inconsistent with Brown's views stated throughout his role in Carter's administration, Kaplan, *Wizards of Armageddon*, 1983, 382–386; Herken, *Counsels of War*, 1985, 298–302. Regarding Brzezinski's hands-on involvement with crafting PD 59, see Memo, Harold Brown, to the President, Subject: "Nuclear Targeting Policy Review," date unclear, pp. 1–4; "8/78–4/79," Box 35, Brzezinski Collection, JCPL; Special Coordination Committee Meeting Notes, April 4, 1979, pp. 1–9, "8/78–4/79," Box 35, Brzezinski Collection, JCPL; and Brzezinski's memorandum to the President, date unclear, but providing a synopsis and action plan based on the Special Coordination Committee, with specific reference to "(1) stable deterrence? (2) stable crisis bargaining? And (3) effective war management?" "8/78–4/79," Box 35, Brzezinski Collection, JCPL; a memorandum from Vic Utgoff to Brzezinski, April 5, 1979, "8/78–4/79," Box 35, Brzezinski Collection, JCPL, makes clear what guidance Brzezinski's team was providing to Harold Brown. As well, Brzezinski had discussions among his staff on drafts of PD 59 on how to proceed with winning its acceptance by Carter as well as the specific wording to be used in the text of the presidential directive; see William E. Odom to Brzezinski, March 22, 1980, "5/80–1/81," Box 35; note also strategy document by Fritz Ermarth to Jasper Welch and Victor Utgoff, on the subject, "Countervailing Strategy and the Targeting Problem," March 20, 1980, seeking "a concept for dealing with its [strategic competition] worst contingency," with respect to how to get PD-59 past Brown and signed by Carter, "In my view, this more comprehensive approach would move the doctrinal process across a broad front at a time when we are unlikely to get Harold Brown or the President to sign on to a directive that is broad enough and innovative enough to generate real progress. If we take this comprehensive approach now, then we may be ready for a real PD in early 1981." Page 1 of 2, "3/80–4/80," Box 35, Brzezinski collection, JCPL.

[31] Memo from William E. Odom to Brzezinski, July 24, 1980, "Targeting PD Briefing for the President," notes that "flexibility," "targeting categories," and "acquisition policy" sections reflect Brzezinski's nuclear strategy perspectives and not Brown's.

"Brown's view of our defense posture a rudderless ship piloted by a bland [sic] man."[32] Brzezinski's staff clearly drafted PD 59 and strategized to gain Secretary Brown's and the President's final approval for the directive.[33] By the end of Carter's term, offensive realism, in the form of NUTS, would become official policy.[34]

The differences between Secretary of Defense Brown's strategic stance and that of Secretary of Energy Schlesinger and National Security Advisor Brzezinski are vivid and apparent in drafts of the ensuing presidential directive and the US nuclear targeting policy. Brown held that "a full-scale thermonuclear exchange would constitute an unprecedented disaster for the Soviet Union and for the United States," and that there could be no guarantee whatsoever that even a "tightly controlled use of the strategic forces for larger purposes could be kept from escalating to a full-scale nuclear war."[35] Brzezinski's team redrafts the presidential directive to further its goals. It seeks the flexibility and "ability to design nuclear employment plans on short notice in response to the latest and changing circumstances" not limited by (1) stipulated "pre-planned options," (2) prior attack, or (3) potential collateral damage.[36] Thus, it views engaging in nuclear conflict as thinkable and winnable, retains the first-right to engage in nuclear warfare, and seeks leverage to bargain acceptable terms in favor of the United States. It rejects Thomas Schelling's limited nuclear option

---

[32] The entire section reads, "Today I saw for the first time a copy of Brown's posture statement in its final form. I was staggered by it. Last year it marched to the tune of PD-18. From this year's version it is impossible to infer the existence of PD-18. Not only does it lack a coherence which only a national and military strategic [stet] can import, but many of its meandering sections are anti-strategy and anti-doctrine musings. I don't know who cleared the thing on our staff but he did not bring the outlines of PD-18 to bear on it. If I were a member of Congress, I would call Brown's view . . . ," memo, William E. Odom, to Brzezinski, Jan. 29, 1979, NLC-12-21-9-11-4, JCPL, p. 2 of 2. See also, memo, from William E. Odom to Brzezinski, July 24, 1980, "5/80–1/81," Box 35, Brzezinski Collection, JCPL; note that Odom states that Brown's reticence was not revealed to Carter. Furthermore, it is clear that Brzezinski had Odom working on a draft of PD 59; memo from William E. Odom to Brzezinski, March 22, 1980, subject "Draft PD on Nuclear Targeting," see p. 3 of 5 with additional discussion of strategy to bring Brown on board the directive, at p. 5, "3/80–4/80," Box 35, Brzezinski Collection, JCPL.

[33] See Memo from William E. Odom and Jasper Welch to Brzezinski, March 25, 1980, subject, "Nuclear Targeting Policy," and follow-up memorandum by same authors to Bzezinski, March 26, 1980, subject, "Targeting Policy"; "3/80–4/80," Box 35, Brzezinski Collection, JCPL. Indeed, Brown's role was merely to suggest some revisions on Brzezinski's draft of PD 59; see memo from William Odom and Jasper Welch to Brzezinski, April 17, 1980, Subject, "Draft PD on Nuclear Employment Policy," one-page memo, with nine pages of draft and commentary; "3/80–4/80," Box 35, Brzezinski Collection, JCPL.

[34] Kaplan, *Wizards of Armageddon*, 1983, 384–385.

[35] Chapter 5, "The Nuclear Capabilities," draft of "Targeting Policy," p. 69, attached to memo from Brzezinski to the Secretary of Defense, undated, but requests a response by April 4, 1980, and a cover memorandum is dated April 9, 1980, "3/80–4/80," Box 35, Brzezinski Collection, JCPL.

[36] These three points are numbered 3, and 6, 10, pages hand labeled 7E 1 and 2, attached to memo from William E. Odom and Jasper Welch to Brzezinski, dated March 25, 1980, "3/80–4/80," Box 35, Brzezinski Collection, JCPL; April 4, 1980, "3/80–4/80," Box 35, Brzezinski Collection, JCPL.

of achieving controlled escalation through "psycho-political effects," which in its view characterized NSDM-242 rather than to uphold "the First Principle of War ... that is, destroy the enemy's army or its ability to fight" in view of our "scarce [and vulnerable] ... nuclear weapons" over "days, and weeks ... or months" to ensure destruction of our opponent and vie to secure US survivability and recovery."[37] Additionally, whereas the benefit of NUTS, at least from the perspective of prospective public evaluation, had been to take innocent populations out of harm's way, the Brzezinski plan clearly "retain[s] this city-busting opinion in the pre-planned options section," which was intentionally redacted from PD 59's release to make it more palatable for those who would question its superiority to assured destruction.[38]

Carter's indefinite postponement of the neutron bomb project in March 1978 further substantiates the narrative that he came into office supporting Schelling's nuclear strategy of assured destruction relying on retaining counterstrike capability through submarine-based missiles.[39] This action makes obvious Carter's aversion to the militant hard-line position consistent with NUTS strategic doctrine of flexible response, which treats nuclear weapons as conventional weapons and seeks to maintain escalation control by exerting asymmetric advantage. Carter's action went against the advice of all his national security advisors, individually and collectively.[40] Carter shocked and dismayed his national security team by standing against this anti-populace, building-preserving, nuclear warhead. The president's national security corps thought that Carter had no grasp of military strategy, and they felt disrupted, stymied, and embarrassed by what to them seemed to be his uncomprehending and solo intervention.[41] The neutron bomb was integral to Schlesinger's strategy of flexible response that treated nuclear weapons as conventional arms, and it was particularly suited to achieve extended nuclear deterrence to afford Europe protection from a potential Soviet invasion. Here the concepts underlying MAD and NUTS differ on how to

---

[37] Point 1, ibid., 7E.; for explicit rejection of Schelling's style limited nuclear options, and the rejection that political control versus strategic control should oversee nuclear targeting, see memo from William E. Odom to Brzezinski, March 22, 1980, "Draft PD on Nuclear Targeting, p. 2, "5/80–1/81," Box 35, Brzezinski Collection, JCPL; see also pp. 2–4 of this same document.

[38] On the city-busting stipulation in the pre-planned options section, see notes combined with memo from Odom and Jasper Welch to Brzezinski, March 25, 1980, "Nuclear Targeting Policy," "3/80–4/80," Box 35, Brzezinski Collection, JCPL, point 1, on page hand labeled #7E; on the careful redaction of the city-busting feature of PD 59's pre-planned options, see memo from Odom to Brzezinski, January 7, 1981, "Distribution of PD-59," with attached redacted copy of PD 59, NLC-12-37-4-8-6.

[39] Schlesinger makes this point, "Interview with Dr. James R. Schlesinger," 1984, 72. For internal Carter administration discussions, see Jim Thomson to Brzezinski, February 22, 1979, "8/78–4/79," Box 35, Brzezinski Collection JCPL, "The most significant setback would have to be the neutron bomb affair."

[40] Schlesinger, "Interview with Dr. James R. Schlesinger," 1984, 58.

[41] Schlesinger states this in his exit interview, and it is evident in the aftermath of Carter's indefinite postponement of the neutron bomb in Folder "2–4/78," Box 17, National Security Affairs Collection, JCPL.

maintain effective deterrence, with the former looking to manipulating the risk of engaging in nuclear war, and the latter treating force as well calibrated with predictable consequences.[42]

Schlesinger referred to this as "the neutron bomb fiasco" and stated that the president was "kind of blind on natural security problems."[43] Clear about his impact on Carter's administration, Schlesinger states that in his role overseeing the Department of Energy, "ultimately, in November of '78, I got presidential approval – it was announced – of the production of the components of the neutron bomb," meaning that in fact "although you don't have a prompt neutron bomb capability, you are six hours away from having neutron bomb capability."[44] Schlesinger was well aware of his active perpetuation of flexible response, which this enhanced radiation weapon exemplified. He further observes,

I have been a patron of enhanced radiation warheads since my days at RAND, subsequently my days at the Atomic Energy Commission, and I called for deployment when I was Secretary of Defense, and ultimately produced the components as Secretary of Energy, so I have a consistent, although in the eyes of some, a somewhat checkered career on this subject.[45]

Even after Schlesinger left Carter's administration in August 1979, he "worked with [Senator] Sam Nunn to put to use the President's expenditures from the administration on national security."[46]

CARTER'S NUCLEAR SECURITY DILEMMA

Carter's early approach was characterized by classical liberalism and its promise of peace for peace. However, the exercise of either promising to support a

---

[42] Robert Jervis discusses the difficult problem of extended deterrence, and how Schelling's and Schlesinger's approaches differ with the former relying on manipulating risk, and the latter depending on incurring tangible punitive military damage consistent with escalation dominance, see "Security Studies: Ideas, Policy, and Politics," in *The Evolution of Political Knowledge: Democracy, Autonomy, and Conflict in Comparative and International Politics*, ed. by Edward D. Mansfield and Richard Sisson (Ohio State University Press, 2004), 100–126, especially fn 34, p. 126. Jervis notes that Schelling's deterrence via manipulating risk depends on demonstrating the irrational stance of being prepared to go down the slippery slope of engaging in suicidal war which "even if true, 'is a dead end'" because preparing to accept ultimate devastation as a means of securing stable peace signifies abandoning strategic rationality, "Security Studies," 2004, fn 34, p. 126, Jervis quotes Lawrence Freedman, *Evolution of Nuclear Strategy*, 1981, 400. Here Jervis acknowledges the conclusive incoherence of MAD if strategic rationality is one's sole means of addressing security. Jervis himself is aware that a deterrence situation viewed as a Chicken game characterized by "competition in risk taking" is best exited by offering the *reassurance* that "the state will not punish the adversary if it behaves in the desired way" by cooperating; first quote is from Robert Powell's commentary on Jervis's essay "Security Studies," called "Nuclear-Deterrence Theory: Where We Left Off When the Berlin Wall Came Down," 2004, in the same volume, 131–136, at 133; second quote is from Jervis "Security Studies," 2004, at 111.
[43] Schlesinger, "Interview with Dr. James R. Schlesinger," 1984, 62.
[44] Ibid., 63.
[45] Ibid., 63.
[46] Ibid., 71–72.

nuclear deterrent counterstrike or effectively engaging in Schelling-style bluffing seemed out of reach for Carter. In Robert Jervis's words, "Making either threats or promises credible is difficult enough, doing both simultaneously is especially demanding . . . President Carter probably succeeded in convincing the Soviets that he would cooperate, but he also tempted them to exploit him."[47] Thus, President Carter faced the dilemma of how to credibly threaten the USSR with a devastating counter strike that would serve no purpose besides killing millions of hapless Soviet citizens. As a devoted man of conscience, maintaining the credibility of this deterrent threat and immoral promise was outside Carter's reach.

Behind-the-scenes conversations offer one insight into what led President Carter to sign Presidential Directive 59, which put the US military on a footing treating nuclear weapons as conventional forces, planned to fight a protracted nuclear war, and incorporated the offensive MX missile system. However, understanding the broader intellectual and political climate is also important. Thomas Schelling and Robert McNamara terminated their active engagement with arms control by the late 1960s at the same time that Albert Wohlstetter, Herman Kahn, Colin Gray, Paul Nitze, and Edward Teller initiated a vocal public campaign to promote a pro-nuclear-use policy.[48] Jervis defended mutual assured destruction, initially in his 1976, 1978, and 1984 publications, and then more effectively in *The Meaning of the Nuclear Revolution* in 1989.[49] Carter was counseled by hawkish Secretary of State Zbigniew Brzezinski; his Secretary of Defense Harold Brown; and his Secretary of Energy James R. Schlesinger.[50]

---

[47] Robert Jervis, *Meaning of the Nuclear Revolution* (Cornell University Press, 1989), 58. James R. Schlesinger makes this direct observation of Carter, "Interview with Dr. James R. Schlesinger," 1984, 60–61.

[48] George Kennan was a prominent arms control advocate who was not a defense rationalist. Jervis, by contrast, fully engaged rational deterrence theory, even if he ultimately pushed beyond its confines in finding it limited for leaving the debate stuck with unilateralist and escalating deterrence; for the best statement of this acknowledgment, see Charles Glaser, "The Security Dilemma Revisited," *World Politics* (1997), 50:1, 171–201, at 193; see also Achen and Snidal's comment on Jervis's position, "Rational Deterrence Theory," 1989, at 155. Freedman, too, though thorough in his knowledge of rational deterrence theory, ultimately concludes that it was not able to defend the saner policy of MAD (see especially pp. 377–400). Note that Schelling continued to publish in the area of arms control, e.g., Thomas Schelling, "A Framework for the Analysis of Arms-Control Proposals," *Daedalus* (1975) 104:3, 187–200.

[49] See Robert Jervis, *Perception and Misperception in International Politics* (Princeton University Press, 1976; "Cooperation under the Security Dilemma," *World Politics* (1978) 30:2, 167–214; *The Illogic of American Nuclear Strategy* (Ithaca: Cornell University Press, 1984); and *The Meaning of the Nuclear Revolution*, 1989). For perhaps the most succinct and effective overview of the strategic debate, see Glaser, "Why Do Strategists Disagree," 1989, 109–171. It is clear that Glaser, too, though sympathetic to MAD (see especially p. 161 acknowledging that the "punitive retaliation [MAD] school holds the strongest positions on the disputed issues," yet fails to carry the debate).

[50] James R. Schlesinger, the author of the "Schlesinger Doctrine," and longtime RAND researcher, served in Carter's cabinet alongside Harold Brown from October 1977 to July 1979. The Schlesinger Doctrine promoted limited nuclear options (LNOs) from small tactical nuclear

Carter would face considerable foreign policy challenges, most notably the Iranian hostage crisis and the Soviet invasion of Afghanistan. At the same time, MAD was facing increasing scrutiny for holding innocent civilians hostage for the good behavior of their government. Its common features with NUTS in this regard were overlooked as the United States shifted its focus from assuring the USSR of its peaceful intention unless provoked into war as a last resort to securing the capability and resolve to prevail at all levels of armed conflict. This seemed to be the only means to solve the Prisoner's Dilemma riddle of avoiding mutual defection by having the wherewithal to maintain a credible punitive threat.

All are agreed that the nuclear arms race was angst ridden.[51] Carter had entered office amenable not only to arms control but also to disarmament. He made his intention clear in Presidential Directive 18, which directed that the United States should "'take advantage of our relative advantages in economic strength, technological superiority and popular political support' *both* to seek Soviet cooperation in resolving conflicts, renegotiating arms control agreements, and constructively dealing with global problems *and* to counterbalance adverse Soviet influence in key areas of the world."[52] His apparent lack of adequate concern for defense frightened the pro-nuclear-use contingent of policy analysts. He wrote in his personal diary in August and October of 1977:

Met with the Committee on Present Danger, Paul Nitze, Gene Rostow, and others. It was an unpleasant meeting where they insinuated that we were on the verge of catastrophe, inferior to the Soviets, and that I and previous presidents had betrayed the nation's interests. I told them I'd like to have constructive advice, balancing all factors with at least the possibility considered that the Soviets did want a permanent peace and not suicidal nuclear war

In Congress, Senator [Scoop] Jackson was the core around whom the most vitriolic anti-Soviet forces coagulated. Their premise was that the Soviets were enormous ogres who were poised to take over the world. This group looked on me as weak and naïve because I argued that the Soviet Union was rotten to the core and that over time our promotion of peace, human rights, and accommodation on arms control would be detrimental to the Soviets and beneficial to our nation.[53]

---

weapons to weapons of catastrophic destruction and sought to "control ... the level of violence in any conflict"; see Lawrence Freedman, *Evolution of Nuclear Strategy* (New York: St. Martin's Press, 1981), 377–392, quote from 384. There can be no doubt that the Schlesinger Doctrine seeks defense in maintaining the relevance of violence to control outcomes in conflict situations by maintaining the asymmetric policy of escalation dominance.

[51] The *Bulletin of American Scientists* kept a constant barometer on their members' estimation of the likelihood of nuclear war; for discussion, see Herken, *Counsels of War*, 1985, 105, 125, 185, 192, 247, 303.

[52] Brzezinski to President Carter, subject, "Capitalizing on Our Economic Advantages in U.S.-SU Relations," undated, NLC-29-11-2-3-3, JCPL, Brzezinski Collection, declassified 2008/04/09, p. 1 of 2.

[53] President Jimmy Carter, *White House Diary* (New York: Farrar, Straus, and Giroux, 2010), August 4 and October 25, 1977, 76, 123–124.

Carter worried some US defense analysts because he seemed to accept that the Soviets had benign intention, and that the United States and the USSR could work together to ensure peaceful coexistence. The Soviet's 1979 invasion of Afghanistan strengthened their belief that the Soviets intended to extend their empire using military force.[54] Notwithstanding that MAD was a fact and not a policy, defense rationalists themselves were hard pressed to defend it against the NUTS plan to prepare to fight and win a nuclear war.[55]

Since analysts conceded that the Prisoner's Dilemma best characterized nuclear security dilemma and arms race, a policy of mutual assured destruction could no longer be rationally sustained. Whereas classical liberals offered the assurance of cooperation, MAD offered the assurance of destruction as a punitive threat to unlock the perceived Prisoner's Dilemma, which was derived from an Assurance Game (Stag Hunt) under conditions of significant risk. Not only did they agree that the United States most preferred unilateral defection in a nuclear standoff, coercive bargaining, and an arms race but more importantly, the signature feature of applied PD logic stipulated that these openly hostile preferences were wholly required for self-defense, even though the United States really preferred to get along amicably. By 1988, even Jervis, perhaps the most prolific and steadfast supporter of MAD throughout the 1970s, observed that "a central question for the work on anarchy is how cooperation is possible when actors are in a Prisoner's Dilemma."[56] However, the Prisoner's Dilemma model in particular clearly signifies that each actor hopes to gain by exploiting the other. Jervis makes this point in no uncertain terms: "Each is driven by the hope of gaining its first choice – which would be to exploit the other."[57]

Schelling introduced the initial ambiguity of accepting that a Stag Hunt Assurance Game transforms into the more virulent Prisoner's Dilemma as a function of uncertainty about others' intentions. Hence, he gave rise to a characteristic Prisoner's Dilemma pedagogy that sanctioned the idea that a predatory stance is wholly legitimated by and consistent with benign intent. Schelling's analysis was accepted by strategists who felt compelled to address the "worst contingency" security dilemma, which everyone seemed to concede

---

[54] John Gaddis has since concluded that the Soviet invasion of Afghanistan followed from their characteristic support of an internal security risk to the Marxist regime; *The Cold War: A New History* (London: Penguin, 2006), 220.

[55] On the status of MAD as a fact and not a policy, see Jervis, *The Meaning of the Nuclear Revolution*, 1989, 46–73. On the inability to provide a rational defense of MAD as a policy, see, e.g., Patrick Morgan, "New Directions in Deterrence," in *Nuclear Weapons and the Future of Humanity*, ed. by Avner Cohen and Steven Lee (Totawa, NJ: Rowman & Allanheld, 1986), 169–190. See also Glaser, "Why Do Strategists Disagree," 1989, 162.

[56] Robert Jervis, "Realism, Game Theory, and Cooperation," *World Politics* (1988), 40:3, 317–349, at 322; throughout his writings, Jervis reserves the PD game for contexts in which actors have predatory intent and reserves the Stag Hunt, or Assurance Game, for actors who have the first preference of cooperating with others, see, e.g., "Was the Cold War a Security Dilemma," *Journal of Cold War Studies* (2001), 3:1, 36–60.

[57] Jervis, "Realism, Game Theory, and Cooperation," 1988, at 318.

resembled the disconcerting PD.[58] The general acceptance of the widespread applicability of the Prisoner's Dilemma and PD logic of gain without regard for others is a direct result of the development of rational deterrence theory entangled with evolving nuclear policy. Consider security analyst Charles Glaser's defense of Prisoner's Dilemma logic and asymmetric deterrence. In the PD model of the security dilemma, derived from an assurance situation, the United States adopts the preferences of a predator in self-defense. Glaser notes that even though the United States adopts a predatory stance, its leaders still assume that other nations recognize that it is actually a peace seeker: "This line of argument plays a central role in the 'deterrence model,' which rejects the security dilemma completely, albeit implicitly, by assuming that the adversary knows the state [United States] is a pure security seeker." The science of deterrence opposes aggression, hence combining the US reflection of predatory preferences "with the assumption that the adversary is greedy, the deterrence model calls for highly competitive policies and warns against the dangers of restraint and concessions." The upshot of the Prisoner's Dilemma approach to superpower security entailed that "in describing the cold war competition between the United States and the Soviet Union, the deterrence model held that the Soviets were bent on expansion for entirely greedy reasons and knew that they had nothing to fear from the United States."[59] Whereas the competitor is viewed as an aggressor, one's own action, although directly opposing the other's interests, is viewed as inherently peaceful. Thus, analysts continuously tended to insist that the United States represented the "good guys" with upstanding values, failing to recognize the deepening chasm between their resolution of nuclear security and classical liberalism.[60]

It was standard to view the high-stakes nuclear superpower standoff in terms of a single-play Prisoner's Dilemma, which is the default in game theory as a result of emphasis on tangible outcomes, which even in an assurance standoff (Stag Hunt) with sufficient uncertainty necessarily transforms into the intransigent PD.[61] However, the PD formalization of the security dilemma and

---

[58] "Worst contingency" is quoted from Brzezinski's staff member Fritz Ermarth to Jasper Welch and Victor Utgoff, "Countervailing Strategy and the Targeting Problem," March 20, 1980, "3/80–4/80," Box 35, Brzezinski Collection, JCPL, p. 1.

[59] This and preceding two quotes from Glaser, "The Security Dilemma Revisited," *World Politics*, 1997, at 193. Note that in his 2001 "Was the Cold War a Security Dilemma?" Robert Jervis points out that during the Cold War, the United States "sought to thwart any potential rivals and open the world to American capitalist penetration" (43) and had the officially stated aim "To reduce and power and influence of the USSR," quoting NSC 20/4, in *Foreign Relations of the United States* (Washington DC: US Government Printing Office, 1948) 1:2, 662–672 at 667. Hence, Jervis speaks of a "deep security dilemma" suggesting that fear for security drives one to have essentially expansionist goals.

[60] See, e.g., Michael Doyle's contrasting understanding of the post–World War II nuclear peace, *Ways of War and Peace* (New York: W. W. Norton, 1997), 301–311.

[61] See, e.g., quote by Don Ross at the head of the Chapter 3, Assurance; "Game Theory," *The Stanford Encyclopedia of Philosophy*, 2006. For example Jack Snyder, *The Soviet Strategic Culture: Implications for Limited Nuclear Options* (RAND Corporation, September 1977);

wholesale adoption of strategic rationality without a doubt shifted the challenge from assuring the other of one's benign intent to motivating cooperation through one's wherewithal to issue credible threats of harm. Thomas Schelling had found the PD game useful for capturing a security dilemma in which each actor prefers peace to conquest yet adopts the preferences of an aggressor as a function of uncertainty. Schelling reasoned that even in this worst-case scenario in which actors pursue goals inconsistent with each other's security, peaceful coexistence could be achieved if each actor could threaten the other from beyond the grave using devastating retaliation. If both nations have nuclear-armed submarines that can withstand a first strike, then each nation has the power to strike back, and it is in neither nation's interest to marshal a first strike.[62]

Schelling's strategically rational defense of MAD looks plausible but was found to have three fatal flaws attributable to its PD structure: immorality, incredibility, and irrationality. Were the United States to be hit by a Soviet all-out first strike and the only recourse was to strike back to wreak similar damage on the Russians, not only would this counterattack be immoral, but it stood indicted for lacking any causal power to serve US interests after deterrence had failed and, thus, any credibility to deter in the first place.[63] Defense rationalists, consequently, reasoned that MAD rested on an immoral, incredible, and therefore irrational threat to strike back when such an act can only seal its own doom: "It is perhaps a central tension in deterrence ... that its ultimate threat is to engage in a senseless act of total destruction."[64] Without any contingency plan in place to fight rather than admit defeat, MAD further seemed effete.[65] The immorality of the threat of massive retaliation was the undoing of MAD because it signified the incredibility of following through, thus rendering deterrence inconsequential. Additionally, MAD could be accused of being suicidal if the act of following through on a counter strike would provoke additional Soviet missile strikes on America.[66]

Jervis, "Security Under the Security Dilemma," 1978; Freedman, *Evolution of Nuclear Strategy*, 1981. Of course, Thomas Schelling's *Strategy of Conflict* (Cambridge, MA: Harvard University Press, 1960) had started this practice.

[62] The question of whether the Soviets had the capability to detect US submarines was raised in the famous Team B Report that concluded that the fact there was no evidence of such technology provides sufficient reasoning that it may exist; Anne Hessing Cahn, "Team B: The Trillion-Dollar Experiment," *Bulletin of the Atomic Scientists* (1993), 49:3, 22, 24–27.

[63] Jervis addresses this point, following Patrick Morgan in noting that "if people were totally rational, deterrence in a world of mutual assured destruction would not work. To carry out your threat would mean the destruction of your own society; so, if the other side thinks you will retaliate, it assumes you are less than rational Robert Jervis, "Deterrence Theory Revisited," *World Politics* (1979) 31:2, 289–324, at 299).

[64] Quoted from ibid., 300.

[65] Discussed by Freedman, *Evolution of Nuclear Strategy*, 1981, 395.

[66] This was Schlesinger's stated concern; see his US Senate testimony, "Uses and Abuses of Analysis," 1968, 340.

Moreover, in the continual contest of wills between the Soviets and Americans all too evident in Carter's daily log of White House events, MAD suggested a posture of "better Red than Dead" and did not provide a strong position from which to bargain.[67] The nuclear security dilemma modeled as a Prisoner's Dilemma transformed into a Chicken game once both sides faced the fear of potential mutual devastation yet still vied for supremacy.[68] Without continually maneuvering to at least achieve mutual cooperation rather than ignominious submission, it seemed that even if MAD did prevent a nuclear war, it would grant the Soviets a victory in the Cold War.

It was a signature belief of the defense rationalists that the threat of violence could be calibrated and applied to either compel or deter actions of the other side.[69] Both Schelling and Kahn advanced this view. For Schelling, the idea had been to manage risk in mobilizing threats, whereas for Kahn, the plan was to manage military application of force to achieve escalation dominance. In either case, politics and war became indistinguishable.[70] The recalcitrant Prisoner's Dilemma game, in which each person in pursuing his best interests mires both in a suboptimal outcome, was accepted in deference to the need to prepare for the worst case in which one's own defense threatens the security of the other.[71] According to the PD analysis of the nuclear security dilemma, nuclear weapons signify that defense must take the form of offensive action from which no one can be invulnerable. Even though mutual vulnerability is inescapable, the voices that clamored for proactive preparation to wage war, rather than those counseling the acknowledgment of reciprocal vulnerability, prevailed.[72]

NUTS seemed suited to address each of the signature weaknesses of MAD.[73] First, it signals the unwavering intention to counterattack if attacked at any

[67] Discussed by Jervis, "Deterrence Theory Revisited," 1979, 301–302; on the difficulty of applying nuclear strategy in its MAD or NUTS form to strategic bargaining, see Glaser, "Why Do Strategists Disagree," 1989, 168.

[68] On strategic bargaining in a nuclear Chicken game even in the context of MAD, see Jervis, *Meaning of the Nuclear Revolution*, 1989, 40–41.

[69] Schelling, *Strategy of Conflict*, 1960; see how the ability to control the outcome of violence is necessary to defend the strategic policy of NUTS and escalation dominance; Glaser, "Why Do Strategists Disagree," 1989, 150–151.

[70] "In particular, Schelling's ideas on tacit communication and the manifestation of signals make it clear that the players are involved in bargaining as much as fighting"; Martin Hollis and Steve Smith, *Explaining and Understanding in International Relations* (New York: Oxford University Press, 1990), 173–174.

[71] This is a central theme in Jervis, "Cooperation Under the Security Dilemma," 1979.

[72] For an alternative resolution of the nuclear security dilemma see, e.g., Edward F. McClennen, "The Tragedy of National Sovereignty," in Cohen and Lee, eds., *Nuclear Weapons and the Future of Humanity*, 1986, 391–406.

[73] Note that Carter's team working under Brzezinski was clear that "the Republican platform includes a lot of nuclear war-fighting doctrine," and that part of the mission of PD 59 was to clarify their policy "and leave no room for confusion." Memo from William E. Odom to Brzezinski, July 24, 1980, "Targeting PD Briefing for the President," "5/80–1/80," Box 35, Brzezinski Collection, JCPL, p. 1 of 1.

rung of engagement.[74] Second, it has a plan if deterrence fails: to fight for victory no matter what.[75] Third, it recommends "firing demonstration shots to show resolve."[76] Fourth, it accepts the challenge of a nuclear Chicken contest of wills, providing the strongest position from which to bargain.[77] Nevertheless, NUTS rests on the fallacy that it is possible to meaningfully engage in nuclear conflict, and it ignored the Soviets' promise to retaliate against *any* use of nuclear weapons and lost sight of the fact that "the primary objective of nuclear strategy is to avoid wars, not to fight them."[78] NUTS openly adopts a one-sided posture on defense in the full knowledge of the fact that achieving strategic dominance is more important than reassuring the other actor of one's benign intent.

President Carter had entered office exemplifying a classical liberal security posture. The classic liberal resolution of the security dilemma for both international relations and civil society, articulated in some form by Samuel Pufendorf, Hugo Grotius, Thomas Hobbes, Benedict Spinoza, John Locke, Adam Smith, Immanuel Kant, John Stuart Mill, Isaiah Berlin, and Friedrich Hayek, rests on a several key pillars.[79] Since self-preservation is basic for every actor, and the

---

[74] Escalation control is linked to flexible response under the reasoning that deterring, or preventing further, acts of Soviet aggression depends on having the flexible capability to prevail at any level of conflict. Of course, the debate is arrested on the point of whether introducing nuclear weapons into a conventional conflict would entail "escalation . . . [that] would still become uncontrollable"; pointed discussion of this debate is in "Senate Foreign Relations Committee Paper on PD 59"; the paper is attached to a memo from Jasper Welch to Brzezinski, September 11, 1980, report is dated September 9, 1980, from San Sienkiewicz, p. 3 of 8, "5/80–1/81," Box 35, Brzezinski Collection, JCPL.

[75] The plan is "To assure the survival of the US as a functioning independent nation, capable of political, economic, and military recovery," stated in "Countervailing Strategy for General War," attached to a memo from Ermarth to Welch and Utgoff, March 20, 1980, "Countervailing Strategy and the Targeting Problem," memo two pages, report p. 1 of 4, "3/80–4/80," Box 35, Brzezinski Collection, JCPL.

[76] Quoted from William Odom memorandum to Brzezinski, "Draft PD on Nuclear Targeting," March 22, 1980, p. 4 of 5, "5/80–1/81," Box 35, Brzezinski Collection, JCPL (note that document is out of temporal sequence in its placement in the file folder).

[77] The ability to bargain, especially in crisis setting, is mentioned in the memorandum leading up to PD 59, e.g., Special Coordination Committee Meeting, April 4, 1979, direct reference to "crisis bargaining," as a key topic for discussion, "Summary of Conclusions"; "8/78–4/79," Box 35, Brzezinski Collection, JCPL. Hollis and Smith provide a helpful discussion of the paradoxes embattling MAD from the perspective of defense rationalism, *Explaining and Understanding in International Relations*, 1990, 173–174. For a thorough analysis of the Schlesinger Doctrine's contribution to the puzzles of deterrence via MAD, see Freedman, *Evolution of Nuclear Strategy*, 1981, 374–395; Glaser, "Why Do Strategists Disagree," 147–148. Note that Schlesinger accepted that the Soviets' behavior would be based on their perception of the credibility of the US deterrent, which he interpreted as a rationale to further buttress US credibility to engage in nuclear war because he worried that the Soviets perceived the United States as benign; hence, Jervis's *Perceptions and Misperceptions in International Politics*, 1976.

[78] For discussion, see Freedman, *Evolution of Nuclear Strategy*, 1981, 379, 391, 385.

[79] Richard Tuck's *The Rights of War and Peace: Political Thought and the International Order from Grotius to Kant* (Oxford: Oxford University Press, 1999) explains how liberal

motive of self-preservation pertains to all actors, actors can peaceably coexist only if they concede to each other the right to exist and therefore voluntarily desist from harming others. This way of interpreting amicable relations among nations or individuals reduces to the pledge of "peace for peace" and the threat of "war for war."[80] Where liberalism views a state of nature as the return to the unconstrained natural right to all things, and civilization as a rarefied state organized by self-adopted rules and commitments, neoliberalism views the achievement of order as a function of equilibria arising from actors' unconstrained aspirations. Liberalism views the recourse to violence as a breakdown of social order; neoliberalism views social order as derived not from promises, but from credible threats of violence.[81]

To understand the transformed approach to mutual security, we must grasp how the Prisoner's Dilemma was used to motivate MAD, and identified as an inescapable logical puzzle miring MAD in the inevitable outcome that deterrence relies on an irrational threat.[82] The incredibility of the MAD deterrent threat was subject to ongoing attention by defense rationalists.[83] The strategic analyst Lawrence Freedman captures the dilemma of nuclear deterrence: "Yet the question of how nuclear weapons could be used in war remained and continued to nag at responsible officials as well as academic strategists. Once one openly admitted that the nuclear arsenal was unlikely ever to be activated then the deterrent lost all credibility."[84] MAD seemed arrested by paradox: if

international relations theory predicated on the no-harm principle anteceded the civil model for liberal governance. Michael Doyle is particularly insightful on the classic liberal tradition in international relations, *Ways of War and Peace*, 205–314. Obviously, classic liberalism would come under attack on many fronts in the late twentieth century; see, e.g., Samuel Moyn, *The Last Utopia* (Cambridge, MA: Harvard University Press, 2010), but its eclipse did not need to result in Prisoner's Dilemma logic of aggressive self-defense regardless of others.

[80] For a succinct discussion of this tradition as it was initially articulated by Grotius and Hobbes, see Richard Tuck, *Hobbes: A Very Short Introduction* (Oxford: Oxford University Press, 1989), 26–29; note how far neoliberalism is from liberalism given that in its day the latter was considered "illiberal" for condoning voluntary slavery and absolute monarchy; at least it established a normative order by uniting might with right instead of permitting might to establish right; on the latter, see Russell Hardin, "Does Might Make Right," in J. Roland Pennock and John William Chapman, eds., *Authority Revisited* (New York: New York University Press, 1987), 201–217.

[81] This is a central theme of Thomas Schelling's research, see *Strategy and Conflict*, 1960; the idea is that both threats and promises are levied to achieve an end that would rather be obtained without needing to take the act as either threatened or promised.

[82] For an emphatic statement of this, see Jervis, "Deterrence Theory Revisited," 1979, 300.

[83] This concern had been articulated by Brennan and the pro-nuclear-use strategists as early as the late 1960s as assured destruction was renamed mutual assured destruction, or MAD; James R. Schlesinger refers to the "suicidal implications" of assured destruction; "Uses and Abuses of Analysis," 1968, 340; Harold Brown admits assured destruction's incredibility deriving from the fact that "it is at least conceivable that the mission of assured destruction would not have to be executed at all in the event that deterrence failed," although it is important that any "potential enemy" not be led to believe this possible; Harold Brown, "Report to Congress 1979 Budget, FY 1980 Authorization Request, and FY 1979–1983 Defense Programs," January 23, 1978, 57.

[84] Freedman, *The Evolution of Nuclear Strategy*, 1984, 392–393.

nuclear armaments would only be used to seal the end of civilization, then there could be no conceivable plan for their use unless one embraced mass destruction and reciprocal suicide. However, if one held the nuclear arsenal with no intention of ever deploying it, then it could not stand as a credible deterrent threat.

Freedman thus goes on to explain, "If weapons had to be designed for operational use then some sort of guidance was necessary, which required stating a preference for one form of nuclear employment against another."[85] Freedman identifies a puzzle over what makes deterrence work, capturing the standard application of the Prisoner's Dilemma model to represent the puzzle of deterrence.[86] Without second-strike ability, each side was vulnerable to the other's initiation of a first strike; the introduction of second-strike capability neutralized the other's unilateral advantage, but only if one would actually follow through on a massive counterattack, or at least was believed that it would do so. Insofar as the strategic policy of MAD kept weapons in their silos until devastation was already certain, American nuclear arms would serve no function. To strategic planners, it seemed necessary to stipulate an operational use for nuclear weapons so that they could serve a constructive causal purpose furthering national goals.[87] If one started with the premise of striving for strategic dominance, even if ultimately the fact of MAD results in a game of nuclear Chicken, at least one clearly signals the intent to prevail rather than settle for submission.

Whereas MAD took seriously the inability to constructively wage nuclear war, and the Soviets' continual assertion that any use of nuclear force would lead them to counter with massive retaliation, NUTS was wholly dedicated to developing the meaningfulness and possibility of waging nuclear war and to acquiring weapons accordingly.[88] The difference between the two perspectives is clear in a brief exchange between Brzezinski and Brown. Brzezinski points to three major points of discussion in moving forward with PD 59: (1) the requirements of stable deterrence, (2) "requirements of stable crisis bargaining," and (3) "requirements of effective war management." Brown, following the logic that escalation control and war management are extremely unlikely, especially in prolonged conflict, said that "it is important to have our planning for all out

[85] Ibid.
[86] See ibid., 392.
[87] Clearly stated in Joseph S. Nye Jr., *Nuclear Ethics* (London: Collier Macmillan, 1986).
[88] On strategists' acknowledgment of the Soviet statements to this effect, see "Information Memorandum," Council of Foreign Relations, September 11, 1980, "It is also doubtful that the Soviets have only a massive strategic nuclear attack option in their war plans, although they often imply that by asserting the inevitability of their massive retaliation or of controlled escalation should they be attacked," p. 6 of "5/80–1/81," Box 35, Brzezinski Collection, JCPL. On the escalation control, flexible response weapons acquisition policy that was designed to link budgeting outlays directly to strategic planning beyond the limits of weapon employment and acquisition necessary for MAD, see pp. 4 and 7 of this document in addition to William E. Odom's clear statement to this effect in his memo to Brzezinski, March 22, 1980, p. 4 of 5, in "5/80–1/81," Box 35, Brzezinski Collection, JCPL.

nuclear war well in hand because all out spasm war is the most likely possibility, given the unlikely possibility of nuclear war in the first place." In other words, nuclear deterrence is sufficient to prevent the escalation into nuclear conflict, and the emphasis should thus be on preventing conflict in the first place. Once nuclear conflict is initiated, he reasons, fighting meaningfully misses the main point of deterrence to avoid war altogether. However, holding out the hope of being able to successfully prevail in prolonged nuclear combat, Brzezinski offers the counter-ing thought that "the very likelihood of all out nuclear war is increased if all out spasm war is the only kind of nuclear war we can fight."[89]

Additional discussion makes clear that the Carter administration abandoned the MAD footing both as an acquisition policy and as an employment policy, notwithstanding the overall recognition, as Jervis repeated throughout his career, that "MAD as a condition with which we and the Soviets are stuck, has obtained at least since the late 1960s."[90] Thus, it is impossible to exit the reality of mutual assured destruction. Nevertheless, the flexible response, coun-tervailing policy was gradually and continually introduced both as the guideline for purchasing weapons systems and for their employment. The MX missile system controverts MAD, which is based on accepting mutual vulnerability. Flexible response plans to employ nuclear weapons as a natural escalation from conventional warfare, with the plan of capping escalation; however, in reality, it cannot guarantee escalation control any more than MAD can guarantee deter-rence. In 1980, as PD 59 was moving through the approval process, US govern-ment defense analysts observed that "MAD as an employment doctrine has never really been in force, thus PD 59, which would be a dramatic departure had that been so, is rather just another step in a gradual and long-run policy evolution."[91] This is because the so-called Schlesinger Doctrine had been inher-ent in strategic rationality since the late 1960s. From McNamara onward, the SIOP had targeted almost every Soviet concern worth targeting. Still, of course, Carter's endorsement of the policy to procure and deploy powerful first-strike weapons and his commitment to having the power to engage in lengthy nuclear war was wholly unprecedented.[92]

---

[89] This exchange is in the minutes to the Special Coordination Committee Meeting of April 4, 1979, p. 2 of 9, "8/78–4/79," Box 35, Brzezinski Collection, JCPL.

[90] "Information Memorandum for United States Senate Committee on Foreign Relations," September 9, 1980, attached to memo to Brzezinski from Jasper Welch, September 11, 1980, quote from p. 6 of 8, "5/80–1/81," Box 35, Brzezinski Collection, JCPL.

[91] Information Memorandum for United States Senate Committee on Foreign Relations, September 9, 1980, attached to memo to Brzezinski from Jasper Welch, Sept. 11, 1980, p. 7 of 8.

[92] Strategists agree that escalation control requires escalation dominance to be successful; see Freedman, *Evolution of Nuclear Strategy*, 1981; Glaser, "Why Do Strategists Disagree," 1989. For the smooth continuity between the Schlesinger Doctrine and Harold Brown's devel-opment of PD 59, see Glaser, "Why Do Strategists Disagree?" 1989, 139, 147–148, 155. Glaser repeatedly argues that escalation equivalence must be escalation dominance for it to make any sense, see his "Why Do Strategists Disagree," 1989, 153, 163, 167. It is clear that Brzezinski and

THE INESCAPABLE IRRATIONALITY OF MAD

In the late 1970s through the mid-1980s, some philosophers and strategists tried to counter the argumentative ground and policy stature gained by NUTS.[93] However, strategists had widely come to accept game theory as a statement of orthodox instrumental rationality, and the puzzle of nuclear deterrence as isomorphic to the Prisoner's Dilemma.[94] Even moral philosophers were not able to successfully defend MAD despite its steadfast commitment to avoiding nuclear war because the moral agent necessarily views the pointless killing of noncombatants that would occur after deterrence fails to be unconscionable. The Prisoner's Dilemma model of nuclear security, which refused resolution by MAD, was only logically solvable by NUTS.[95]

President Carter was especially susceptible and accountable to moral reasoning, a manifest fact amplified by his solo disapproval of the neutron bomb.[96] In 1978, Gregory Kavka, one of the first moral philosophers to become captivated by game theory, argued that deterrence in various forms of punitive retaliation, including massive nuclear retaliation, must be inherently immoral because it depends on taking an action that is evil, the wanton destruction of innocent people.[97] Kavka poses three questions. First, given the immorality of the contemplated retaliatory act, can it be reasonable to act on such an intention at the moment when deterrence has failed and all that remains is gross carnage?

like-minded strategists fully viewed PD 59 to be consistent with the gradual and persistent movement to NUTS from MAD mainly overseen by Schlesinger; for contemporary acknowledgment of this point, see "Information Memorandum for United States Senate Committee on Foreign Relations," Memorandum from the Committee of Foreign Relations Meeting, September 9, 1980, p. 1 of 8, memorandum attached to memo from Jasper Welch to Brzezinski, Sept. 11, 1980, "5/80–1/81," Box 35, Brzezinski Collection, JCPL. To see how procurement could be divorced from employment, see David Lewis, "Buy Like a MADman, Use Like a NUT," in his *Papers in Ethics and Social Philosophy* (Cambridge: Cambridge University Press, 2000), 219–228.

[93] Jervis, "Realism, Game Theory, and Cooperation," 1988; *Meaning of the Nuclear Revolution*, 1989; see also Paul M. Kattenburg, "MAD (Minimum Assured Deterrence) Is *Still* the Moral Position," in Charles W. Kegley Jr., and Kenneth L. Schwab, eds., *After the Cold War: Questioning the Morality of Nuclear Deterrence* (Westview Press, 1991), 111–120.

[94] See, e.g., David Gauthier, *Morals by Agreement* (New York: Oxford University Press, 1986), and especially his earlier "Deterrence, Maximization, and Rationality," *Ethics* (1984), 94:3, 474–495; on the relationship between the puzzle of nuclear deterrence and the Prisoner's Dilemma, see, e.g., Jervis, *The Meaning of the Nuclear Revolution*, 1989, 129–133; and Gregory Kavka, *Moral Paradoxes of Nuclear Deterrence*, 1987, especially 46–47.

[95] This historical record is clear that the moral scrutiny MAD received from philosophers who also used game theory, most prominently Gregory Kavka, first pointed out the immorality of even forming the intention to retaliate through a massive counter strike, see his "Some Paradoxes of Deterrence," *Journal of Moral Philosophy* (1978) 75:6, 285–302.

[96] James R. Schlesinger makes clear both Carter's uncompromising moral character, which took a Kantian position on promises made, and his discomfort with nuclear weapons in "Interview with Dr. James R. Schlesinger," 1984, 60–61, 72.

[97] Kavka, "Some Paradoxes of Deterrence," 1978, 288, see also 286; one of the earliest expositions on game theory through the lens of moral theory is R. B. Braithwaite, *Theory of Games as a Tool for the Moral Philosopher* (Cambridge: Cambridge University Press, 1955).

Second, if it is clear that following through on such an intention must be immoral at the time of its enactment, then is it not the case that the mere formation of the deterrent intent of massive punitive harm is itself immoral? Third, in recognition of the patent immorality of both following through on the act and even the mere formation of the intention to implement the action, must it not be impossible for a moral agent to form such a deterrent intention?

According to this reasoning, deterrence via the threat of assured destruction is immoral. At the behest of President Carter, this deterrent threat must then be contrary to moral judgment, and as it contravenes Carter's bona fide moral preferences, it must be irrational. Enacting MAD once deterrence had failed would be patently immoral and contradicted the commander in chief's humane values. Hence, acting on this threat both failed to further US interests and entailed irreverence for human life and must therefore be irrational, and hence incredible. An incredible deterrent threat is less than worthless.

Schlesinger offered NUTS, or deterrence via strategic dominance, as the ideal antidote to this worrisome forfeit of national security. Despite its mismatch with strategic realities underlying superpower parity, NUTS demonstrated the willingness and ability to wage not only nuclear war but conflict at any level through the introduction of limited nuclear options (LNOs) that blurred the distinction between conventional and nuclear weapons. It circumvented the immorality of MAD by threatening escalation control for any military engagement, proposing all forms of military action as means to prevail, rather than as the final desperate act of a defeated nation. The demonstrated intention was thus considered to be crucial, even more important than the actual feasibility in maintaining effective deterrence or winning a nuclear war.

The tall order of escalation control, however, faces two challenges. First, not only are there no guarantees of capping nuclear confrontation, but "the amount of damage from a 'small' nuclear war might be so great that the damage caused by a small nuclear war might approach tha[t] expected in a full scale nuclear war."[98] Second, the pursuit of supremacy itself has destabilizing implications ensuring a Prisoner's Dilemma arms race at best and all-out war at worst. This point was raised at the Special Coordination Committee Meeting addressing a Memorandum for the President on the 1979 report "Nuclear Targeting Policy Review." With respect to the pursuit of strategic dominance, David Aaron noted that "stability at one level can be the enemy of deterrence at another level." He explained, "For example, overall strategic superiority may create a very stable situation with respect to deterring Soviet military initiatives, but be very destabilizing in the degree to which it encourage[s] Soviet efforts to improve and expand their forces."[99] Countervailing strategists viewed that worst-case scenarios entailed enemy aggression and not accidental or erroneous

---

[98] Lackey, "American Debate," 1987, 37.
[99] Special Coordination Committee Meeting, April 4, 1979, Detailed Minutes, along with the Top Secret Secretary of Defense Memorandum to the President on "Nuclear Targeting Policy Review," "8/78–4/79," Box 35, Brzezinski Collection, JCPL.

misapplication of nuclear devices, proliferation, or the sheer destabilizing impact of mimicking a military posture consistent with preemptive attack.[100]

Carter's dilemma was distinct from that of the counterforce supporters who came to predominate in his administration. Whereas they were content to offer war for war and war (or strategic dominance) for peace, Carter sought to maintain the stance of war for war and peace for peace, as was consistent with classical liberal bilateral assurance of cooperation, and mutual deterrence against pathological hostility. During his early days in office, Carter said that "a single Poseidon boat was enough retaliatory power, that it really can by itself destroy the Soviet Union, and we really don't need any more."[101] Thus, while those promoting the Schlesinger Doctrine accepted the Prisoner's Dilemma model of nuclear security and arms race, Carter's position reflected Schelling's original question: how does a classical liberal or prospective cooperator deter a predator?[102] Here is Carter's riddle as assessed by Kavka:

> Let us call situations of the sort that nation N perceives itself as being in, Special Deterrent Situations (SDSs). More precisely, an agent is in an SDS when he reasonably and correctly believes that the following conditions hold. First, it is likely he must intend (conditionally) to apply a harmful sanction to innocent people, if an extremely harmful and unjust offense is to be prevented. Second, such an intention would very likely deter the offense. Third, the amounts of harm involved in the offense and the threatened sanction are very large and of roughly similar quantity (or the latter amount is smaller than the former). Finally, he would have conclusive moral reasons not to apply the sanction if the offense were to occur.[103]

Carter initially sought to maintain the openness of offering cooperation in exchange for cooperation and demonstrating the unequivocal greatness of Western institutions of market freedom and democratic self-governance. The challenge before him was to deter a predator without becoming one and to maintain the attitude of seeking mutual assurance of cooperation while still having the wherewithal to deter invasion.

Writing in 1978, Kavka had little wisdom for Carter because he finds that "in an SDS, a rational and morally good agent cannot (as a matter of logic) have (or form) the intention to apply the sanction if the offense [military attack] is committed."[104] According to Kavka, the only way around this conclusion is to tie one's decision making to a mechanical device, adopt a corrupted character, or defer to those actors whose character is morally ambiguous. Neither of the first two was possible for Carter. Therefore, NUTS advocates embodied the

---

[100] For pointed discussion, see Lackey, "American Debate," 1987, 36–37.

[101] Statement by Schlesinger, "Interview with Dr. James R. Schlesinger," 1984, 72.

[102] Accepting the PD model of the nuclear security dilemma, of course, entails all the steps from the initial Stag Hunt modified by uncertainty, viewing the threat of MAD incredible, yet also accepting the Chicken game model of brinkmanship in which each prefers to be the sole defector.

[103] Kavka, "Some Paradoxes of Deterrence," 1978, 286–287.

[104] Ibid., 292.

third possibility, and they conceded the PD preference ranking for unilateral defection.[105] Schlesinger clearly insisted that the United States should maintain dominance among competitors and hegemony over allies.

The victory of NUTS over MAD is logically unassailable if one accepts the Prisoner's Dilemma representation of mutual security and the comprehensive reach of strategic rationality. As a result, President Carter exchanged the opportunity to emphasize the mutual assurance of amicable coexistence in favor of the United States demonstrating the predilection and capacity for unilateral defection in striving for ascendance. This posture seemed to solve the nuclear Prisoner's Dilemma game, if the main worry were issuing a credible deterrent threat and having a plan to engage in violence for every conceivable Soviet action. However, if one's main anxiety were to reduce chances for potential conflict as a shared responsibility or if one worried that the capability to dominate, even if not exercised, still maintains the potential to exercise asymmetric advantage, NUTS is unsatisfactory because it risks inviting the adversary to preemptive engagement and flirts with multiplying the dangers of accident and proliferation, not to mention that it stockpiles more weapons than could reasonably fulfill a meaningful destructive purpose.[106]

THE TACIT ALLIANCE BETWEEN OFFENSIVE REALISM AND GAME THEORY

It may seem like a bold claim to assert that applying game theory to solving the problem of nuclear deterrence, and to the more general international relations challenge of anarchy, not only sets up MAD to lose to NUTS but also results in a predatory defense posture. However, if it is generally accepted that game theory is the gold standard of instrumental rationality, then it becomes apparent that the boldness of the assertion does not lie in claiming too much authority for game theory. Instead, it resides in the implication that game theory structures decision making in a manner that restricts agents from utilizing resources and logics of action outside its scope. When stated in this way, most rational choice theorists would likely agree that decision theory represents the state-of-the-art approach to sound action, and that rationales for judgment outside its scope are not only unsubstantiated but also invalid.[107] Jervis observes that "In micro-economics, SEU [subjective expected utility] theories can be both descriptive

---

[105] Lackey reads Kavka to conclude that it is OK to issue the deterrent threat, so long as one does not, as a moral actor, implement it in the case that deterrence fails, but Kavka is clear that this is not a possible position for the rational and moral actor; Lackey, "American Debate," 1987, 40.

[106] Philip Pettit argues that even if another actor has the power to dominate but does not act on it, this still implies the less powerful agent is subject to domination; "Keeping Republican Freedom Simple," *Political Theory* (2002), 30:3, 339–356.

[107] See, e.g., Robert Nozick, *The Nature of Rationality* (Princeton, NJ: Princeton University Press, 1994); for an opposed view, see Jean Hampton, *The Authority of Reason* (New York: Cambridge University Press, 1998).

and prescriptive because of the argument that only those who behave in accordance with them can flourish."[108] Game theory buttresses offensive realism by equating sound judgment with the eschewal of "irrational" or metaphysical considerations consistent with idealism, social constructivism, or naivety.[109] Insofar as offensive realism can boast of being consistent with game theoretic strategic rationality, then its pedigree and policy proposals seem credible beyond dispute.

In his recent defense of defensive realism against its offensive alternative, Tang directly states that his argument rests on acknowledging that "ontological reason must take priority over instrumental reason" in appraising the security dilemma.[110] This idea of straying from instrumental rationality to marshal a defensive realist national security protocol is not reassuring to those who view the world from the perspective of *realpolitik*. Yet, ahead I show how the originators of a classical liberal perspective that rests on a reciprocal stance of not harming others believed they were advancing a strict pragmatism to solve the security dilemma. Recall the elementary theoretical commitments structuring game theory. Only ends, and not the means by which they are obtained, register in payoffs. This assumption rejects the no-harm principle, fair play, the internalization of norms, and commitment to agreements made, unless decision rules not limited to individual optimization are introduced.[111] Payoffs in many game contexts, specifically those that are repeating or involve multiple actors, are assumed to reflect an intersubjectively obvious resource over which actors vie as a criterion of success within their environment.[112] Solidarity and team reasoning are contrary to the individualistic maximization deemed consistent with instrumental rationality. Altruism, although possible in principle, is either too difficult to track mathematically or even more likely too costly.

Each of these assumptions is consistent with a *realpolitik* approach to international relations. From this perspective, the *raison d'état* supersedes principled action and norm-governed conduct.[113] It stands to reason that instrumentally astute states will acquire the scarce natural resources necessary to project power through causal efficacy. Maximin logic of strategic independence demands acting so as not to be dependent on any other actor. Solidarity and altruism

[108] Jervis, "Rational Deterrence: Theory and Evidence," *World Politics* (1989), 41:2, 183–207, at 188.

[109] In his *Nuclear Ethics* (1986), Nye argues that a "realist-Cosmopolitian" synthesis is necessary to ensure the incorporation of humane values into international relations beyond pure instrumental means, 34–41; see also Michale W. Doyle, *Ways of War and Peace* 1997, 205–300.

[110] Shiping Tang, *A Theory of Security Strategy for Our Time: Defensive Realism* (New York: Palgrave Macmillan, 2010), 50.

[111] Nye argues that to be effective nuclear deterrence should encompass both a purely rational set of considerations and nonrational considerations that provide the overall significance of national identity and existential ethos; *Nuclear Ethics*, 1986, 106–107.

[112] On realism's deferral to fungible value, see Doyle, *Ways of War and Peace*, 1997, 47.

[113] Tuck argues that this concern was fundamental for the originators of classical liberalism; see *The Rights of War and Peace*, 1999, especially 1–15.

are easily rejected by the realist, the first because it is contrary to reason, and the latter because only the altruism that pays is worthwhile.

To understand the contrast between strategic rationality and offensive realism on one hand and defensive realism and classical liberalism on the other, let us consider four questions. First, how is self-preservation defined, and what are its imperatives? Second, what is the source of value that affords instrumental power? Third, what is the source of power that enables purposive action among other purposive actors? Fourth, what steps are necessary to deter a predator? These questions afford insight into how strategic rationality offers a perspective on the character of purposive action that forecloses on the possibilities for cooperation anticipated by classical liberalism and defensive realism. Awareness of game theory's predisposition toward offensive realism provides actors with a vantage point from which to evaluate which position makes more sense.

Self-Preservation

Game theory shares with offensive realism the view that an actor's identity is defined by the actor's preferences and opportunities. From the perspective of rational choice, no natural boundaries to personal or national identity define a perimeter, which if transgressed, represents harm. From the perspective of offensive realism, "there is no possibility of drawing a sharp line between the will-to-live and the will-to-power."[114] This means that the survival of a state in anarchy depends on the continual augmentation of power, regardless of the effect on other states. Hence, resonating with Schlesinger's strategic wisdom, offensive realist John Mearsheimer recommends the policy of hegemony to secure the immediate survival and long-term prospects of a powerful state.[115]

It is not immediately obvious that rational choice theory tacitly endorses the pursuit of hegemony among states, or the exercise of domination among individuals.[116] Certainly, the PD preference matrix accepts that every actor will cheat the other, even if that party already cooperated. Both offensive

[114] This is quoted from Reinhold Niebuhr, *Moral Man and Immoral Society: A Study in Ethics and Politics* (New York: Charles Scribner's Sons, 1960 [1932]), 42 by Tang, *Theory of Security Strategy for Our Time*, 2010, 19. Note that Tang suggests that all realists endorse Niebuhr's view, but Tang subsequently goes on to explain why the defensive realist does not view that self-preservation depends on expansionist or revisionist goals, essentially suggesting that there is some boundary condition, or threshold condition, demarcating the dictates of security that is not equivalent to unconstrained maximization of power, see especially pp. 43–58.

[115] John Mearsheimer, *The Tragedy of Great Power Politics* (New York: W. W. Norton, 2001); for discussion, see Tang, *Theory of Security Strategy for Our Time*, 2010, 58.

[116] However, it is possible to surmise this from the Prisoner's Dilemma analysis of international relations hegemonic stability theory, Duncan Snidal, "The Limits of Hegemonic Stability Theory," *International Organization* (1985), 39:4, 579–614; and state of nature underlying social contract theory, Michael Taylor, *Anarchy and Cooperation* (New York: John Wiley, 1976).

realism and game theory accept that actors would if they could without negative repercussions seek resources without limit, despite the impact on others, and therefore will tend to express PD preferences that track fungible rewards. Strategically rational individuals have preference orderings that by design rule out side constraints, fair play, and commitment, which evaluate the appropriateness of action choices based on procedural considerations. They therefore do not recognize or uphold a threshold demarcating self-preservation distinct from the conquest of others. Hence, the theoretical structure of game theory presupposes an offensive realist orientation to interactions among states and people.[117]

The classical liberal who views the security of selfhood, whether of a corporeal person or physical territory, in terms of maintenance of the status quo, is able to differentiate between harming others and self-defense. Acting on the principle of no-harm requires both striving to understand how to enact the concept and constraining one's actions accordingly.[118] Rational choice theory renders it difficult to express the idea that perpetuation of selfhood is not equivalent to unconstrained preference satisfaction. Consequently, the game theoretical perspective of agentive identity is at odds with the classical liberal commonsensical intuition that it is possible to define self-preservation without encumbering it with expansionist intention or unavoidable cross purposes against others.

Source of Value Enabling Instrumental Agency

Offensive realism and game theory both presuppose the existence of raw value, or sources of power in the world that are both necessary for manifesting purposive action and have value prior to and independent from institutions and social practices.[119] This approach to value is appealing in international relations because the idea of anarchy among nations builds on the assumption that instrumental efficacy is independent of qualities of relationships between

---

[117] For discussion, see Philip Green, *Deadly Logic: Theory of Nuclear* (Ohio State University Press, 1966), 213–253.

[118] Note that international relations neoliberal institutionalists and defensive realist theorists declare that the bright-line distinction between their position and that of offensive realists is their focus on "absolute gains" versus "relative gains." Robert Powell, "Absolute and Relative Gains in International Relations," in Baldwin, ed., *Neorealism and Neoliberalism*, 1993, 209–233; Duncan Snidal, "Relative Gains and the Pattern of International Cooperation," in ibid., 170–208. However, I dispute that this is a sufficient criterion to defend classical liberalism. Even a focus on absolute gains in a PD game accords to actors the predilection to seek unilateral gain at others' expense, and the determination of the PD payoff structure will be deduced from preferences tracking fungible sources of value, e.g., Andrew Kydd, *Trust and Mistrust in International Relations* (Princeton, NJ: Princeton University Press, 2007), 1–12.

[119] This game theoretic presupposition for standard operationalization is most evident in Roger Myerson, *Game Theory: Analysis of Conflict* (Cambridge, MA: Harvard University Press, 1991), 3, 22–26; it is outrightly stated by Shaun Hargreaves Heap and Yanis Varoufakis, *Game Theory*, 2nd ed. (New York: Routledge, 2004), 209.

states. Therefore, the source of value underlying the power of nations by definition preexists sociability.[120]

This standpoint is also incorporated into the foundations of game theory. The original understanding of games anchoring the mathematics of its prodigious founder, von Neumann, was that payoffs reflect an ontological property of the world. Interpersonally transferable utility is necessary for zero-sum games, the original focus in *Theory of Games and Economic Behavior*.[121] Von Neumann's perspective presumes that payoffs reflect a quality such as temperature in physics, suggesting that utility as the source of value is similar to energy states that are the underlying cause of temperature. In zero-sum games, the payoffs are intersubjectively precise and are not a function of individual's subjective interpretation.[122] Many games of interest in international relations and political economy identify payoffs relying on a direct statement of tangible resources.[123] If actors compete over scarce resources conveying the instrumental power to achieve goals, then any encounter with another actor potentially has the Prisoner's Dilemma structure in which every actor has the first preference for suckering the other and taking all the disputed resources.

The other avenue of defining utility, permitting it to be subjective, such that the payoffs to games are a psychological property of actors that does not necessarily track any specific ontological feature of the world, renders that uncertainty about the intentions of the other actor is perpetual. However, expected utility theory must be applied, and this theory will default to tracking fungible value.[124] Here, again, offensive realism and game theory reach the same conclusion.[125] Even if there is a good objective reason, based on interpersonally transferable utility, to model a security dilemma as an Assurance Game, it is wise for the security strategist to accept that actors may interpret the significance of outcomes according to some idiosyncratic means of judgment

[120] This position is inconsistent with Thomas Hobbes's *Leviathan*, ed. Richard Tuck (Cambridge: Cambridge University Press, 1996), a point I explore in the next chapter.

[121] For discussion, see Nicola Giocoli, "Do Prudent Agents Play Lotteries: Von Neumann's Contribution to the Theory of Rational Behavior," *Journal for the History of Economic Thought* (2006), 28:1, 95–109.

[122] It is not without irony that the only way to obtain precise value for von Neumann in most contexts was to default to relying on cash value, which is the product of complex networks of interaction, yet he treated it as though it were like temperature in physics; John von Neumann and Oskar Morgenstern, *Theory of Games and Economic Behavior* (Princeton, NJ: Princeton University Press, 1944), 23. For an alternative view see John R. Searle, *The Construction of Social Reality* (New York: Free Press, 1995).

[123] Von Neumann Morgenstern utilities that track an interpersonally transferable source of value can be subjectively modified by an affine translation without slipping into a more individualistic and encompassing subjectivity, von Neumann and Morgenstern, *Theory of Games*, 1944, 23–29; for the assumption of interpersonally transferable utility, see Thomas Schelling, who employs this vocabulary in "Hockey Helmets, Concealed Weapons, and Daylight Saving: A Study of Binary Choices with Externalities," *Journal of Conflict Resolution* (1973) 17:3, 381–428.

[124] Myerson, *Game Theory*, 1991, 3, 22–26.

[125] Doyle, *Ways of War and Peace*, 1997, 47.

that is forever opaque to each actor. This is Schelling's argument that permitted the nuclear security dilemma, which he assumed best resembled a Stag Hunt, to be translated into a Prisoner's Dilemma game on the basis of irreducible uncertainty, which he quantified as 80 percent risk.

The initial and crucial step in understanding the difference in approach between offensive realism and defensive realism is to see that the latter is confident that there are potential sources of value and power to act that transcend those raw resources that are available in an anarchic state of nature. The argumentative move taken by classical liberals is to note the common ground all actors have in confronting the proverbial state of nature; this offers as much epistemological confidence as does instrumental rationality's acceptance of cause and effect.[126] Whereas for the rational strategist, the only intersubjectively accessible and causally efficacious value is interpersonally transferable utility, for the classical liberal, value is created on the basis of finite resources by establishing practices predicated on mutual toleration and shared expectations that depend on relinquishing the intent to harm others.[127]

The first of these sources of value is the prospect dividend, or the value of resources to their owner who can be confident in their possession, which results from the mutual tolerance facilitating self-preservation. Whether or not it is agreed that mutual toleration is a viable practice, it is at least possible to agree that, in principle, if actors did not threaten each other, then the classical liberal world carved into distinct property rights would yield value over and above the value of the extant raw resources.[128] Security in possession is worth more than simple possession. Actors would pay more to keep what they already own than its fungible value, which means that no actor has the ability to pay for the conditions of his livelihood as subjectively valued with all of his resources actually on hand. Prospect theory reveals that goods in hand are worth more than their replacement cost. Therefore, security in possession is worth more than the fungible value of all the resources available to purchase replacement and supersedes its base material resource value.

Already then the classical liberal sees a source of value that is not available to the offensive realist: security of possession. Another source of value is that gained through sociable forms of interaction, either in institutions or through normative conduct. Maritime law, finance, aviation, science, and technological innovation provide sources of value that are not possible in a state of nature.[129]

---

[126] Hobbes, *Leviathan*, Chapter 13.

[127] Note the temptation for international relations theorists to presume that a devotee of realism and *realpolitik* must track sources of value commensurate with the laws of physics; see, for example, Randall Schweller, *Maxwell's Demon and the Golden Apple: Global Discord in the New Millennium* (Johns Hopkins University Press, 2013).

[128] Daniel Kahneman and Amos Tversky, "Prospect Theory: An Analysis of Decision under Risk," *Econometrica* (1979) 47:2, 262–292.

[129] This is a central point for Hobbes in *Leviathan*, Chapter XIII in passage with phrase, "life of man solitary, poore, nasty, brutish, and short."

The neoliberal school of international relations makes the case for the value of institutions and the possibility of cooperation but tries to derive this social capital from the purely strategic considerations of rational choice.[130] However, as I argue, it is ultimately impossible to rescue classic liberalism from within the confines of game theoretic strategic rationality. On this point, the offensive realist school agrees, arguing that liberalism necessarily tends toward some form of idealism or passivism.

Source of Power to Act in World with other Agents

For the offensive realist and rational strategist, there is one source of power to act in the world: marshaling the instrumental means to threaten other actors.[131] Domination is the means of successful exploitation. In the nuclear era, as Schelling recognized and emphasized, threatening to harm others is often sufficient, even without actually following through. Within international relations, for the offensive realist and according to the game theoretic approach pioneered by Schelling, it follows deductively from the view of self-preservation and of value that applied violence is the currency of purposive action. Tang describes this position: "[Glen] Snyder explicitly argues that an imperialist state will pursue both conquest and intimidation (Snyder 1985, 165), and Mearsheimer emphasizes that an offensive realist state is not a 'mindless aggressor' (Mearsheimer 2001, 37), thus also implying that intimidation is a tool of an offensive realist state."[132] Conveniently, as Schelling observed in 1960, the threat of applying pain to the opposition is often sufficient, without actually needing to perpetrate violence. Tang continues,

Hence, seeking to establish hegemony with intimidation, although perhaps more palatable for the victim, is still a form of intentionally threatening other states. As a matter of fact, intimidation is an especially "wise" strategy when the aspiring hegemon still lacks the material power to impose its will, or it is simply too costly to impose hegemony by force, because it allows the hegemon to be viewed as not "excessively" aggressive.[133]

Violence must be calibrated and prescribed in doses to achieve its greatest effect. The proponents of the nuclear war plan of NUTS stood united with game theorists in perceiving that social order, understood to be the achievement of an actor's way in the world among other strategic actors such that favorable stability arises, is the product of knowing the oppositions' preferences and leveraging the threat of punishment to influence their choices. Rewards also

---

[130] This approach is articulated by Robert Axelrod and Robert O. Keohane, "Achieving Cooperation under Anarchy: Strategies and Institutions," in Baldwin, ed., *Nerealism and Neoliberalism*, 1993, 85–115.

[131] Technically, positive incentives may be used as well, but coercive threats are more typical in the arena of national security; Thomas Schelling, "War without Pain, and Other Models," *World Politics* (1963), 15:3, 465–487.

[132] Tang, *Theory of Security Strategy*, 2010, 58.

[133] Ibid.

function as incentives and can be perceived as the removal of harm or the threat to harm.

The ability to rationally threaten harm in precisely measured doses is both what enables actors to realize goals and the force producing the social order that exists. Social order, such as it exists, is produced out of anarchy by vying for sources of raw power that are independent from sociability. Under the best of circumstances, strategic competition results in regular patterns, as opposed to the unpredictable chaos that could be expected in a perpetual state of war. Where strategists understand perceptions to influence judgment, the most cautious interpretation of others' intentions is employed in strategic analysis.

The classical liberal and the defensive realist do not dispute that natural resources, subject to the laws of physics, convey the instrumental power to act in the world. However, they depart from strategic rationality in holding the opinion that the ability to act among other purposive agents relies on shared expectations and norms and a tacit or explicit system of rights. These arise from proposing a different reconciliation of the security dilemma that perplexes the offensive realist and the game theoretic approach to security.[134] This alternative resolution rests on viewing the requirements of self-preservation to be built up from reproducing patterns of sustainability that respect the status quo rather than reflect an unbounded maximization of expected gain. It pivots on retracting the threat of harming others, in direct contrast to offensive realism and strategic rationality. Classical liberalism proposes pursuing mutual viability and even cooperative exchange through reassurance that one's intention is not to exploit or dominate others.[135] According to this view, agents bear responsibility for not compromising others' security to pursue their own. A classical liberal could in self-defense act on strategic imperatives, but this would be an aberration rather than the rule.[136]

By adopting this approach to the security dilemma, the classical liberal and the defensive realist are poised to capitalize on three sources of value that transcend the raw power the offensive realist and rational strategist pursue. These are the prospect dividend, social capital resulting from shared ventures, and unbounded positive-sum good will and civility that arise from sociability and mutual respect.[137]

---

[134] For Kydd, whether a security dilemma is a Prisoner's Dilemma in which every actor seeks to sucker others is strictly determined from a calculation of fungible rewards; *Trust and Mistrust*, 2007, 6–7.

[135] See, e.g., Tang, *Theory of Security Strategy*, 2010, 70.

[136] For discussion, see Michael W. Doyle, "Kant, Liberal Legacies, and Foreign Affairs," *Philosophy and Public Affairs* (1983) 12:3, 205–235, notice Immanuel Kant's allowance for asocial sociability.

[137] This source of value is achieved by the manifestation of circumstances conveying value to their experiencers that far outpaces the value of the scarce and tangible resources giving them corporeal existence. See, e.g., Seth Godin, *Lynchpin: Are You Indispensable?* (New York: Penguin, 2010). It builds on the insights of Amartya Sen, who argues that with respect to the social capital of human resources, the experiential value and productive power of capability and

Like the prospect dividend, social capital, or commonly held institutions and infrastructure, is dependent on resources. Yet it too embodies value beyond the raw power to instrumentally achieve goals or strategically intimidate others, through yielding the right to harm others by maintaining a formal or informal set of mutual expectations. This system of tacit or explicit rights is not the product of violent threats and de facto possession. As Hobbes argues, such a system requires the threat of force to assure participants and to thwart rouge predators but otherwise relies on voluntary compliance.[138] It is imperative to demonstrate the willingness to cooperate when others do so.

The neoliberal, offensive realist approach flagrantly contradicts this understanding of a shared world of purposive agency by insisting that the rational actor recognizes no limitations to preference satisfaction and the will to power other than incentives. From this perspective, both oneself and others pursue self-defense and self-gain in a self-help system without exit. The classical liberal calls on two modes of action that the rational strategist disregards for being irrational. The first is voluntarily constraining one's actions so as not to harm or threaten others. The second is following the norms of a system that provide the extra social capital value over and above raw resources valuable in anarchy. The first may be thought of as side constraints to action that facilitate a universal system of individual, or national, rights to self-preservation.[139] The second has the characteristics of fair play, pointed to by both Adam Smith and John Rawls as fundamental to enacting a capitalist system predicated on realizing exchange value beyond the raw power of natural resources that served as the criterion of value under mercantilism.[140]

For the purposes of classical liberalism and defensive realism, it is worth pointing out, but not dwelling on, the other additional source of value available once the security dilemma has been resolved through a system of mutual expectations and rights built up from reciprocal respect rather than the perpetual deployment of threats to harm. Whereas the prospect dividend and social capital discussed earlier are linked to scarce resources, and therefore remain a finite source of value, there is the potential for unbounded positive sum value that may accrue from amicable social relations in the form of friendship, aesthetic beauty, good will, understanding, healing, trust, and esteem that of course remain dependent on the incarnation of ontologically existing properties yet convey potentially unlimited subjective value. The offensive realist and

---

functioning outpace the sheer financial wherewithal to capture value that must be inherently finite; see *Development as Freedom* (New York: Knopf, 1999), and *Rationality and Freedom* (Cambridge, MA: Belknap, 2002).

[138] Discussed in Chapter 5.

[139] Doyle, *Ways of War and Peace*, 1997, 205–314; this has its parallel in civil society, articulated as side constraints that permit the practice of property rights; see Robert Nozick, *Anarchy, State, and Utopia* (Oxford: Basil Blackwell, 1974).

[140] Doyle, *Ways of War and Peace*, 1997, 230–250; see also Adam Smith, *Theory of Moral Sentiments* (New York: Cambridge University Press, 2002); for commentary, see S. M. Amadae, *Rationalizing Capitalist Democracy* (Chicago: University of Chicago Press, 2003); John Rawls, *Theory of Justice* (Cambridge, MA: Belknap, 1970), chaps. 5 and 6.

neoliberal approaches, which both instrumentalize and commodify all value, cannot recognize unbounded sources of value because the single criterion metric useful for strategic manipulation cannot stray too far from raw power without taking on the aura of transcendent or metaphysical value.[141]

Classical liberals and defensive realists stand confident on three points. First, they perceive that the sources of value and the power to act among purposive agents derive from mutual respect over and above raw resources. Second, whereas strategic rationality treats others as mere means to achieve goals, purposive action for the classical liberal is characterized by self-knowledge and the acknowledgment of like-minded agents who see relationships as a way to achieve mutual goals. To prosper is to create constructive channels of mutual exchange based on the right to exist as an end in oneself, rather than a mere means.[142] Third, classical liberals see the dictates of their own self-preservation as compatible with others' like imperatives, and they have the self-knowledge to be certain that they will cooperate if guaranteed or assured that others will do so as well.

Of course, no reassurance can be offered to an offensive realist to obtain mutual cooperation because the offensive realist views the requirements of self-perpetuation as inherently mutually antagonistic. A strategic rationalist could, however, accept that the security dilemma is a Stag Hunt. Let us suppose with game theorist Andrew Kydd that a state calculates that it is in its best interest to cooperate.[143] If this calculation tracks raw resources over which actors compete, then the classical liberal must always be wary that the rational strategist may at any moment calculate that aggressive expansion is the best policy.[144] If this calculation is predicated on subjective factors, then perceptions remain forever opaque. Furthermore, the rational strategist does not admit or act on the perception that mutual respect, rather than a combination of raw power and calibrated incentives some violent, is what enables purposive actors to be the most effective in realizing value. The offensive realist and the rational strategist thus remain perplexed by the Prisoner's Dilemma model of the national security

[141] Georg Simmel, *The Philosophy of Money* (New York: Routledge, 2011, reprint edition), 283–291, presents a clear argument for how bourgeois capitalism permits the development of value beyond both zero-sum division of tangible resources into allotments bounded by de facto or legal possession and security of possession to the aesthetic products of culture in the arts and humanities.

[142] Doyle, *Ways of War and Peace*, 1997, 205–214.

[143] See how Kydd defines states' security prerogatives based on whether they view their requirements with respect to a PD or Stag Hunt sheerly as a matter of calculation, *Trust and Mistrust*, 2005, especially 7–8. Note that with respect to resource dilemmas and the social contract, it is generally accepted that the PD model applies with Brian Skyrm's *The Stag Hunt and the Evolution of Social Structure* (Cambridge: Cambridge University Press, 2004) being the exception; on the game theoretic analysis of whether Hobbes's state of nature is best represented by a PD or Stag Hunt, see Chapter 5.

[144] Kydd, *Trust and Mistrust*, 2007, 7–8, makes clear that forming a relationship with a rational strategist depends on always identifying when it may be in that actor's interest to exploit one's interests.

impasse. For them, self-help and anarchy cannot be exited or transcended. By making the concessions to the PD payoff matrix and the exhaustive reach of strategic rationality, the offensive realist ends up trapped in the Prisoner's Dilemma. The focal remedy then is the hope that the situation is indefinitely and exactly repeated, so that incentives are supplied by actors themselves.[145]

So far, the difference between the two realisms and two liberalisms hinges on a difference of opinion about what makes action intelligible and meaningful, with the offensive realists and neoliberal political economists placing their reliance on a single model of instrumental agency, and the defensive realists and classical liberals having a broader scope.[146] In looking to mutual respect instead of credible threats to bootstrap out of a state of nature, classical liberals endorse the no-harm principle as the source of the prospect dividend. They understand that fair play is the basis of social capital. They also comprehend that commitment to agreements made is not only fundamental to the prospect dividend and to social capital but further provides the basis for realizing unbounded sources of value in friendship, justice, trust, good will, and fulfillment.

The classical liberal further rejects the idea that all value can be measured on a single scale. A scale tethered to tangible resources of central interest to realists is insufficient to adequately capture the manner in which cooperative ventures that grow out of the no-harm principle, mutual respect, and commitment generate more value than the total allotment of physical attributes permitting them to exist. Game theory purports to acknowledge the same insight by claiming to offer a comprehensive science of decision making. However, the mathematically tractable default of using cash value for payoffs makes it in principle impossible to reflect even the prospect dividend. Agents' appraisal of value must ultimately be demonstrated by their willingness and *ability* to pay for goods. As actors' de facto ability to pay for goods cannot exceed their goods on hand, there is no practical way for any agent to express, or accordingly register, that the value of their lives or property exceeds the raw resources granting their existence.

THE ROAD NOT TAKEN

The paradox of nuclear deterrence, like the Prisoner's Dilemma, seems to require that conditions necessary for a satisfactory solution contradict the aims and preferences of the actor seeking to deter. Whereas the United States is a nation whose public image is based on the fundamental values consistent with classical liberalism, deterring the potentially aggressive Soviet Union seemed to require abandoning a position of offering cooperation for cooperation and military

[145] Axelrod and Keohane, "Achieving Cooperation under Anarchy," 1993; the repeated Prisoner's Dilemma and its idealized unstable equilibrium is discussed in Chapter 11, "Tit for Tat."
[146] Note, for example, how Nye, *Nuclear Ethics*, 1986, 27–41, suggests that realism could be compatible with a cosmopolitan approach to world politics; however, his discussion implies that cosmopolitanism adds considerations above and beyond a strictly realist approach.

engagement for aggression. Even though moving to the footing of NUTS contradicted the material fact that the world was not large enough to accommodate useful detonation of all the accumulated thermonuclear warheads, still strategists concluded that maintaining deterrence must rely on pursuing the bid for nuclear supremacy and credibly sustaining the intention to wage nuclear war if necessary.

However, in taking this path, strategists underplayed the threat of nuclear mishap through error, accident, or escalation and ignored the common ground of Earth's inhabitants shared not just by the Soviet Union and the United States, but also by other nations, of living in a habitable environment. Leaving aside for the moment the difficulty of the puzzle of solving nuclear deterrence, we can conclude that by the 1980s the game theoretic approach to nuclear security stood as "normal science," to invoke Thomas S. Kuhn's paradigmatic language.[147] The game theorist Steven J. Brams published a comprehensive treatment of superpower conflict that modeled deterrence as a Chicken game and the arms race as a Prisoner's Dilemma.[148] He concludes that "we may wish the strategic problems the superpowers face were not so obdurate, but, in a curious way, their obduracy forces the players to come to grips with the haunting dilemmas, especially involving the use of nuclear weapons, to which these game give rise."[149] The puzzle of nuclear deterrence and the need to exercise sovereignty by deploying nuclear weapons coevolved with coming to view game theory as the canonical statement of purposive rationality. Quoting the political theorist David Gauthier, Brams finds, along with the nuclear strategists sustaining credible deterrence and game theorists more broadly that "the alternative to eschewing all threat behavior, in the bluntest terms 'can only be the willingness to accept victimization, to suffer passively a nuclear strike, or to acquiesce in whatever the potential striker demands as the price of its avoidance.'"[150] Thus Brams finds that both the exigencies of nuclear deterrence and the recommendations of strategic rationality necessitate moving to a stance of threatening harm as a vital means of self-defense.

In 1980, the moral philosopher Kavka revisited the morality of deterrence and identified a firmer deference for MAD that denied both the Prisoner's Dilemma structure of the problem and the all-inclusive reach of strategic rationality.[151] In his essay, Kavka provides a glimmer of clarity for the defensive

---

[147] Thomas S. Kuhn, *Structure of Scientific Revolutions*, 50th anniversary edition (Chicago: University of Chicago Press, 2012).

[148] Steven J. Brams, *Superpower Games: Applying Game Theory to Superpower Conflict* (Yale University Press, 1985).

[149] Ibid., 152.

[150] Ibid., 36, quote from David Gauthier, "Deterrence, Maximization, and Rationality," *Ethics* (1984) 94:3, 99–120, 494. Brams's quotation from Gauthier shows the tight relationship between the philosophy of deterrence, rational choice theory, and game theoretic analysis of superpower strategy.

[151] Gregory Kavka, "Deterrence, Utility, and Rational Choice," *Theory and Decision* (1980), 12, 41–60. Kavka's earlier work, discussed in "The Inescapable Irrationality of MAD" section "Some Paradoxes of Deterrence," 1978.

realist and classical liberal who seek to assure others of their cooperative orientation yet have sufficient influence to deter predators. Dovetailing his analysis with rational decision theory and John Rawls's appropriation of von Neumann's mimimax principle to secure his second principle of justice, Kavka performs a rigorous utilitarian analysis of whether rational choice most supports (1) unilateral disarmament, (2) counter value deterrence via MAD, or (3) counterforce deterrence via NUTS.[152] In this second paper, Kavka concludes that, indeed, the position most consistent with nuclear strategic realities and utilitarian ethics is that of minimal deterrence via MAD. Disarmament is too likely to lead to Soviet domination. Counterforce targeting is too destabilizing for mimicking preemptive weaponry and granting such capability, and it perpetuates arms racing.[153] Kavka solves his earlier problem, which President Carter faced, of issuing a credible deterrent threat by focusing on the overall beneficial implications of maintaining deterrence. Kavka argues that his worries articulated in "Some Paradoxes of Deterrence" may be overcome along the lines of Schelling's invocation of chance: so long as the Soviet Union is not 100 percent certain that there will be no retaliatory consequences for its attack on the United States, deterrence is sufficiently plausible given the catastrophic nature of nuclear warfare.[154]

The most interesting feature of Kavka's proposed resolution of the nuclear security dilemma with MAD is that he takes up Schelling's burden of answering the offensive realist's challenge of how to deter an aggressor signified by accepting the PD model. Kavka observes, "One view of the balance of terror is that it results from each side selfishly pursuing its national interests, rather than adopting a moral posture and seeking to promote the interests of mankind as whole."[155] In other words, no adequate resolution of the nuclear security dilemma can ignore this worry that either the United States or the Soviet Union seeks success through exploitation and risks annihilating both nations and more countries besides. Kavka concludes that given the impossibility of sufficiently reducing uncertainty about the other's intentions, maintaining deterrence as opposed to disarming, for the time being, is essential for national security. Not only does Kavka reject NUTS in favor of MAD, but he also proposes that the long-term solution to dissolving the nuclear standoff will only arise from altering the other's perceptions by assuring that party of one's own cooperative intention. Thus, he recommends "changing U.S. and Soviet *perceptions* of each other and gradually building mutual trust between the two nations and their governments."[156] Here, Kavka

---

[152] John Rawls, *Theory of Justice*, 1970.
[153] Kavka, "Deterrence, Utility, and Rational Choice," 1980, 55–56.
[154] Even in reading and rereading Kavka's "Some Paradoxes of Deterrence," 1978, and "Deterrence, Utility, and Rational Choice," 1980, essays, the philosopher remains unclear on how deterrence can be defended by the purely moral and perfectly rational actor; this is my most charitable reading to render his overall conclusion intelligible.
[155] Kavka, "Deterrence, Utility, and Rational Choice," 1980, 58.
[156] Ibid.

rejects offensive neorealism by suggesting that neither nation is inherently aggressive and proposes that each build trust, and he invites us to question the overarching reach of strategic rationality.[157]

Thus, Kavka's proposed reconciliation of Cold War nuclear tensions offers a means to solve the problem of issuing a credible threat, accepts the challenge of a PD model for the security dilemma and arms race, yet recommends building trust over time by assuring the other of one's reciprocal cooperation. His philosophical work, which gained acknowledgment in the 1980s, seems to suggest that one can successfully defend MAD against NUTS even within the paradigm of strategic rationality.[158] However, in his 1980 defense of MAD, Kavka circumvents Schelling's two admissions. He first insists on acknowledging uncertainty about the intentions of the other, ultimately suggesting that the demeanor of cooperation befits a superpower with appropriate moral values. Of course, the problem here is that once a superpower stipulates that it prefers mutual cooperation over mutual defection, then it will be flat footed in both the Chicken game of nuclear deterrence and bargaining in the overarching arms race. However, given that the common interest in avoiding nuclear war outpaces that of maintaining a bid for nuclear ascendance, the exercise of nuclear sovereignty invites placing national integrity consistent with defensive realism and classical liberalism on the highest level of priority. Hence, second, Kavka opens the door to recognizing trust as a category of action on par with reciprocal no-harm that exceeds the standard operationalization of strategic rationality.

CONCLUSION

In conclusion, several points are evident:

1. By 1980, the Prisoner's Dilemma fulfilled multiple functions: it captured the toughest case security dilemma as well as more mundane ones besides, it reflected the arms race, and it thus came to represent the state of anarchy and the puzzle of emerging from it.
2. The PD is a core logical puzzle at the heart of noncooperative game theory and it is reinforced by game theorists' assumption that actors pursue

---

[157] This is a key point that will be discussed in full in the introduction to Part II: "Government." Is moral agency wholly consistent with strategic rationality, or must strategic rationality displace some views of moral conduct? The philosopher who addresses this question most directly in conjunction with nuclear deterrence, no less, is David Gauthier in "Deterrence, Maximization, and Rationality," 1984; Joseph S. Nye Jr., in *Nuclear Ethics*, 1986, concludes that the United States must have motives that transcend strategic rationality to maintain the integrity of its values. Brams, *Superpower Games*, 1985, suggests wholly working within the confines of strategic rationality, but it is difficult to surmise if he takes a position that would support MAD over NUTS; no rational deterrence theorist appears to have offered a sound defense of MAD over NUTS within the context of the 1970s nuclear security debate.

[158] Lackey, "American Debate," 1987, 40–41. It is improbable that Carter was aware of Kavka's papers, but of all the contemporaneous literature discussing nuclear security, Kavka's reasoning was most pertinent to Carter's orientation.

fungible scarce resources in competition with each other, and the fact that insecurity in an Assurance Game makes it appear indistinguishable from a Prisoner's Dilemma to external observers.

3. The PD representation of the paradox of nuclear deterrence reinforced the view that deterring an aggressor requires threatening punitive damage that mimics suckering the other actor by defecting, recommends coercive bargaining to ensure one's most favorable settlement, and transposes the exercise of ensuring one's own cooperation into threatening sanctions on others to secure their compliance.

4. Within the paradigm of strategic rationality, NUTS must win the nuclear security debate because it sustains the practice of issuing credible threats, engaging in coercive bargaining, and using flexible nuclear forces to exert escalation dominance. MAD relies on the manifest fact of essential equivalence, building a common recognition of the futility of full-scale thermonuclear war and the high probability of uncontrollable escalation should nuclear confrontation break out, and it deters via counterattack instead of flexible escalation using evolving nuclear options.[159]

5. Noncooperative game theory carries the claim of exhaustive validity over all decision making and renders focal the Prisoner's Dilemma by its own theoretical structure. Thus, viewing relations through its lens necessarily negates categories of action that have alternative means of legitimation. President Carter's conversion from MAD to NUTS demonstrates how offensive neorealism consistent with noncooperative game theory secured a logical victory for escalation control and flexible response that attempted to defy the "life condition" of mutual assured destruction by demonstrating the intent and capability to prevail in prolonged nuclear conflict and to secure flexible nuclear military capability to address varying levels of conflict.[160]

James Schlesinger stands witness to President Carter's moral stance evident in his affirmation that, "Why, of course, if we made the promise to them we have to enter into an agreement to fulfill another administration." Even at Schlesinger's prompting that reneging on an agreement with allies could give the United States extra bargaining leverage, Carter replied that, "Of course we have given our word."[161] However one evaluates the exigencies of international

---

[159] These distinctions are clear from comparing Secretary of Defense Harold Brown's "Report to Congress 1979 Budget, FY 1980 Authorization Request, and FY 1979–1983 Defense Programs," January 23, 1978, see especially pp. 56–59, to National Security Advisor Zbigniew Brzezinski's team "Draft PD on Nuclear Employment Policy," materials attached to memo from William Odom and Jasper Welch, April 17, 1980, and "PD on Nuclear Force Employment Proposed Revision," April 15, 1980, "3/80–4/80," Box 35, Brzezinski Collection, JCPL, especially p. 7E (hand marked).

[160] The acknowledgment of the inescapable "life condition" of MAD is in the "Information Memo" from Stan Sienkiewicz to All Members of the Committee on Foreign Relations, September 9, 1980, "5/80–1/81," Box 35, Brzezinski Collection, JCPL, p. 5 of 8.

[161] "Interview with Dr. James R. Schlesinger," 1984, 59–60.

politics, we can recognize that strategic rationality contradicts the practice of making agreements and keeping them. It recommends the building of social order out of threats, coercive negotiation, and individual maximization without regard for others instead of reciprocal respect, assurance of one's cooperative intention, and the good will to make at least one person better off and no one worse off.

PART II

GOVERNMENT

Introduction

At work here is the law of the instrument [that is, game theory and the Prisoner's Dilemma]: give a small boy (or a researcher) a hammer and he will find things that need hammering. As Kaplan (1964, 29) points out, often the problem is not that some techniques are pushed to the utmost, but that others may, in consequence, be ignored.

Robert Axelrod, 1970[1]

Because it captures the structure of a recurring sort of social predicament, there can be no doubt but that the prisoner's dilemma model is bound to be of use to policy-makers ... This complexity in co-operative predicaments probably means that the state will often be necessary to select and then reinforce one of the available resolutions. However much we relish the invisible hand, we may still require the strong arm. The lesson is as old as Hobbes, but there is no reason here for surprise. So, after all, is an appreciation of the prisoner's dilemma. Due to the inherent failure of individuals' ability to cooperate, third-party enforcement must be invoked.

Philip Pettit, 1985[2]

It is easy to appreciate how an American president, even the most conscientious, would be impaled on the horns of the intractable nuclear security dilemma. By all indications, the nuclear security dilemma should be resolved by assuring recipro-cal cooperation backed by potential devastating retaliation, in keeping with the classical liberal adage of war for war and peace for peace. However, rational strategists considered the threat to harm millions of innocent people once deter-rence had already failed to be both immoral and irrational. In 1978, Gregory Kavka, who contributed the most rigorous philosophical analysis of this problem at that time, had little to offer President Jimmy Carter to place the strategy of

---

[1] Robert Axelrod, *Conflict of Interest: A Theory of Divergent Goals with Applications to Politics* (Chicago: Markham, 1970), 6; Abraham Kaplan, *The Conduct of Inquiry* (Chandler, 1964).

[2] Philip Pettit, "The Prisoner's Dilemma and Social Theory: An Overview of Some Issues," *Australian Journal of Political Science* (1985) 20:1, 1–11, at 1.

mutual assured destruction on firm ground, save that he should leave the role of issuing credible deterrent threats to agents who were either less than moral or less than rational with the hope they would never be carried out.[3] In 1980, Kavka renewed his efforts to defend the minimum deterrence of MAD against the countervailing war-fighting posture of NUTS. He recommended that the United States should disambiguate its intentions from those characterized in the Prisoner's Dilemma model of the nuclear security dilemma and arms race by clearly demonstrating the intention to cooperatively avoid nuclear war. He recommended building trust by offering assurance of intentions to follow through on cooperation if assured reciprocity rather than demonstrating the preference to secure dominance consistent with PD logic.

Leaving aside for the moment how best to resolve the paradox of nuclear deterrence, in the 1980s philosophers and social scientists came to accept that much more mundane contexts – most prominently the social contract, collective action, and market exchange – were best modeled with the Prisoner's Dilemma. This model no longer was reserved for worst-case scenario planning; instead, the fact that the Prisoner's Dilemma had proven to be useful in analyzing nuclear deterrence and the potentially apocalyptic arms race was used as an advertisement for its relevance.[4] One of the greatest proponents of the game theoretic approach to studying social interactions, Robert Axelrod, moved swiftly from the Cold War security quandary to individual behavior among friends and business partners and the riddle of how to exit the state of nature associated with Thomas Hobbes's anarchy. He begins his highly acclaimed and influential book *The Evolution of Cooperation* with the following questions:

When should a person cooperate, and when should a person be selfish, in an ongoing interaction with another person? Should a friend keep providing favors to another friend who never reciprocates? Should a business provide prompt service to another business that is about to be bankrupt?[5]

Axelrod prepares to offer advice to anyone confronting these questions and bases the credibility of his wisdom on its relevance to nuclear security. He next asks, "How intensely should the United States try to punish the Soviet Union for a particular hostile act, and what pattern of behavior can the United States use

[3] This is what Kavka concludes in "Some Paradoxes of Deterrence," *Journal of Moral Philosophy* (1978) 75:6, 285–302 and what Douglas Lackey reports in that other moral philosophers had gravitated toward accepting the minimum deterrence characteristic of MAD, yet remained stymied by the fact that only those manifestly willing to use nuclear weapons in line with NUTS could maintain credible deterrence by the 1970s, "The American Debate on Nuclear Weapons Policy: A Review of the Literature, 1945–1985," *Analyse and Kritik* (1987) 9, 40–43. This Lackey literature review is essential reading in the nuclear security debate and makes clear the portal both international relations theory and political theory went through conjointly in tackling the problem of nuclear security with the preeminent tool of rational deterrence theory, which is equivalent to game theory.

[4] This is evident in one of the most important texts discussing the PD, Richmond Campbell and Lanning Sowden, *Paradoxes of Rationality and Cooperation* (Vancouver: University of British Columbia Press, 1985).

[5] Robert Axelrod, *Evolution of Cooperation* (New York: Basic Books, 1984), vii.

to best elicit cooperative behavior from the Soviet Union?"[6] Building on his 1970 *Conflict of Interest*, Axelrod's application of strategic rationality spans the Cold War security dilemma, the broader sphere of international relations, market exchange, governance and the social contract, interpersonal relations, and even evolutionary biology.

Axelrod, along with frequent coauthor and neoliberal institutionalist Robert Keohane, was a key figure in establishing the neoliberal approach to political economy. Neoliberalism provides an explanation for how narrowly self-interested rational actors may achieve cooperative outcomes under the assumption of strategic rationality. Axelrod uses the issue of international relations security to introduce the significance of strategic rationality and the Prisoner's Dilemma more generally. He observes, "The most important problem is the security dilemma: nations often seek their own security through means which challenge the security of others."[7] In his enthusiasm to identify situations to which the PD model is applicable, Axelrod overlooks international relations theorists' original analysis of the conditions that must pertain for a security dilemma composed of hopeful cooperators to devolve into the state in which each actor prefers unilateral defection and a Stag Hunt transforms into a Prisoner's Dilemma because "either offensive weapons exist and are superior to defensive ones, or that weapons systems are not easily distinguishable."[8] Axelrod, who relies on interpersonally transferable utility that he directly relates to viability and successful purposive agency, popularizes the indefinitely repeated Prisoner's Dilemma, with two individuals encountering each other in exactly the same circumstances over prolonged periods, as the remedy for avoiding the suboptimal outcome. Thus, he accepts that actors often exhibit the Prisoner's Dilemma predilection for unilateral defection yet shows how the ill consequences of mutual defection can be avoided if the environment is appropriately structured to ensure indefinitely repeated dyadic interactions with actors who have perfect memory recall.[9]

Axelrod links his research to studies of cooperation and altruism in evolutionary biology and made connections between traditional Western political theory and game theoretic modeling. He is quick to draw attention to the apparent confluence of three centuries of thought under the single rubric of the Prisoner's Dilemma: "Hobbes regarded the state of nature as equivalent to what we now call a two-person Prisoner's Dilemma, and he built his justification for the state upon the purported impossibility of the sustained cooperation in such a situation."[10] Throughout his work, Axelrod articulates the intellectual

---

[6] Ibid.

[7] Ibid., 4.

[8] Art Stein, "Cooperation and Collaboration," in *Neorealism and Neoliberalism*, ed. David Baldwin (New York: Columbia University Press, 1993), 49.

[9] Axelrod's cooperative solution to the PD through indefinitely repeated play is the topic of Chapter 11, "Tit for Tat."

[10] Axelrod, "Emergence of Cooperation Among Egoists," 1985, 320–338, 321.

foundations of neoliberal markets and governance in the Prisoner's Dilemma social contract.[11]

The reinvention of the social contract in keeping with strategic rationality and the Prisoner's Dilemma model of the security dilemma have been coextensive with the emerging practice of neoliberal economics, and viewing behavior to be programmed into agents, therefore modeling presocial organisms in evolutionary contexts on par with human subjects (Part III). Many game theorists, social scientists, and philosophers contributed to this shift in consciousness. Some were attracted by the scientific promise and utter novelty of game theory, most prominently Duncan Luce and Howard Raiffa, and others, such as James M. Buchanan, Gary Becker, and Richard Posner, appreciated the power to reexamine stale debates and produce arguments they found appealing. David Gauthier and Gregory Kavka were dedicated to investigating the implications of strategic rationality and the PD for political theory.[12] The fact that they found the Prisoner's Dilemma analytically identical to the challenge of practicing nuclear deterrence invites us to investigate the significance of their approach and its litany of implications.[13] The stakes are nothing less than these researchers' simultaneous design and ratification of the blueprint for the neoliberal world order that entails accepting the comprehensive reach of game theory and displacing alternative logics of action including appropriateness, incommensurable and non-finite valuation, and solidarity. Researchers using game theory were not the only ones to effect this change in consciousness. Publication editors; referees; academic hiring and tenure committees; and the interplay of actors and networks, institutions, and financial support advanced this new approach.[14]

[11] This had been prevalent until the early twenty first century. Even though this analysis remains the default, Brian Skyrms has argued that the Stag Hunt is more appropriate for modeling the social contract. In *The Stag Hunt and the Evolution of Social Structure* (New York: Cambridge University Press, 2003), Skyrms relies on interpersonally transferable utility, 65–81, assumes indefinitely repeating play between the same two actors to transform a PD into a Stag Hunt, 4–5, and moves quickly to encompass both humans and *E. coli* bacteria into his understanding of the social contract, 15; Herbert Gintis finds that the Chicken game is most germane, *Bounds of Reason* (Oxford: Oxford University Press, 2009).

[12] Lackey, "American Debate," 1987, provides the most comprehensive analysis and bibliography of political theorists' engagement with the paradox of cooperation under the nuclear security dilemma modeled as a PD.

[13] The paradox of nuclear deterrence, the generalized PD game (in all its forms, two to *n*-individuals, single shot, repeating, and indefinitely repeating), the Newcomb Problem, and the Toxin Puzzle are all classed as the same decision problem. Gauthier, "Deterrence, Maximization, and Rationality," *Ethics* (1984) 94:3, 474–495, makes the first comparison evident. On the Newcomb problem, see David Lewis, "Prisoner's Dilemma Is a Newcomb Problem," in *Paradoxes of Rationality and Cooperation*, 1985, 251–255; for Lewis on nuclear deterrence see his "Finite Counterforce" in *Nuclear Deterrence and Moral Restraint*, ed. Henry Shue (Cambridge: Cambridge University Press, 1989). On the Toxin Puzzle, and its relationship to nuclear deterrence, see Gregory Kavka, *Moral Paradoxes of Nuclear Deterrence* (Cambridge: Cambridge University Press, 1987), 47.

[14] On the complexity of these considerations see John McCumber, *Philosophical Excavations: Reason, Truth and Politics in the Early Cold War* (Chicago: University of Chicago Press, forthcoming).

Professionals implemented the approach in law, public policy, and the design of institutions.[15]

Before moving into the chapters in this Part II, readers are invited to first consider a central question: how did the worst-case planning that was deemed suitable to, if not the factual realities, then the logical intricacies of rational deterrence become transferred into the interiority of civil society to produce nuclearized sovereignty? How did strategic rationality, which typically assumes consequentialism (only outcomes matter), realism (value exists prior to social relations), and hyper-individualism ("other-regarding" signifies viewing others as strategic maximizers like oneself), come to be the only approach to coherent action available to individuals throughout their lives?[16] The pivotal point of paradigmatic shift occurred when the analytic philosopher and political theorist David Gauthier extended the anxiety over the incredibility of immoral threats underlying mutual assured destruction to question the coherence of moral promises and commitments. With this theoretical move, Gauthier mires social contract theory in the same quagmire as that which sunk MAD. Assuring others of one's intention to cooperate in a situation with Stag Hunt rewards devolved into a Prisoner's Dilemma because deterring a potential aggressor demanded credible threats relying on one's impersonation of a predator. Thus, individual autonomy and any collective sovereignty derived from it came to exhibit the same rationale for action as that characterizing NUTS: the pursuit of asymmetric success despite others through coercive bargaining and unilateral defection to sucker others when possible.

Gauthier's 1984 essay "Deterrence, Maximization, and Rationality" marked the point at which analytic political theory fully embraced rational decision theory and endorsed the PD view of governance, thereby fully breaking with the classical liberal world of reciprocal respect for one another's right to exist, enjoyment of property as an extension of personhood, and autonomy to enter into private agreements without the need for invasive government mandating compliance. Gauthier's work accepts that rational individuals must comport with the axioms of decision theory; no other logics of action are available to them. The neoliberal subject can expect to be caught in Prisoner's Dilemmas at any moment and will need to devise means to resolve them using the tools of noncooperative game theory. The classic liberal world of reciprocal no-harm and voluntary compliance with agreements made recedes because strategic

---

[15] Leading texts establishing each respective field are Gary Becker, *The Economic Approach to Human Behavior* (Chicago: University of Chicago Press, 1978); Edith Stockey and Richard Zeckhauser, *Primer for Policy Analysis* (New York: W. W. Norton, 1978); and Douglas Cecil North, *Institutions, Institutional Change, and Economic Performance* (Cambridge: Cambridge University Press, 1978).

[16] Nicola Giocoli points out the hyper-individualism of game theory requiring that actors choose actions that render their fates independent of others' choices despite the radical interdependence of social life, "Nash Equilibrium," *History of Political Economy* (2004) 36:4, 649–666; for an engaging analysis of how strategic rational actors encounter each other, see David Lewis, *Convention* (Cambridge, MA: Harvard University Press, 1969), 5–36.

rationality does not condone behavior that does not augment each individual's expected utility in consideration of achieved outcomes. This neoliberal alienation of selfhood is even more complete than in the classic liberal's proposal that freedom is equivalent to selling one's labor on the open market.[17] No domain of relations or experiences can transcend the single criterion imperative to trade off every eventuality against another and make choices accordingly throughout the supergame of each individual's life. Individuals are free to do whatever they please, subject to the constraint of their willingness and ability to pay for their choice. Instead of market relations being privileged practices contained within the rule of law, building on security to achieve positive-sum growth and experiential cultural value that may exceed limited physical resources by nonlinear measures, expected utility theory and game theory are the rules that all life forms must successfully navigate to survive and propagate.[18] Individuals are free to act on any choice contained in their set of opportunities. The volitional quality of choice is eroded in favor of strategic rationality, which was originally designed as a logic of action that could provide a complete set of instructions needing no conscious implementation.[19] Stress is perpetual because one's status is never secure as a result of uncertainty, risk, and other actors' combative stances and likely adoption of coercive bargaining tactics.

Gauthier's work represents a milestone because he definitively explains how the puzzle of achieving credible deterrence is equivalent to satisfactorily resolving a "whole range of situations, including most prominently generalized Prisoners' Dilemmas."[20] He concludes that the PD mirrors not only the nuclear security dilemma but also both the challenge of maintaining social order in a civil society and the problem of conducting amicable exchange: "The state of nations and, more especially, of nuclear powers is our nearest analogue to the state of nature, and Hobbes's advice applies to it."[21] In this single densely argued article, Gauthier takes Thomas Schelling's twin admittance of the complete validity of strategic rationality and its central analytic problematic Prisoner's Dilemma from the domain of worst-case preparedness

---

[17] This is Karl Marx's critique of the type of freedom offered by modern capitalism, "Communist Manifesto," *The Marx-Engels Reader*, 2nd ed. (New York: W. W. Norton, 1978).

[18] In Georg Simmel, *Philosophy of Money* (London: Routledge and Kegan Paul, 1978), 291 and 446–467, it is possible to discern how neoliberal capitalism represents not so much an alienation from objective material reality so far as from a subjectivity of endless possibility of individual and cultural innovation to one mandatory logic of action tracking ontologically existing presocial value necessary for survival and propagation.

[19] John von Neumann and Oskar Morgenstern, *Theory of Games and Economic Behavior* (Princeton, NJ: Princeton University Press, 1953), 33; Ken Binmore, *Natural Justice* (New York: Oxford University Press, 2005) amplifies this point, 210; for a comprehensive discussion, see Philip Mirowski, *Machine Dreams: Economics Becomes a Cyborg Science* (Cambridge: Cambridge University Press, 2002).

[20] Gauthier, "Deterrence, Maximization, and Rationality," 489.

[21] Ibid., 492.

into the heart of the social contract. Thus, relationships among individuals and between individuals and the state all fall under the umbrella of strategic rationality. Its appellation "rational choice" softens its strategic and combative character, instead emphasizing individual's choice between being rational or irrational.

In his work, Gauthier attempts to solve both the paradox of nuclear deterrence and the problem of Hobbesian anarchy, which he claims spans international relations and civil society, by finding a way out of the Prisoner's Dilemma. He credits Schelling for first identifying that the PD best models the nuclear security dilemma. However, rather than following Schelling's and Kavka's recommendations to exit the PD into the more favorable Stag Hunt game, Gauthier turns his attention on cutting through the Gordian knot of the PD's inevitable suboptimal, mutually impoverishing outcome. Gauthier's research is at the epicenter of the rupture between classical liberalism and neoliberalism because he fully embraces both strategic rationality and its entailed paradox that self-interest leads to mutual disaffection in the Prisoner's Dilemma.

Gauthier's analysis identifies two means of exiting the PD. On the one hand, if one can issue a credible threat to retaliate should the other fail to cooperate, then the other will cooperate, and the PD is solved. The other way out of the PD, so Gauthier reasons, is to credibly promise that one will cooperate if the other does as well. Neither solution is complete without both parties adopting the proposed solution symmetrically: each would need to deter the other, and each would need to credibly promise to cooperate if the other does. Deterrence and assurance have the same puzzling structure because in both cases individuals offer to take an action that at the time of its execution fails to have any causal import in realizing their preferences. From the perspective of strategic rationality, because the act serves no instrumental function in promoting the actor's goals at the time it is performed, the act must be irrational. Therefore, neither should the rational actor implement the act, nor is it credible that either the deterrent threat or assuring promise has any weight in averting the mutual defection outcome of the Prisoner's Dilemma.

Gauthier's analysis follows from equating solving the PD game by assuring the other of one's intent to cooperate in solving it by compelling cooperation via the threat of retribution. Thus, he equates a credible promise to cooperate with a credible threat to retaliate. In both cases, the PD structure of the problem accepts that each most prefers to sucker the other. The Prisoner's Dilemma reinforces the credibility of threatening the opposition in the case they fail to cooperate because, as with nuclear deterrence, the United States adopts preferences consistent with defecting and punishing the opposition *regardless* of the Soviets' cooperation or defection. However, following through on the punishment in the case of nuclear deterrence signifies both murder and suicide. This parallel construction of promising and threatening underscores the question of how, in either case, the commitment to follow through on a statement of intent can be credible if it lacks the casual power to realize the agent's expected gain at the time the action is taken. Specifically, if I threaten to retaliate if you fail to cooperate, but the act of retaliation does not further my goals or maximize my gain, then after you fail to cooperate, punishing you serves no purpose for me.[22] Similarly, so Gauthier argues, if I make a promise to exchange money for a good, once I have the good, and delivering the money serves no causal purpose in maximizing my expected gain, then I have no instrumental reason to fulfill my commitment. In both cases, the utility maximizing agent will not follow through on either the threat or the promise.

Both Kavka and Gauthier agree that the United States must be perceived as first and foremost a moral nation that would eschew acting on the initiative of gross brutality, specifically in the case of the failure of deterrence. However, their analyses hone in on different aspects of the paradox of deterrence. Kavka persistently emphasizes that *moral scruples* prohibit actors from issuing credible deterrent threats. Gauthier stresses that the *causal inefficacy* of the punitive retaliation means that perpetrating punishment fails to maximize the expected utility of the perpetrator and thus is irrational to enact and impossible to credibly threaten.

Whereas Kavka suggests that the assurance solution of the Prisoner's Dilemma may be possible because of the integrity of choice consistent with the moral institution of promising and the underlying hope to achieve mutual cooperation despite actors' presumed temptation to sucker the other, Gauthier concludes that both immoral threats and moral promises are contrary to strategic rationality. In game theory, expected utility only pertains to outcomes, which typically track an instrumentally salient property of the world evident to intersubjective inspection. Thus, if expected utility in games only registers the value of outcomes, and every consideration of worth must be reflected in the agents' expected utility function, then exiting the Prisoner's Dilemma by promising the other agent of one's intention to cooperate if assured reciprocity must fail. After the other actor cooperates, the PD preference matrix, which contains all the information relevant to individual choice, still requires the rational actor to defect.

Thus, Gauthier and the philosophers working in the rational choice tradition concurred that contracts from dyadic exchange through to complex multi-actor social contracts are best modeled as Prisoner's Dilemmas and resist solution by rational actors. Agents who comport with game theoretic instrumental rationality are caught in an undertow because they "cannot reason themselves into being rule followers [who maintain commitments entered into either by agreement or tacit consent] and yet rule following seems absolutely critical to social life."[23] The adoption of strategic rationality, as a result, contradicts social practices heretofore deemed sensible and fundamental to social concourse. Thus, in confronting the nuclear security dilemma wherein initially deterrent

---

[22] Footnote 13 applies equally here.
[23] Gerald Gaus, *The Order of Public Reason* (Cambridge: Cambridge University Press, 2011), 146–147.

threats that compromised the moral character of the United States were deemed wrong for intending harm, hence incredible and irrational, subsequently moral assurances were ensnared because they contradicted the structure of strategic action.

The nuclear security dilemma opened the door to a world in which the rational autonomous decision maker obeys the axioms of expected utility and rules of strategic choice just as the nuclearized superpower exercises sovereignty through compelling and deterring other nations. Agents accept that the same *modus operandi* of incentives and outcomes, as opposed to also considering legitimacy and procedures, delimit all intelligible relations. Part II investigates the implications of viewing the rationale for government in terms of strategic rationality and Prisoner's Dilemma. Chapter 5, "Hobbesian Anarchy," discusses how the rational choice reading of Hobbes's *Leviathan* and its PD model of anarchy leads theorists to consider that coercive force is the necessary and sufficient condition for achieving civil society instead of as a necessary condition to be complemented by consent and self-restraint. Chapter 6, "Social Contract," follows James M. Buchanan's and John Rawls's debate over fair play and the motivational weight of hypothetical consent to terms of distribution. Buchanan's neoliberal account of government privileges status quo power relations and cannot exit from the realm of coercive bargaining. Arguing that all people have the same temptation to transgress law independent of their means or inclusion in the social contract, Buchanan suggests a justification for a maximal security state consistent with mass incarceration. Chapter 7, "Unanimity," argues that Buchanan's attempt to derive constitutional legitimacy from unanimous agreement is undermined by the lack of motivational content of strategically rational consent. Chapter 8, "Consent," follows Richard Posner's further demolition of consent by his equation of the legal weight of *ex post* and *ex ante* consent. Posner's theoretical position has the implication of permitting that better-off individuals can acquire resources from less well-off individuals by paying compensation after harm without prior consent. Lastly, Chapter 9, "Collective Action," analyzes how the Prisoner's Dilemma model of joint action fails to distinguish between the challenges of the temptation to free ride and resignation over the causal negligibility of contributing, in situations classified as large-scale collective actions or "tragedies of the commons." The conflation of the hope to cheat with the concern that one's actions cannot make an appreciable difference makes it difficult for either political theorists or institutional designers to establish effective regulatory regimes.

Together, these chapters articulate the implications of a consistent rational choice approach to political theory that yields a basis for neoliberal political economy in stark contrast to the classical liberal state of mutual respect and minimal intervention. Viewing rational agency in terms of expected utility and strategic rationality creates a tendency to measure all value in terms of a zero-sum finite resource and to accept as nonnegotiable the imperative to maximize gain individualistically as the primary means of survival. Free riding is not one potential outcome, but the rationally condoned result.[24] Coercive bargaining too is consistent with rational choice and cannot be surmounted. The strength of neoliberal political theory is that it embraces value pluralism and rejects paternalism.[25] However, accepting that individuals *might* be narrowly self-interested, unprincipled, and uncooperative is different from teaching that rationality implies this behavior.

[24] Russell Hardin makes this point explicit in "The Free Rider Problem," *Stanford Encyclopedia of Philosophy*, 2003, available online, http://plato.stanford.edu/entries/free-rider/, accessed January 30, 2015.
[25] It is not clear that the rational choice approach rules out paternalism in cases wherein individuals' behavior falls short of strategic rationality; see Richard Thaler and Cass Sunstein, *Nudge* (London: Penguin, 2009).

5

Hobbesian Anarchy

I do not know what greater contribution human industry could have made to human happiness. For if the patterns of human action were known with the same certainty as the relations of magnitude in figures, then ambition and greed, whose power rests on the false opinions of the common people about right and wrong, would be disarmed, and the human race would enjoy such secure peace that (apart from conflicts over space as the population grew) it seems unlikely that it would ever have to fight again.

Thomas Hobbes, 2013[1]

In Hobbes' ... [*Leviathan*], each individual in the state of nature can behave peacefully or in a war-like fashion. "Peace" is like ... ["cooperate"] because when everyone behaves in this manner it is much better than when they all choose "war" ("confess") ... However, bellicosity is the best response to both those who are peaceful (because you can extract wealth and privilege by bullying those who choose peace) and those who are bellicose (because might can only be stopped by might). In short, "war" is the strictly dominant strategy and the population is caught in a *Prisoner's Dilemma* where war prevails and life is "nasty, brutish and short."

Shaun Hargreaves Heap and Yanis Varoufakis, 2004[2]

Game theory seemed aptly suited to warfare because it assumes that without threat of coercive sanctions, no agreement, consensual arrangement, or voluntary obligation is sacrosanct. The Prisoner's Dilemma, accordingly, seemed to capture the essence of the nuclear security dilemma. International relations theorist Alexander Wendt describes this most tense security dilemma in terms of the

---

[1] Thomas Hobbes, from his "Epistle Dedicatory," quoted in Richard Tuck's introduction to Thomas Hobbes, *Leviathan*, ed. by Richard Tuck, revised student edition (Cambridge: Cambridge University Press, 2013), xxvi.

[2] Shaun Hargreaves Heap and Yanis Varoufakis, *Game Theory*, 2nd ed. (London: Routledge, 2004), 174.

"logic of Hobbesian anarchy," which is well known: "the 'war of all against all' in which actors operate on the principle of *sauve qui peut* and kill or be killed."[3] Wendt explains that Hobbes's anarchy "is the true 'self-help' system ..., where actors cannot count on each other for help or even to observe basic self-restraint."[4] In this case, survival depends on military power, and when actor A increases security, those gains reduce B's security because B is unsure of whether A's intentions are defensive. Harkening back to the nuclear security dilemma, and the condition of uncertainty, which is only enhanced when actors deploy weapons with offensive value, Wendt concludes that "security is a deeply competitive, zero-sum affair, and security dilemmas are particularly acute ... because of the intentions attributed to others."[5]

Notwithstanding the numerous differences between nation-state actors and individual human actors, theorists routinely apply Hobbes's argument about the achievement of social order from anarchy to cases across domestic and international relations.[6] That is because, for many rational choice theorists, there is a direct parallel structure between the concerns of individual security and national security. In the words of David Gauthier, "war is the consequence of national insecurity, and the natural desire to preserve oneself."[7] According to Gauthier, "the natural condition of mankind is inherently unstable"; the competition for increased security through increasing power necessarily makes all actors increasingly insecure. Gauthier directly compares individuals' and states' pursuit of security: "Just as two nations seeking to strengthen themselves to prevent conflict with the other, find themselves locked in an arms race which tends to bring on that conflict, so men, seeking to strengthen themselves to prevent being overcome, find themselves locked in a race which ensures that most are overcome."[8] Gauthier reaches the glum conclusion that actors must

---

[3] Alexander Wendt, *Social Theory of International Politics* (Cambridge: Cambridge University Press, 1999), 265; Wendt also considers that the Prisoner's Dilemma underlies international relations anarchy in his "Collective Identity Formation and the International State," *American Political Science Review* (1994), 88:2, 384–396.

[4] Wendt, *Social Theory*, 1999, 265.

[5] Ibid.; Wendt's discussion is consistent with the game theoretic view that if actors care about others' relative power, then the Prisoner's Dilemma payoff matrix closely approximates a zero-sum game; for discussion, see Michael Taylor, *The Possibility of Cooperation* (New York: Cambridge University Press, 1987), 143–148.

[6] For another example of Hobbesian anarchy analyzed from the perspective of international relations, see Charles R. Beitz, *Political Theory and International Relations* (Princeton, NJ: Princeton University Press, 1979), 27–50; and Michael Doyle, *Ways of War and Peace* (New York: W. W. Norton, 1997), 111–136. Important differences between the domestic and international case are the relative power differentials and longevity of the actors. There is debate over the significance of Hobbes's theory; Campbell Craig argues that for Hobbes, self-preservation was the most pressing concern; *Glimmer of a New Leviathan: Total War in the Realism of Niebuhr, Morgenthau, and Waltz* (New York: Columbia University Press, 2003), ix–xviii.

[7] David Gauthier, *Logic of Leviathan* (Oxford: Clarendon, 1969), at 17.

[8] Ibid.

pursue power as a condition of their survival, but that this "perpetual and restless desire of power after power" entails "impotence ... [and] death." Gauthier's conclusion resonates with neoliberal political philosophy, which concurs that the challenge of achieving order under a social contract ultimately comes down to solving the Prisoner's Dilemma game. He relies on the PD model to capture the essence of his worry that actors will not find it rational to, and therefore will not, follow through on agreements they made. [9]

Rational choice theorists now regularly suggest that Hobbes's state of nature is a Prisoner's Dilemma in which each individual at all times seeks to take advantage of others. [10] The solution, accordingly, is Prisoner's Dilemma governance, or govermentality, if the target meaning refers to how techniques of governing and subject's rationales for action are inseparably developed: the strong state is introduced to keep everyone in line with the law. [11] This Prisoner's Dilemma logic is extended to the idea that individuals only follow through with agreements because of the threat of third-party enforcement. This rational choice reading of Hobbes suggests that coercive force is a necessary and sufficient condition to achieve the rule of law. As with deterrence theory, the question of legitimacy, or consent to mutual self-restraint granting to each the right to exist, is superfluous.

Comparing a traditional reading of Hobbes with a contemporary game theoretic reading reveals multiple clear differences between these two approaches spanning almost four centuries. It is simple to observe that Hobbes did not view the exercise of leaving the state of nature in terms consistent with the Prisoner's Dilemma game because he clearly asserted the overwhelming mutual benefit derived from living in a commonwealth. Thus, if one were forced to rely on a game theoretic model to make this point, exiting the state of nature is better regarded as an Assurance Game or possibly an Assurance Dilemma because the upside of achieving a state of mutual cooperation is fantastic. However, Hobbes further argues that not only is it difficult for men to reach sound conclusions of judgment in logic and cause-effect relations in a civil society, but that the state of nature is characterized by fundamental uncertainty and unpredictability stemming from its configuration and actors' inability to have systematic knowledge. Additionally, Hobbes's entire focus is on security as a threshold condition, and not, as rational choice theory suggests, the pursuit of preference satisfaction without constraint. Thus, whereas Hobbes's construal of the proverbial state of nature at a minimum more resembles a Stag Hunt than a Prisoner's Dilemma, even more pointedly his actors are not limited to strategic rationality and may

---

[9] Ibid., 79, 97; see also his *Morals by Agreement* (Oxford: Oxford University Press, 1987).

[10] Michael Taylor has a clear exposition of this perspective in his *Anarchy and Cooperation* (New York: John Wiley, 1976); and *The Possibility of Cooperation*, 1987).

[11] "Governmentality" is Michel Foucault's concept articulated in chap. 4, "Governmentality," in *The Foucault Effect: Studies in Governmentality*, ed. Graham Burchell, Colin Gordon, and Peter Miller (Chicago: University of Chicago Press, 1991), 87–104.

grasp natural laws and self-made agreements as rationales for action required for living in harmony.[12]

The rational choice interpretation of Hobbes's *Leviathan* pushes past its classical liberal roots to reach the neoliberal conclusion that a maximum security state with invasive monitoring and surveillance powers will be necessary to prevent all citizens from acting on their overriding prerogative to defect from cooperation whenever the opportunity presents itself without penalty. Hobbes's three fundamental laws, resting on analytic consistency, are irrelevant to the neoliberal Hobbes. These state that actors must seek peace as a condition of their self-preservation, that this demands relinquishing the right to every form of impulse or desire satisfaction available, and requires following through on agreements made. The pages ahead present the contemporary reading of Hobbes in game theoretic terms, introduce the classical liberal Hobbes, and discuss how the former celebrates Hobbes's notorious Foole while the latter holds this character in contempt.[13] The chapter concludes by exploring how a society populated by strategic rational actors may be best exemplified by the example of Hobbes's Foole, who, as the logic of the Prisoner's Dilemma recommends, each defect even after his partner in exchange pursuing the fruits of collaboration has cooperated.

HOBBES'S LEVIATHAN AND THE PRISONER'S DILEMMA

Many theorists tackling the rationale for government through the worldview of rational choice propose that Hobbes's state of nature is best represented by a Prisoner's Dilemma game.[14] There are many possible reasons for such an argument. The intense insecurity characterizing a state of nature resembles that described by Schelling in "Mutual Fear of Surprise Attack."[15] Competition over scarce resources requisite for agency presents the characteristic Prisoner's Dilemma payoff matric in which all actors pursue more power for themselves. Also, if agents' solely act on myopic preferences consistent with egoism and jockeying for preeminence, and view making agreements as cheap talk, then

---

[12] Brian Skyrms reviews Hobbes's state of nature in terms of a Stag Hunt, but he derives this view from a repeating Prisoner's Dilemma (discussed in Chapter 11, "Tit for Tat") and moreover reinforces a limited view on human agency discussed later in this chapter; *Stag Hunt and the Evolution of Social Structure* (Cambridge: Cambridge University Press, 2004), 1–6.

[13] See Jean Hampton, "The Knavish Humean," in *Rational Commitment and Social Justice*, ed. Jules L. Coleman, Christopher W. Morris, and Gregory S. Kavka (Cambridge: Cambridge University Press, 1998), 150–167; Gauthier, *Logic of Leviathan*, 1969, 79, 97.

[14] The best known are Gauthier, *Logic of Leviathan*, 1969; Gauthier, *Morals by Agreement*, 1987; Gregory S. Kavka, *Hobbesian Moral and Political Theory* (Princeton, NJ: Princeton University Press, 1986); Jean Hampton, *Hobbes and the Social Contract Tradition* (New York: Cambridge University Press, 1988); Daniel Farrell, "Hobbes as Moralist," *Philosophical* Studies (1985), 48, 257–283; Daniel Farrell, "Reason and Right in Hobbes' *Leviathan*," *History of Philosophy Quarterly* (1984), 1:3, 293–314; see also Taylor, *Possibility of Cooperation*, 1987.

[15] Chapter in *The Strategy of Conflict* (Cambridge, MA: Harvard University Press, 1960).

any resource dilemma characterized by strategic competition over finite goods seems best modeled by the PD.

Regardless of which motive propels actors – fear, gain, or glory – game theory seems to capture the problem of cooperation. Game theorists collapse these three motives, which Hobbes distinguishes, into a single criterion payoff matrix.[16] Additionally, game theorists must assume that actors are aware of all conceivable outcomes, that they have consistent preferences over these potential outcomes across their lifetimes, and that they can accurately estimate relative probabilities of outcomes coming to pass. Once one concedes these assumptions, and that the scenarios that actors repeatedly face reflect the paradigmatic Prisoner's Dilemma, then Hobbes's argument for Leviathan seems appropriate. The Prisoner's Dilemma best represents the state of nature in which individuals confront one another. The state must step in to alter actors' incentives so that they can achieve the mutually desirable outcome of cooperation versus mutual attack. According to this reading of Hobbes, the state is particularly required because even if actors agree to cooperate, each still has the most favored temptation to defect. Hence, according to Hobbes:

And covenants, without the sword, are but words, and of no strength to secure a man at all. Therefore notwithstanding the laws of nature [self-preservation, and the predilection to seek peace once self-preservation is assured] ... if there be no power erected, or not great enough for our security; every man will, and may lawfully rely on his own strength and art, for caution against all other men.[17]

Game theorists concur. Agreements without binding force behind them are mere words or in the parlance of game theory, cheap talk. The primary problem of social order then is that given the myriad situations throughout civil society in which agents' preferences and choices are surmised to resemble that of a Prisoner's Dilemma, everyone prefers mutual cooperation to mutual defection, yet everyone prefers sole defection to unanimous cooperation.[18] An enforcement body must therefore be introduced to alter the otherwise tempting payoff for attempting unilateral defection.

[16] Hobbes, *Leviathan*, 1996, 88; Taylor explains his transferal of Hobbes's fundamental motives of human agents into the single criterion metric; *Possibility of Cooperation*, 1987, 138–144, note that he argues that Hobbes's motive of glory can be captured by relative versus absolute gains, whereas others might attempt to capture it by reputation (viewing actors' past performance as indications of their future performance, disregarding the fact that past correlation between a situation and an outcome cannot prove or guarantee future correlation).

[17] Hobbes, *Leviathan*, 1996, 199; this passage is quoted by David Gauthier, *Logic of Leviathan*, 1969; Gauthier goes on to observe that this passage is as relevant to "nations as to individual men" because "agreement requires enforcement; enforcement requires power," 210.

[18] Recall that this preference matrix is considered characteristic of an intense security dilemma, arms races, disarmament, social trust, free riding, domestic labor, public goods, market exchange, bargaining, joining a trade union, corruption, the increasing inefficacy of antibiotics, and deciding to stand in a crowded stadium Shaun Hargreaves Heap and Yanis Varoufakis, *Game Theory: A Critical Introduction*, 2nd ed. (New York: Routledge, 2004), 175–180.

This straightforward reading of Hobbes may be referred to as Rational Choice Hobbes. Many theorists have developed this interpretation of Hobbes's *Leviathan* and the challenge of achieving social order.[19] It now stands as a canonical reading of Hobbes as well as a standard game theoretic argument explaining the need for government. In their authoritative game theoretic text, Hargreaves Heap and Varoufakis move directly from introducing the Prisoner's Dilemma game to using it to provide the rationale for government. The difficulty is that in every situation characterized by the Prisoner's Dilemma preference matrix, all would agree to mutual cooperation but then would look to exclude themselves from carrying through as promised.[20] Hargreaves Heap and Varoufakis conclude that "there needs to be a mechanism for enforcing agreement. Hobbes's 'sword', if you like." Crucially, they add, "It is this recognition which lies at the heart of a traditional liberal argument, dating back to Hobbes, for the creation of the State (or some enforcement agency to which each submits)."[21]

The same problem of how to achieve compliance with an agreement made in a state of nature resurfaces within a nation-state.[22] Individuals must pay taxes to enable the state to function. Yet, each prefers to free ride on others' tax payments and is only motivated to pay into the state's coffers because of the

[19] In addition to those mentioned in note 14, see James M. Buchanan, *Limits of Liberty: Between Anarchy and Leviathan* (Liberty Fund, 2000); Hargreaves Heap and Varoufakis, *Game Theory*, 2004, 34–38.

[20] David Lewis addresses this characteristic predicament in his discussion of conventions vs. social contracts: in a social contract, individuals may prefer for others to cooperate while they themselves engage in lone disobedience, whereas under conventions all individuals prefer mutual cooperation to sole defection as in an Assurance Game; *Convention*, 1969, 90–91. Lewis offers a game theoretic interpretation of Hobbes's resolution of the social contract, suggesting that (consistent with preferring peace to war), actors' preferences will reflect mutual cooperation over unilateral defection, and that therefore, "these accepted obligations [in the form of "tacit consent or fair play" to honor the sovereign's commandments] will be counted as a component of preferences, not as an independent choice-determining force"; *Convention*, 2002, 94. However, Lewis's resolution of Hobbes's state of nature into the social contract misses that in orthodox noncooperative game theory using expected utility theory over outcomes not only gives competition over scarce resources the payoff structure of a Prisoner's Dilemma, but moreover leaves no latitude "to think of someone's preferences as the resultant of *all* the more or less enduring forces that go into determining his choices," which would permit "the modification of our preferences by obligations" that exist because individuals prefer peace over war, 93 and 94. Thus, Hobbes's three analytic laws of nature prescribe obligation, which then transforms preferences contradicting orthodox noncooperative game theory. For a clarifying discussion, see Amartya K. Sen, who observes that "the rejection of self-goal choice reflects a type of commitment that is not able to be captured by the broadening of the goals to be pursued ... and it has close links with the case for rule-based conduct, discussed by Adam Smith; "Goals, Commitment, and Identity," in his *Rationality and Freedom* (Cambridge, MA: Harvard University Press, 2002), 206–224, at 219–220.

[21] Hargreaves Heap and Varoufakis, *Game Theory*, 2004, 174; note that Phillip Pettit concurs with this argument for law enforcement, "The Prisoner's Dilemma and Social Theory," 1985.

[22] Taylor, *Possibility of Cooperation*, 1987, 125–179, clearly articulates this, and he argues that conditions under the state will only exacerbate the PD structure of cooperating, see 167.

threat of sanctions. Thus, running the state is likewise conceived as a multi-person Prisoner's Dilemma game.[23] Gregory Kavka articulates the PD proportions of the problem. Security is a public good, which is defined as a resource supplied as a result of private contributions that everyone has access to regardless of whether that individual actually contributed. Kavka explains, "Most any individual faces a situation similar to multiparty prisoner's dilemma . . . Because contributing is not necessary for receiving the benefit, relatively few will contribute relatively little, and the good in question will be undersupplied." According to this PD analysis, although everyone seeks peace, everyone prefers to forgo contributing while others pay. The solution is taxation at the point of the sword. Kavka concludes, "financing of the good by enforced taxation can eliminate the prisoner's dilemma aspects of the situation and produce the socially desirable outcome. Hobbes realized this as regarded the public goods of national defense and safety from attack by one's fellows."[24] In his analysis of Hobbes's argument for absolute sovereignty, Michael Moehler agrees that the "problem of compliance is best modeled by a one-shot PD game."[25] Economists also join in on this understanding of the rationale for government.[26] It is difficult to overemphasize the extent to which contemporary theorists informed by rational choice theory accept that the social contract and government are challenges best represented by the Prisoner's Dilemma model.[27]

The rational choice understanding of Hobbes is easy to present. The state of nature is a Prisoner's Dilemma game, often in the context of two actors who may or may not meet each other again. Certainly there may be spoils from cooperating, but each has the incentive to seek unilateral security. Figure 10 presents how game theorists portray the problem of cooperating. It holds among individuals attempting to cooperate in a state of self-help. If we consider the choice between living in a state of nature and living in a civil society, everyone prefers living in a civil society. Yet, when it comes down to living under the social contract, everyone prefers to make an exclusion for himself or herself. So the choice between "obey the sovereign" and "disobey the sovereign" takes on the character of a Prisoner's Dilemma game, again.[28] Sanctions are introduced to resolve this Prisoner's Dilemma. The state must be sufficiently robust to have the surveillance and punitive powers to entice everyone to obey private contracts and public laws (Figure 11).

The game theoretic analysis of the problem underlying the achievement of social order identifies a Prisoner's Dilemma at three points: in the state of nature

---

[23] Taylor explains this, ibid., 127–148; see also his *Anarchy and Cooperation*, 1976.

[24] Kavka, *Hobbesian Moral and Political Theory*, 1986, at 246 (and preceding quote).

[25] Michael Moehler, "Why Hobbes' State of Nature Is Best Modeled by an Assurance Game," *Utilitas* (2009), 21:3, 297–326, 311.

[26] See Dennis Mueller, *Public Choice III* (Cambridge: Cambridge University Press, 2003), 9–12.

[27] Brian Skyrms is an important exception; see *The Stag Hunt*, 2004.

[28] Taylor observes that "individual preferences in Hobbes's state of nature have the structure of a Prisoner's Dilemma game *at any point in time*" [his emphasis], and that actors are faced by two choices, to cooperate or defect, *The Possibility of Cooperation*, 1987, at 134.

FIGURE 10. Rational Choice Concept of Hobbes's State of Nature
Left individual is row player, right individual is column player.

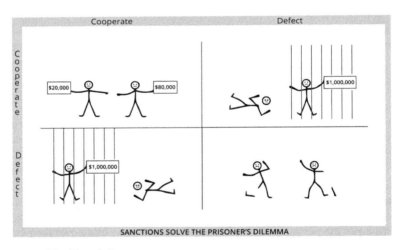

FIGURE 11. Neoliberal Governance

prior to the establishment of the sovereign state, motivating the rationalization for the state, and within the state because actors will seek to defect from both private agreements and contributing to the tax base of the state. The enforcement power of the state is fundamental to establishing and maintaining the social order game theorists envision, and they pinpoint Hobbes as a key pioneer reaching this conclusion. This opinion has prevailed despite the fact it both contradicts Hobbes's classification as an early liberal political theorist and places neoliberal political theory on a bleakly authoritarian footing connoted by this heavy-handed solution to the problem of achieving social order.[29] Game theorists are aware that somehow the rational choice diagnosis of the challenge of maintaining governance is a more onerous task than Hobbes foresaw. One textbook acknowledges that for Hobbes the state would be absolute, but "its interventions ... quite minimal." In contrast, the game theoretic PD analysis "seems to suggest that the State ... will be called upon to police a considerable number of social interactions ... the boundaries of the State ... will be drawn quite widely."[30]

TRADITIONAL HOBBES

The divergence between the expected utility maximizing view of game theory and the traditional reading of Hobbes makes explicit the extent to which the logic of game theory provides a rationale for government that is at odds with classic liberalism. Hobbes, we recall, is the first great liberal author who emphasizes that the achievement of personal security is more important than which particular form of government is established.[31] Hobbes observes that in the state of nature – an intense security impasse in which all individuals fend for themselves – every individual has the right to all things, including one another's bodies.[32] The reason for leaving the state of nature and entering into a civil society is thus to achieve comfortable living. Where there is no assurance that others seek peace, no individual has the guarantee of security. It is the guarantee of security that makes possible civilized life. The state of perpetual war is to be avoided at all costs less than the toll of living in this unenviable state:

> In such a condition [state of nature], there is no place for Industry; because the fruit thereof is uncertain; and consequently no Culture of the Earth; no Navigation, nor use of the commodities that may be imported by Sea; no commodious Building; no Instruments of moving, and removing such things as require much force; no Knowledge of the face of the Earth; no account of Time; no Arts; no Letters; no Society; and which is worst of all, continuall feare, and danger of violent death; And the life of man, solitary, poore, nasty, brutish, and short.[33]

Hobbes is clear that the benefits from living together in a civil society are far greater than any hoped-for gains in a state of nature. To use the lexicon of game theory, this puzzle of achieving a commonwealth is that of a Stag Hunt, and not a Prisoner's Dilemma game. Hobbes thus urges us to take a long-term perspective to gain an accurate sense of the conditions that lead to peace. Knowledge of arts and sciences grows over decades and centuries. It can be lost quickly in war.

---

[29] On Hobbes as an early liberal, see Tuck, introduction to Hobbes's *Leviathan*, ed. Richard Tuck (Cambridge: Cambridge University Press, 1996), ix–lvi.

[30] Hargreaves Heap and Varoufakis, *Game Theory*, 2004, 175.

[31] Note that Richard Tuck views Hobbes's *Leviathan* (1996) as developing a representative conception of government, xxxiv–xxxviii.

[32] Hobbes, *Leviathan*, 1996, 91.

[33] Hobbes, *Leviathan*, 1996, 89.

Hobbes is adamant that even under the best of circumstances, the human mind is prone to errors in logic and judgments of cause and effect.[34] Individuals have their own opinions, and moreover typically consider themselves the most clever. Obviously then in a state of nature, with everyone at war with everyone, unpredictability is preeminent. With respect to actors' attempts at achieving collaboration, each will have his or her own opinion about which way is best, so that not only pure reason and science but also concepts of morality and the content of law will be subject to endless dispute.[35] The world of people Hobbes describes is the inverse of that studied by game theorists because Hobbes suggests that knowledge and predictability arise from life in a commonwealth, whereas game theory supposes that rational actors have good knowledge of possible outcomes and a completely consistent set of preferences over those outcomes prior to inhabiting civil society.

Hobbes identifies the sources of conflict among men as competition, diffidence, and glory, which follow from the hopes of self-gain, security, and reputation.[36] Even though Hobbes adheres to a materialist account of causation, he still holds to a subjectivist account of sensation, and hence of individuals' propensity to judge phenomena as good or bad, and just or evil. Differences in opinion could erupt even over mathematics, and weights and measures.[37] Thus, the state of nature is a war of beliefs and opinions, a "war of minds."[38] All people prefer to trust their own counsel and to live in accordance with their own judgment of right and wrong: "They will hardly believe that there be many so wise as themselves."[39]

Hobbes lays out nineteen natural laws, self-evident to reason, that establish the conditions by which peace may be obtained. The first three of these laws, identified earlier, are the most salient: to seek peace when possible, to forfeit the right to all things, and to keep covenants made.[40] Hobbes tells us that the sum of the laws of nature add up to the Christian golden rule: "Do not that to another, which thou thinkest unreasonable to be done by another to thy selfe."[41] It is on interpreting Hobbes's laws of nature that a debate erupts between traditional readers of Hobbes and rational choice theorists. The latter insist that Hobbes steadfastly argues that it is the power of the sword that upholds conduct in keeping with Hobbes's laws of nature. Otherwise, if individuals were to abide voluntarily by the laws of nature, there would seem to be no need for civil government backed by coercive power. If it were rational for humankind to cooperate without the sword, then why the need for a Leviathan?

---

[34] Ibid., chap. 5, "Of Reason and Science," 31–37.
[35] Tuck discusses this in his introduction to Hobbes, ibid., xxix, 74.
[36] Ibid., 88; see also 54, 199.
[37] For commentary, see Tuck's introduction in ibid., xxv.
[38] Quote is from Hobbes, cited by Tuck, in ibid., xxx; for commentary see, xxviii–xxx.
[39] Ibid., 87.
[40] Discussed in ibid., 92, 100.
[41] Ibid., 188, see also 117.

When considering the means necessary to establish peaceable governance and amicable living, Hobbes differentiates between social animals such as bees and ants, which form a confederacy without the introduction of artificial organization, and humans, for whom social organization depends on the introduction of a common power.[42] We can ask whether the precise role of the sovereign is simply to introduce sanctions on defectors. Alternatively, the sovereign could provide a standard for a practice so basic as weights and measures on which allegiance may form as a joint product of individuals' consent and the sovereign's assurance that any predators will be dealt with.[43] Whereas game theorists stress the former possibility, a close reading of Hobbes suggests the latter.

Hobbes lists five aspects unique to humans who must resort to artificial governance in contrast to living creatures such as ants and bees that can cooperate by natural instinct. First, people compete with one another for honor and dignity or, in other words, compete for value produced through relationships. Second, social animals naturally work for their common good, whereas people only do so under duress as they seek their personally defined good first. Third, many people have a tendency to "thinke themselves wiser, and abler to govern the Publique, than the rest ... and thereby bring it to Distraction and Civill warre."[44] These individuals, at cross-purposes with one another, create confusion over what is good and what is evil. Social animals cannot distinguish between damage and injury and hence take no offense to fellows' actions, whereas people excel at offending and taking offense. Finally, Hobbes concludes,

The agreement of these creatures is Naturall; that of men, is by Covenant only, which is Artificiall; and therefore it is no wonder if there be somwhat else required (besides Covenant) to make their Agreement constant and lasting; which is a Common Power, to keep them in awe, and to direct their actions to the Common Benefit.[45]

Here, Hobbes presents his central idea of the social contract: that agreements are the primary vehicle for people to achieve joint endeavors.

Whereas game theorists focus on the joint preference for mutual cooperation over mutual defection in a Prisoner's Dilemma game, a traditional reading of Hobbes focuses on the role of the sovereign in establishing a common direction and common standard for action.[46] Moreover, in the social world Hobbes envisions, actors freely license their own future conduct to coordinate with others by giving their word. The game theoretic expected utility maximizer

---

[42] Ibid., 119.

[43] In his discussion of agreement, conventions, and social contracts, David Lewis observes that, "Given sufficient interest in conforming to any standard, a convention ... might result," 85, and suggests that this outcome could arise without a Leviathan-like enforcement body, yet he also acknowledges the power of a standard-bearer to exclude from participation nonconforming individuals, pointing to the importance of enforcement, 103; *Convention: A Philosophical Study* (Oxford: Blackwell, 2002).

[44] Quote is from Hobbes, *Leviathan*, 1996, 119, list goes on to 120.

[45] Ibid., 120.

[46] Discussed by Tuck, ibid., xxxiii.

constantly calculates personal gain, and the payoff matrices presume that the basis of value exists and can be appraised prior to the establishment of social institutions.[47] However, from the perspective of the classical liberal Hobbes, it is precisely the right to select any actions based on momentary action following from their passions that agents foreswear to unite in a commonwealth. For Hobbes, the overriding and essential goal individuals must have in pursuing self-preservation is to achieve allegiance to government. Individuals achieve this by agreeing among themselves to put peace first and self-promotion second. Recall that Hobbes's second law of nature is to yield the right to all things. This implies both respecting others' persons and yielding the propensity to demand that one's opinion prevail over others'. Hence, at the point where game theorists deem that maintaining the commonwealth is an *n*-person Prisoner's Dilemma because each seeks to have others bound by the law but to make a personal exemption, the classic liberal Hobbes insists that to attain peace, men yield their private authorship over their actions to the sovereign.[48] Thus, individuals consenting to governance agree to abide by the sovereign's code of right and wrong, and not their private judgment. This choice between anarchy and civil society is not a Prisoner's Dilemma. Living in a state of mutual cooperation is superior to living in a state of nature, and the condition for living at peace is precisely that of not making an exclusion for oneself. To seek exclusion for oneself is to perpetuate the conditions comprising a state of nature.

HOBBES'S FOOLE AND THE RATIONAL ACTOR

The contrast between these two opposite interpretations of Hobbes becomes manifestly apparent in the rational choice discussion of Hobbes's Foole, who cannot see the reason for complying with an agreement after the other agent has already delivered, unless there is a direct reward or penalty forthcoming. Hobbes's Foole is a minor character in *Leviathan*, yet his disposition so accurately captures that of the rational actor that many rational choice theorists have devoted considerable attention to explaining Hobbes's reasoning in condemning the Foole's logic. Whereas the Foole was a peripheral actor for Hobbes, he is the textbook rational actor of game theorists who seek to provide a rationale for governance. What counts as prudential reasoning for the classic Hobbes, game theorists consider to be a breach of prudence and instrumental rationality.[49]

---

[47] Monetary rewards are deemed to have value in and of themselves regardless of individuals' propensity to follow the norms on which money is dependent for its value. For Tuck's observation that the original Hobbes does not presume that actors are expected utility maximizers, see his introduction, ibid., xxxiii. The rendering of Hobbes conveyed in this chapter resembles that of Tuck.

[48] Ibid., chap. 16, "Of Persons, Authors, and things personated," 111–115.

[49] For discussion of whether the Foole may be persuaded merely by prudential concerns, see Beitz, *Political Theory and International Relations*, 30–31. Daniel Farrell takes the position that the

Rational choice theorists see Hobbes's Foole as the prototypical rational actor because they interpret both the state of nature and life in a civil society to be best modeled by a single-shot Prisoner's Dilemma among two actors, and therefore they question the logic by which it is best to cooperate if the other person does first.[50] Game theoretic logic dictates that in the Prisoner's Dilemma, it is always best to defect, regardless of what the other actor chooses to do. Thus, either Hobbes's reasoning must be made consistent with prudential considerations or it must be admitted that Hobbes transcends strategic logic in making his argument. The standard rational choice strategy is to consider that no one really has any reason to carry through in a single-shot Prisoner's Dilemma, unless repeated encounters and personal reputation can be invoked. In the absence of either, it would be necessary to invoke either some long-term indirect prudential or moral considerations.[51] Hobbes states his natural laws, which provide the rationale for abiding by promises made in distinction to actors' pursuit of momentary desire.

Hobbes addresses two cases in which an individual is bound to carry through on a promise made that falls outside the social contract defended by the sovereign. One is the situation of the Foole, in which the other party has already complied with his side of an agreement in a state of nature.[52] The other is the case of ransom in a state of nature wherein a conquering party releases a captive on the ground of his promising to pay a ransom.[53] The two situations are similar in that in each case, one party has already delivered on his side of an agreement. Hobbes thus makes clear that the difficulty of entering into and following through with agreements is the problem of assurance that the other will comply because personal compliance is a product of one's own discretion. The sovereign offers a guarantee by backing up others' commitments with the threat of coercive sanctions. However, Hobbes is equally clear that a person, even the Foole, has a reason to fulfill his part of an agreement if the other party has already done so, independent of the sovereign's sword. In other words, deferring momentarily and anachronistically to the vocabulary of game theory, Hobbes considers agreements to have more in common with an Assurance Game than a Prisoner's Dilemma. He is explicit that after one's partner in an

---

traditional reading of Hobbes is naïve; see "Hobbes as Moralist," *Philosophical Studies* (1985) 48, 257–283.

[50] Russell Hardin, "Hobbesian Political Order," *Political Theory* (1991), 19:2, 156–180; Hardin, "Utilitarian Logic of Liberalism," *Ethics* (1986), 97:1, 47–74. Skyrms deviates from this consensus by viewing the state of nature and the situation of Hobbes's Foole as an indefinitely repeated PD that has a payoff matrix that can be viewed as a Stag Hunt; *Stag Hunt*, 2004, 4–5. I leave the repeated PD for discussion in Chapter 11 and leave aside Skyrms's treatment of the Foole's situation as indefinitely repeating because it is clear that according to Hobbes, there is no reason to anticipate that the Foole will indefinitely encounter the same partner in repeating circumstances.

[51] On this point see Hampton, "The Knavish Humean," 1998.

[52] Hobbes, *Leviathan*, 1996, 101–102.

[53] Ibid., 98.

exchange has delivered, then it is consistent with reason to follow through on the terms one consented to.[54]

Rather than presenting the Foole with a calculated tally of why he should obey in the interest of direct gain, Hobbes makes two points. First, although one cannot know whether or not an individual exchange will lead to a longer-term relationship, it is wise to follow through with agreements in case they do. But second, individuals who would make an agreement and not fulfill their part after others have done so are simply not fit for society. Hobbes thus reemphasizes his points that civil society is far superior to a state of constant insecurity and that the condition for joining into civil society is obeying the requirements for peace by relinquishing the latitude to act on every passion and interest, and carrying out agreements made. The predicament is not a Prisoner's Dilemma and would only appear to be so in the eyes of the Foole.

Hobbes provides additional reasoning that builds on what one can surmise is the perspective of either long-term prudence or the realization that through social interdependence, actors are better off behaving like the naturally cooperative creatures that act as though they viewed realizing individual gain as a joint exercise.[55] Hobbes additionally states his laws of nature, which he offers as an analytic set of truths anchored by empirical facts.[56] As such, the laws of nature cannot provide a direct motive for action because they more resemble mathematical statements than passions including avarice, ambition, or lust that can animate people[57]; rather, these laws stipulate the principles individuals should comply with to satisfy their overriding pursuit of secure livelihood.[58] Any acts in accordance with these laws then, although not following from passionate motives derived from hope for direct gain, fear, or prestige are nonetheless recommended both because they will indirectly satisfy individuals' goals and because they are consistent with the demands of natural law that stipulate the conditions that must obtain for peace to characterize relations.[59]

Game theory reasons differently than Hobbes. Nash argues that any agreement is necessarily meaningless without the threat of coercive sanctions.[60] The traditional liberal argument is that agreement or consent signifies communication that one finds the terms of trade acceptable, and thus intends to voluntarily comply.

---

[54] Ibid., 102.
[55] Hobbes is clear that any existential quality about terse anxiety over one's self-preservation can only be had through interdependence; ibid., 72–73; Hobbes suggests joint maximization in ibid., 87.
[56] Ibid., 91–111.
[57] Ibid., 99.
[58] International relations theorist Campbell Craig argues that part of Hobbes's challenge, and that of others concerned that individuals and nations achieve peace rather than war, is to keep the contrast between peace and the conditions necessary to maintain it in actors' minds, and that is particularly pressing when considering thermonuclear war; *Glimmer of a New Leviathan*, 2002, xv–xvii.
[59] See Tuck's discussion, in his introduction to *Leviathan*, 1996, xxxii.
[60] For discussion see Ken Binmore's introduction to John Forbes Nash Jr., *Essays on Game Theory* (London: Edward Elgar, 1997), i–x.

There is a crucial distinction between being wary that the other may have ill intent and having ill intent oneself. This distinction is often missed in the quick application of the Prisoner's Dilemma to model a security dilemma such as that of Hobbes's state of nature. The Prisoner's Dilemma model of a bargain, while following logically from game theoretic payoff matrices that emphasize tangible rewards, does not permit the usual intuition underlying free and voluntary exchange: that trade evidently makes both parties better off and does not require a totalitarian police state, private vigilantism, or a society of nosey neighbors to work.

Furthermore, being the author of one's own action – that is, making promises and serving as one's own guarantor for them – is a key aspect of what distinguishes a person from the "Children, Fooles, and Mad-Men" who do not have the faculty of reason and cannot serve as their own guardians.[61] Hobbes admits that making a promise or contract can only provide a weak motive in comparison to the hope for material gain. However, the practice of entering into an agreement, once out of the state of nature, signifies having the intention to abide by terms and that one consents that they comport with one's overarching desire to live in a civil society rather than a state of perpetual war. In taking this step, an individual acknowledges the self-incurred obligation, and thus offers to voluntarily comply or to face punishment for reneging.[62] Hobbes builds his argument for the feasibility of civil society up from the plausibility that two people can enter into agreements, so it is crucial for Hobbes that it is rational to carry out one's own side of an agreement even after the other party has already done so.[63]

It is radical that rational choice theory departs from Hobbes over the two-person bargain rather than the agents' consent to absolute allegiance to the sovereign. Game theory undermines the least controversial case of consent by failing to recognize that basing all interactions on momentary pursuit of subjective self-gain according to a pre-social metric locks individuals into a perpetual state of war. Furthermore it assumes the unrealistic premises that actors have effective foreknowledge, consistent preferences over all possible world states, and that the situations they find themselves in have one structure, that of the Prisoner's Dilemma because fungible resources with absolute value independent of social relations and institutions or relative value when comparing others' rewards inform actors' insatiable desires.

---

[61] Hobbes, *Leviathan*, 1996, 113.

[62] Ibid., 117–121.

[63] It is not clear that putting Hobbes's *Leviathan* in the terms of rational choice theory is helpful because rational choice renders every decision a calculation of gain, as though this were possible in view of a clear value metric and certainty (which I argued earlier are fruits of civil society according to Hobbes). However, if one were trying to understand the implications of the traditional reading of Hobbes in terms of rational choice theory, it is either the case that Hobbes views all agreements as best reflecting a Stag Hunt game or he is invoking some type of extra-rational commitment. If the latter is permitted, it must be the case that Hobbes deems it prudential to cease acting on momentary passions in exchange for adopting the features of self-expression conducive to living amicably in a commonwealth.

However, for the sake of argument, let us accept the rational choice reading of Hobbes that insecurity in a state of nature is best captured by one-play Prisoner's Dilemmas, and that compliance with the sovereign would similarly resemble a one-play Prisoner's Dilemma if sanctions were not threatened to alter the payoffs for defection.[64] Even if agents enter into the terms of a contract, they all have the ever-present incentive to renege on their word. One way to plausibly construct this argument is to view the state of nature as an intense security dilemma in which agents' security depends on both defensive and offensive acts. Allegiance to an absolute sovereign is the means for everyone to exit this precarious existence.[65]

Still, once in the commonwealth, there are plenty of opportunities for a paradigmatic utility maximizer to prey on others, both in one-on-one relationships and in the relationship to the sovereign's law.[66] In rational choice theory, the response to both of these difficulties – that is, of cheating private citizens and breaking public law – is to impose penalties that alter individuals' evaluation of outcomes. The solution becomes a maximal security state with the power to police all interactions and threaten punishment accordingly. Only this would stand as a sufficient deterrent to Hobbes's calculating Foole, who can identify an abundance of opportunities to break the law and to prey on others.

In articulating a response to the game theoretic, incentive-driven Foole who would "be willing to seize any 'golden opportunities' for immoral gain that come one's way," Patrick Neal concludes that the only counter is "the existence of a sovereign so powerful that each and every act of injustice could be found out and punished, and every citizen would know this, and hence never consider such an activity."[67] The deterrent power necessary to keep Hobbes's Foole in line, to Neal, resembles the totalitarian regime described by George Orwell in *1984*. Gregory Kavka, on the other hand, argues that all that is required to counter the Foole's hope for illicit gain is sufficient uncertainty that he might be caught and punished. However, Kavka is quick to acknowledge that in the less-populated and non-anonymous British country life of the sixteenth century, this

---

[64] Not everyone is agreed on whether the state of nature itself is a Prisoner's Dilemma in Hobbes, but all do agree that the problem of everyday interaction is a Prisoner's Dilemma that must be solved by the threat of sanctions. See, e.g., Hardin, "Hobbesian Political Order," 1991, especially 166. Still, even in the most technical introductions of the rational for government, it is standard to introduce it as the solution to a single-shot Prisoner's Dilemma that represents the proverbial state of nature; see Mueller, *Public Choice III*, 2003, 9–10.

[65] There are numerous difficulties in achieving this concerted allegiance, leading some commentators to view the problem of leaving a state of nature in terms of a coordination game (e.g., Patrick Neil, "Hobbes and Rational Choice Theory," *Western Political Quarterly* [1981], 41:4, 635–653, 642) or to view the state of nature as an Assurance Game (e.g., Moehler, "Why Hobbes' State of Nature Is Best Modeled by an Assurance Game," 2009).

[66] These are the primary challenges focused on by Gregory S. Kavka, "The Rationality of Rule-Following: Hobbes' Dispute with the Fool," *Law and Philosophy* (1995), 14, 5–34; and Moehler, "Why Hobbes' State of Nature Is Best Modeled by an Assurance Game," 2009.

[67] Neal, "Hobbes and Rational Choice Theory," 1981, 648.

requirement was far less cumbersome than it would be in the faceless mass society of late modernity. He observes,

This is because in the modern world, people can expect to get away with their violations of core moral rules by simply picking up stakes and moving on after a few violations, and before they are likely to get caught. Thus, our mobile modern society is populated by a variety of robbers, scam-artists, and other blue- and white-collar criminals who seem to make a successful living out of exploiting and cheating others. And their conduct, under modern conditions, is perfectly sensible from a prudential perspective. They are living the life of Hobbes' Foole and getting away with it, using the anonymity of the vast modern city, and the mobility offered by auto, train, bus, and airplane to provide them with an effective cloak of concealment as any offered by Gyges' ring.[68]

Kavka, like Neal, accepts that adopting the premise of rational agency in the contemporary world can only result in a peaceful civil society with the introduction of advanced policing technologies. For game theorists, there is no appropriate response to Hobbes's Foole, or his contemporary counter-parts, besides an effective threat that noncompliance will be detected and punished.[69]

A minimal security state rests on voluntary compliance with Hobbes's first three laws of nature as a prudential measure to realize long-term interests. Agents are to cooperate when assured reciprocity, forgo acting on errant desires, and uphold commitments made. However, this type of agency runs counter to the principles of action formalized by noncooperative game theory. Even if we accept that the state of nature is best understood in terms of an Assurance Game, rather than an Assurance Dilemma or Prisoner's Dilemma, rational choice theorists still tend to interpret living in civil society as a Prisoner's Dilemma.[70] Given the Prisoner's Dilemma structure of cooperation, nothing prevents individuals from

---

[68] Kavka, "The Rationality of Rule-Following," 1995, at 29.

[69] This is the conclusion reached by Hargreaves Heap and Varoufakis, *Game Theory*, 2004, 34–38, 127, 129, 174–175.

[70] Even though contemplating the state of nature in terms of a Prisoner's Dilemma, commentators converge toward the thesis that the preference for establishing civil society over a state of nature makes this problem more of a coordination situation: all prefer to live in a commonwealth over a state of nature (Hardin, "Hobbesian Political Order," 1998; Kavka, *Hobbesian Moral and Political Theory*, 1986, 186; Hampton, *Hobbes and the Social Contract Tradition*, 1988; Moehler, "Why Hobbes' State of Nature Is Best Modeled by an Assurance Game," 2009; Neal, "Hobbes and Rational Choice Theory," 1981). At this point, the greatest challenge is that of choosing a leader because different parties will benefit disproportionately. This is thought of as a battle of the sexes game; see Hampton, *Hobbes and the Social Contract Tradition*, 1988, for extensive commentary, 132–188. Note that part of the distinctive quality of Hobbes's rational actors is that Hobbes presumes their primary goal is self-preservation, Hampton, "*Hobbes and the Social Contract Tradition*," 1988, 35; Bernard Gert, "Hobbes on Reason," *Pacific Philosophical Quarterly* [2001], 82, 243–257) and hence ascribes a realist or objectivist element to his account of individuals' evaluation of utility. Russell Hardin is emphatic on the reading that, although exiting the state of nature is more of a coordination puzzle for Hobbes, living in a commonwealth is characterized by routinely encountered Prisoner's Dilemmas; Hardin, "Hobbesian Political Order," 1991.

calculating that they can take advantage of others and free ride on others' efforts to maintain a government. As no rational actor will voluntarily comply with the law, the government must threaten sanctions on hopeful predators.

The divergence between rational choice liberalism and the classic liberalism exemplified by Hobbes's *Leviathan* over the role of the sword in establishing and maintaining social order becomes evident in John Rawls's defense of fair play in his *Theory of Justice*.[71] According to the principle of fair play, people have a duty aligned with their long-term prudential interests to abide by rules of conduct to which they consent *ex ante*, even if there are occasionally opportunities for self-gain by taking advantage of others.[72] Fair play is similar in form to the commitment to keep one's promises, especially if the other has already delivered. Rational choice theory and noncooperative game theory, on the other hand, adhere to John Nash's view that coercive bargaining will necessarily trump so-called gentlemen's agreements. To John Nash and other game theorists, individuals calculate how to maximize expected utility without artificial constraints, such as respect for others. In fact, noncooperative game theory presumes that individuals' main *modus operandi* is to maximize expected threatening and maximizing against others when warranted by cost-benefit analysis of the personal consequences. As Don Ross incisively observed, noncooperative game theory is well suited to tell us how best to kick a person down a hill.[73]

CONCLUSION

The rational actor model, which eludes the structure of action required by far-sighted preferences, personal restraint, or commitment, is taught in the most elite citadels of learning.[74] It is standard to introduce puzzles of action by asserting that monetary payoffs directly reflect agents' preferences over action and outcomes.[75] Textbook rational actors would accept and even seek golden opportunities for self-gain made by excluding themselves from the common rules guiding conduct. Numerous interactions are viewed as Prisoner's Dilemma: when I sell you my house, I prefer to walk away with your money and the house

---

[71] John Rawls, *Theory of Justice* (Cambridge, MA: Belknap Press, 1971).

[72] Although at first Rawls argued that his theory of justice and concept of duty of fair play were consistent with rational choice theory, he was persuaded of the opposite by critics. For discussion and commentary, see Rawls, "Justice as Fairness: Political, not Metaphysical," *Philosophy and Public Affairs* (1985), 14:3, 223–251; S. M. Amadae, *Rationalizing Capitalist Democracy* (Chicago: University of Chicago Press, 2003), 263–270.

[73] See Don Ross, "Game Theory," *Stanford Encyclopedia of Philosophy*, available online at http://plato.stanford.edu/entries/game-theory/, 1997, revised 2014, accessed January 23, 2014.

[74] Consider the canonical text of Mueller, *Public Choice III*, 2006.

[75] See, e.g., the discussion of Daniel Kahneman and Amos Tversky's presentation of the Allais Paradox, in Hargreaves Heap and Varoufakis, *Game Theory*, 2004, 16; see the standard discussion of the ultimatum game. Most work on bargaining readily presumes that utility can be stated in dollar terms; ibid., 162–163.

rather than complete the exchange.[76] Hardin observes that "Hobbes is perhaps the original discoverer of the fact that ordinary exchange relations are, in other words, a Prisoner's Dilemma problem unless there is some coercive power to back them up."[77] The extent to which a simplistic account of Hobbes's argument for the social contract rests on the Prisoner's Dilemma, in which agents seeks to take advantage of the other as their first choice, cannot be overemphasized.

Amartya Sen has argued that the expected utility maximizer is none other than a "rational fool."[78] He observes that "economic theory has been much preoccupied with this rational fool decked in the glory of his *one* all-purpose preference ordering."[79] This agent is incapable of solving Prisoner's Dilemma predicaments, mainly because he lacks the ability to commit to an action independently of its calculable rewards.[80] Commitment cannot be captured in game theoretic expected utility functions ranking outcomes independent of the means by which they are achieved, yet following through on expressed intention is a feature of action many individuals not only relate to but also practice.[81] In short, members of civil society depend on commitment, in any of its expressions – including fair play, reciprocal respect of rights, political obligation, or side constraints – for their municipalities, states, and nation-state to function.[82] It represents a coherent rationale for action, and agents may choose whether or not to abide by its norms of conduct. Commitment, whether in the form of loyalty, rule following, or principled engagement, is the archetype of behavior that motivates acts independently of agents' calculation of rewards.[83] Countering the neoliberal tendency to view all agents as

[76] Hardin's example, "Hobbesian Political Order," 1991, 175; see also Hardin's car sale example, "Utilitarian Logic of Liberalism," 1986; and Robert Axelrod, *Conflict of Interest* (Chicago: Markham, 1970), 58–60.

[77] Hardin, "Hobbesian Political Order," 1991, 174; see also Hardin's "Exchange Theory on Strategic Bases," *Social Science Information* (1982) 2, 251–272. For further discussion of the perception that routine exchanges are Prisoner's Dilemma games, see Robert Axelrod, *Conflict of Interest: A Theory of Divergent Goals with Applications to Politics* (Chicago: Markham, 1970).

[78] Amartya K. Sen, "Rational Fools: A Critique of the Behavioral Foundations of Economic Theory," *Philosophy and Public Affairs* (1977) 6:4, 317–344.

[79] Ibid., 336.

[80] Ibid., 340.

[81] Ibid., 341; see also Margaret Gilbert, *Joint Commitment: How We Make the Social World* (New York: Oxford University Press, 2013).

[82] See Daniel Hausman and Michael McPherson's commentary on this with respect to Adam Smith's view that market society is fundamentally moral, *Economic Analysis, Moral Philosophy, and Public Policy*, 2nd ed. (Cambridge: Cambridge University Press, 2006), 85.

[83] Much mischief is furthered by the mistaken impression that in the expected utility functions used in orthodox noncooperative game theory, "*all* the more or less enduring forces that go into determining ... choice" can be incorporated, whereas this is patently not the case because these functions track instrumentally relevant salient properties of the decision-context (quote from David Lewis, *Convention*, 2002, 93). If, as David Lewis suggests, individuals incorporate into their preference rankings the overall desire for peaceful coexistence or truth telling (which he also suggests has a PD structure; *Convention*, 2002, 182) and thus act in accordance with obligation

individualistic expected utility maximizers and stipulating that only reward structures can maintain lawful conduct, Sen observes that "to run an organization *entirely* on incentives to personal gain is pretty much a hopeless task."[84] Daniel Hausman and Michael McPherson likewise acknowledge that teaching the rationality of cheating in the form of self-interested utility maximization consistent with Hobbes's rational Foole endorses "a pernicious moral cynicism" and risks "becoming self-fulfilling prophesies."[85]

The axioms of expected utility theory are artificial constructions. The questions driving economic modeling, such as whether it is rational to free ride or whether rationality can condone cooperating in a Prisoner's Dilemma game, have forgone conclusions derived from the analytic structure of the theory itself. It is crucial to understand, in Sen's words, that the "nature of man in these current economic models continues, then, to reflect the particular formulation of certain general philosophical questions posed in the past" by economists and decision theorists.[86] The rational actor model is a wholly abstract, mathematical concept of rationality that distills down to norms of internal consistency.[87] It cannot be empirically falsified, and yet it presents endless room for posing questions and deriving answers.[88] The Prisoner's Dilemma and its structurally related cousins Newcomb's Problem, and Toxin Puzzle are thought experiments derived from asking what would the rational agent, upholding the axioms of rational decision theory, do in these mathematically defined scenarios.[89] *Homo strategicus*, the existential utility maximizer, provides a formal expression of Hobbes's Foole. The exigencies of nuclear deterrence opened the door to relying on purely formal decision theory to provide guidance in pressing real world public policy choices. The application of game theory to concrete lived circumstances demands that only instrumentally relevant fungible, and hence intersubjectively quantifiable, aspects of experience register in what counts in agents' choices.

to the sovereign, this would either have the structure of John Rawls's fair play or would introduce a set of considerations into decision making transcending those that register in establishing the equilibrium of a noncooperative game as the status quo point from which to settle on a bargain as John Nash suggests. On fair play, see John Rawls, "Justice as Fairness," 1985; on an effort to augment orthodox game theories expected utility functions with a parallel decision rule, see Joseph Heath, *Following the Rules: Practical Reasoning and Deontic Constraints* (Cambridge, MA: MIT Press, 2011); see also Gerald Gaus, *The Order of Public Reason: A Theory of Freedom and Morality in a Diverse and Bounded World* (New York: Cambridge University Press, 2011).

[84] Sen, "Rational Fools," 335.
[85] Hausman and McPherson, *Economic Analysis*, 2006, 74–75.
[86] Sen, "Rational Fools," 322.
[87] Ibid.; see also Hargreaves Heap and Varoufakis, *Game Theory*, 2004, 8–14.
[88] On the challenges of falsifying rational decision theory, see Sen, "Rational Fools," 324–325.
[89] See the edited collection by Richmond Campbell and Lanning Sowden, *Paradoxes of Rationality and Cooperation: Prisoner's Dilemma and Newcomb's Problem* (Vancouver: University of British Columbia Press, 1985); on the Toxin Puzzle, see Gregory S. Kavka, "The Toxin Puzzle," *Analysis* (1983), 43:1, 33–36.

This rational actor is also the quintessential neoliberal agent. This agent cannot, in principle, keep agreements made without the threat of external sanctions. A neoliberal society, resting on the premise of rational action, must therefore look to a totalitarian surveillance state to achieve compliance with the law.[90] The society in which every individual is a strategic rational actor is one that relies on rational deterrence theory to mobilize credibility underlying the exercise of nuclear sovereignty.

The following points become clear:

1. Political scientists largely view the problem of anarchy in international relations among nation-states and among individuals without government challenge with a parallel structure.[91] Since the 1970s, rational choice theorists have used the Prisoner's Dilemma to model anarchy in both contexts, in keeping with Thomas Schelling's analysis of the nuclear security dilemma.

2. The broadly accepted neoliberal resolution of anarchy among individuals prior to government is the introduction of a sovereign state that enforces the rule of law by threatening sanctions on citizens who all alike prefer to make an exception for themselves rather than voluntarily participate in upholding the social contract.

3. Therefore, the neoliberal rationale for government is that it provides the necessary and sufficient means to maintain the rule of law; considerations of legitimacy, fair play, and commitment do not motivate actors.

4. This understanding of the achievement and maintenance of national sovereignty breaks away from early modern liberalism because even Thomas Hobbes rejected the idea that actors can have sufficient knowledge to make prudent choices in a state of nature, and that people's motives are sufficiently organized to cohere as a consistent, well-ordered set of desires.

5. For Hobbes, value in the form of meaningful expectation arises from achieving civil society, and certainly not prior to security of personhood. Security is a threshold condition, rather than an unbounded set of interests or passions, to which people owe their highest consideration by seeking peace and, by implication, refraining from threatening other individuals and upholding agreements made to them.

---

[90] Neoliberal theorists strive to soften the harshness of these findings by looking to repeating Prisoner's Dilemmas and Tit for Tat play to achieve endogenous enforcement; this is discussed in Chapter 11.

[91] Steven Krasner discusses how classical liberalism is a theoretical paradigm that encompasses relations among individuals in a nation-state and among nation-states in international relations; "Organized Hypocrisy," in his *Sovereignty: Organized Hypocrisy* (Princeton, NJ: Princeton University Press, 1999), 105–126; Richard Tuck analyzes the development of this theoretical tradition, which studies how order arises in international affairs and domestic state politics; *The Rights of War and Peace: Political Thought and the International Order from Grotius to Kant* (Oxford: Oxford University Press, 2001).

6. By contrast, the neoliberal theory put into practice by accepting the subjective evaluation of action through the lens of noncooperative game theory first endorses individuals' resort to harmful threats to others in violation of the intention to avoid injury to one another as the shared basis giving rise to civil society out of chaos, and second treats agreements as cheap talk without motivational significance. Hence, the world of noncooperative game theory views the achievement of social order in terms of equilibrium states, optimal or not, that arise independent of actors' self-awareness of the role that hypothetical or actual consent plays in reinforcing their compliance with the rule of law because the canonical rational actor maximizes instrumental gain by lone disobedience when possible.

7. Of course, this rational fool is a straw person, a made-up artificial construction. However, insofar as game theory is taught as the gold standard of rational choice without a clear and significant discussion of its caveats, assumptions, and limitations, then neoliberal subjects will be forged on the template of noncooperative game theory by direct pedagogy, the need to negotiate neoliberal institutions, and the encounter with neoliberal agents.

6

Social Contract

In a strictly personalized sense, any person's ideal situation is one that allows him full freedom of action and inhibits the behavior of others so as to force adherence to his own desires. That is to say, each person seeks mastery over a world of slaves.

James M. Buchanan, 1974[1]

The reconciliation between morality and self-interest that mutual advantage theorists seek is hard to achieve, since the rules of justice creates a classic prisoners' dilemma: even if you and I jointly benefit from the rules, I will do still better if you obey them and I allow myself to violate them when it suits me.

Daniel M. Hausman and Michael S. McPherson, 2006[2]

The financial crisis of 2008 provoked a lively debate among economists, investors, and consumers. For some, the collapse of the housing market stemmed from "progressive" attempts in the 1990s to make housing more affordable for the less well-off. For others, it was the result of "neoliberal" efforts to roll back regulation of the free market. Whatever the cause, the government's response was clear: a bailout for the banks nearly equivalent to the earnings of the top 1 percent of American income earners from the prior decade.[3]

Roots of the crisis aside, how did we reach a point where such a bailout could become the consensus response? Property transfer laws, lax financial regulations,

---

[1] James M. Buchanan, *The Limits of Liberty: Between Anarchy and Leviathan* (Yale University Press, 1974), 92.

[2] Daniel M. Hausman and Michael S. McPherson, *Economic Analysis, Moral Philosophy, and Public Policy*, 2nd ed. (Cambridge: Cambridge University Press, 2006), 210–211.

[3] US Senator Bernie Sanders reports in an online op-ed column that the wealthiest 400 American citizens had a wealth increase of $670 billion over the past eight years, www.sanders.senate.gov/news/record.cfm?id=303313, posted September 19, 2008; these figures are consistent with the US Bureau of Census 2002 Current Population Survey figures used by Robert Rector and Rea Henderman Jr., "Two Americas: One Rich, One Poor; Understanding Income Inequality in the US," *The Heritage Foundation*, August 24, 2004; for recent commentary, see "Workingman's Blues," *The Economist*, July 24, 2008.

and neoliberal political philosophy all underlie the practices that led to this eventuality and its relative acceptance. According to the neoliberal Prisoner's Dilemma theory of the social contract, no income disparity is too great. So long as the least well-off are better off than in a state of nature, they have little ground for complaint. In this increasingly individualized and privatized economy, the worth of citizens depends solely on their command of resources and earnings potential. The financial crisis is indicative of late-modern neoliberal economics, which by applying the tool of noncooperative game theory and expected utility theory emphasizes the Prisoner's Dilemma model of the social contract and exchange and therefore inevitably condones profiteering through displacing costs on others. The neoliberal approach to economics also condones coercive bargaining by which negotiator's own profit margin is enhanced by threatening to minimize the best worst-case outcome for rivals.

This chapter and the two that follow examine two leading schools of neoliberal political economy: public choice and law and economics. Public choice promotes a concept of individualized pay-as-you-go responsibility akin to that embraced by the Thatcher and Reagan administrations of the 1980s. Law and economics, which views the role of justice to be that of maximizing the generation of wealth independent from either distributional considerations or individuals' consent, has become a formidable approach to jurisprudence in the United States. These chapters demonstrate how the social contract derived from the Prisoner's Dilemma, which has become the accepted replacement for the classical liberal social contract of mutual benefit, inevitably accepts both the unremitting coercive bargaining of public choice theory and the legally condoned wealth extraction of law and economics.

James M. Buchanan, who received the Nobel Prize in economic science in 1986, is arguably the most influential founder of public choice theory, which has over time been absorbed into conventional economic and public policy analysis.[4] However, Buchanan's neoliberalism marks a sharp break from classic liberalism. The contrast between Buchanan's neoliberalism and John Rawls's classic defense of liberal principles illuminates this shift. Whereas Rawls's *Theory of Justice* draws on consent and voluntary compliance to secure government, Buchanan offers a staunch defense of coercive force. This chapter discusses Buchanan's debate with Rawls over the founding principles of a constitutional order.

Throughout this chapter and into the next two, we can identify four distinct positions on the justification and maintenance of property rights. Classical liberals differ on the extent to which they view the right to private property as self-evident prior to a social contract, with Adam Smith and Robert Nozick arguing that the transparency of property rights claims preexists incorporation

---

[4] For analysis of the theoretical origins and development of public choice, see S. M. Amadae, *Rationalizing Capitalist Democracy* (Chicago: Chicago University Press, 2003); for the mainstream acceptance of public choice theory, see Dennis Mueller, *Public Choice III* (Cambridge: Cambridge University Press, 2003). Note that Mueller opens his 700-page tome with a discussion of the Prisoner's Dilemma foundation of the social contract, 9–11.

into a state, and both John Locke and John Rawls leaving latitude for the state to specify the particulars of ownership if the two caveats articulated by Locke are not met. Locke's limitation on individuals' legitimate ownership of productive means comes into play if either insufficient raw resources (Rawls primary goods) are left in common for those without to sustain themselves or spoilage would occur without permitting legitimate transfers of ownership. Public choice and law and economics similarly differ in their emphasis on whether the content of and claim to property ownership exists prior to or after the constitution of a state. The former stresses that individuals exercise their claim over possessions before entering into a constitutional order but differs from classic liberals in deriving individuals' claims from their de facto possession of goods rather than a normative rationalization. The latter view the practice of property rights to be coextensive with and dependent on the rule of law, yet they differ from classical liberals in treating property rights as arbitrary and at the discretion of the sovereign if a property claim is in the hands of individuals not making efficient use of it.

BUCHANAN'S SOCIAL CONTRACT AND THE PRISONER'S DILEMMA

Buchanan's early work, best expressed in his and Gordon Tullock's 1962 *Calculus of Consent*, demonstrates a commitment to classic liberal philosophy. This book argues that starting with a premise of rational self-interest, individuals can design a constitution that at least minimally serves everyone's interests and secures nearly unanimous agreement.[5] Such a view coheres with the libertarian adage that the government that governs least, governs best. My book *Rationalizing Capitalism Democracy* demonstrates step-by-step the reconstruction of the theoretical underpinning of capitalist democracy as a US Cold War response to communism, Marxism, and totalitarianism in the late 1940s through early 1970s.[6]

However, as the 1960s gave way to the 1970s, the major theme on Buchanan's mind was not so much the threat of communism as it was that of social anarchy, as expressed in student protests, and what he took to be a general breakdown of social decorum. His 1975 *Limits of Liberty: Between Anarchy and Leviathan* is a response to this latter problem of social order. Buchanan articulates his main theme on the first page of the preface: "'Law,' in itself, is a 'public good,' with all of the familiar problems in securing voluntary compliance. Enforcement is essential, but the unwillingness of those who abide by law to punish those who violate it, and to do so effectively, must portend erosion and ultimate destruction of the order we observe."[7] According to Buchanan, we live in a dark world in which the

---

[5] James M. Buchanan and Gordon Tulluck, *The Calculus of Consent* (Ann Arbor: University of Michigan Press, 1962).

[6] Amadae, *Rationalizing Capitalist Democracy*, 2003.

[7] Buchanan, *Limits of Liberty*, 1975, ix.

only force standing between civilization and chaos is the sword. Moreover, the weak link is not only actors' reluctance to comply with social norms and rules but additionally actors' reticence to punish offenders, which welcomes disorder.

The turbulence of the 1960s and early 1970s is thus as important to assessing the origins of contemporary neoliberalism as were the threats of communism and totalitarianism to the reconceptualization of classic liberal principles in the 1950s.[8] The newly articulated Enlightenment-era blueprint for capitalism offered by Adam Smith was predicated on the voluntary renunciation of any claims on others' personhood, property, and contracts, in keeping with sympathetic and impartial judgment of third-party injuries, personal conscience, and the legitimate rule of law.[9] The central unifying theme was that of "negative liberty," or the "no-harm principle": all individuals should be free to do as they please, so long as they does not violate the integrity of persons or their possessions. Smith argued that if the rule of law were restricted to commutative justice, relinquishing any role for redistribution, then prosperity would emerge. Each person, in advancing personal well-being limited only by refraining from injuring and stealing from others, would necessarily contribute to the joint stock comprising a nation. Voluntary consent was seen as the foundation of interactions in the marketplace, and subsequently that of individuals' social contract with government, as the franchise continued to increase throughout the nineteenth and twentieth centuries. By contrast, the neoliberal justification of government is predicated on an analysis of individuals' partially aligned and partially conflictual private preferences over social outcomes. Individuals' identity is formulated in terms of their logically ordered preferences over all conceivable states of the world and their actual opportunities, instead of by the legal and normatively regulative attributes of their personhood, property, and contractual obligations. The predominant operating theme of this new relationship between the individual and government is contained in the Prisoner's Dilemma.

To some extent, the Prisoner's Dilemma corrective to classic liberalism may be viewed as a realistic admission that no one ever really consents to a social contract, and that status quo property rights cannot be metaphysically or historically justified. Hobbes's idea of submitting to the sovereign, regardless of the actual terms for each individual, on a voluntary basis invites the additional assumption that any specific terms will only be agreeable if backed

---

[8] This effort resulted in both a theoretical investigation of and reassertion of fundamental classical liberal principles as in the work of Louis Hartz, *The Liberal Tradition in America* (Mariner, 1961), discussed by Duncan Bell, "What Is Liberalism?" *Political Theory* (2014) 42:6, 682–715, and it also initiated the tradition of rational choice liberalism discussed in Amadae, *Rationalizing Capitalist Democracy*, 2003.

[9] Note that Adam Smith employs empirical methodology to deduce his theory, *An Inquiry into the Nature and Causes of The Wealth of Nations*, 2 vols., ed. R. H. Campbell and A. S. Skinner (Liberty Fund, 1976); and his *Theory of Moral Sentiments*, ed. D. D. Raphael and A. L. Macfie (Liberty Fund, 1982). For discussion, see Amadae, *Rationalizing Capitalist Democracy*, 2004, 205–219.

by sufficient coercive power.[10] This conclusion, as will become evident ahead, would become the position Buchanan articulates in *Limits to Liberty*. However, neoliberal political thought jettisons the Archimedean reference point of classic liberalism of voluntary consent in a two-person agreement that presumes the possibility of uncontested status quo property rights and builds on actors' reciprocal respect for the other's inherent right to exist. Recall that Hobbes built his social contract theory out of the reasonableness that individuals reaching an agreement they directly participate in ratifying will voluntarily comply with it.

Consider a routine single-meeting bargaining situation, for example, on the western frontier with one individual who has a gold coin and another who has a horse. Each hopes to secure both the horse and the gold coin for personal possession but also worries about receiving the "sucker's payoff" of personal injury and no goods. If both would-be traders cooperate, they will achieve an amicable exchange; if both renege, the coin is lost, and the horse runs away. According to John Nash's analysis of bargaining encompassed by noncooperative game theory, any rational individual caught in such a situation will necessarily not cooperate, thereby achieving a worse outcome than would be achieved by cooperating.[11] The PD structure of individuals' preferences that only register instrumental gain of salient scarce resources ensures that regardless of what the other agent does, every individual is better off defecting. If a bargain does lead to a settlement, not only will it require that compliance is secured by punishment for defectors, but also the outcome will likely be the result of threatening to harm as much as is credible if no agreement is reached. Game theorists believe that rational individuals have no means to circumvent the mutually impoverishing outcome of joint defection, unless external penalties are introduced to induce each to cooperate (or if individuals repeat the identical transaction indefinitely, as discussed in Chapter 11). Crucially, even if fully assured of the other's cooperation, the paradigmatic rational bargainer will defect whenever possible unless confronted by fungible costs.

In the transition from neoclassical economics to late twentieth-century neoliberal economics, strategic rationality and by implication the Prisoner's Dilemma became the guide for comprehending many human interrelations. Consider the contrast between Friedrich Hayek's description of trade and that of Buchanan. Hayek presents trade as occurring between individuals on the basis of voluntary consent rather than the threat of coercion.[12] However, to neoliberals like Buchanan, in any exchange transaction, each party has the

---

[10] Note that in the game theoretic ultimatum game, a purely rational actor would accept a 1%–99% split, suggesting that no coercive force need be applied because the actor who receives 1% has an absolute gain that makes him better off than previously; Shaun Hargreaves Heap and Yanis Varoufakis, *Game Theory*, 2nd ed. (New York: Routledge, 2004), 162–163.

[11] For discussion, see Ken Binmore's introduction to John Forbes Nash Jr., *Essays on Game Theory* (Brookfield, VT: Edward Elgar, 1997), ix–xx.

[12] Friedrich Hayek, *Constitution of Liberty* (Chicago: University of Chicago Press, 2011), 208.

incentive to cheat the other. Because individuals are incapable of forming binding agreements by consent, trade itself must be encased within a system of government-enforced sanctions to ensure that no one opts to cheat the other. This leads to a great irony: if we are to have a minimal state in which the unfettered market thrives, then that state needs to be able to broadly monitor and enforce that market. At the same time that the concept of a public was being hollowed of content, individual privacy was as well.[13] Solving the Prisoner's Dilemmas abounding throughout society requires pervasive inspections of individuals' activities. Employees' phone conversations are routinely recorded; trucks are marked "How's my driving, call 1–800-555–1212 to report." No one is trusted to do her job as it is assumed all will cut any corner they can get away with. This is not simply a call for transparency, but for monitoring, surveillance, and carefully calibrated threats and incentives applied privately or publicly. The only way that individuals will comply with legal restraints is through total visibility; neoliberal citizens and consumers will not at any time internalize norms unless the behavior they represent directly pays off.[14] Neoliberalism also deviates from Jeremy Bentham's reformist utilitarianism, relying on technologies resembling his Panopticon prison blueprint to produce useful individuals who eventually may become self-motivated citizens and workers.[15]

To the rational choice theorist, the problem of achieving social order out of anarchy and the problem of maintaining compliance with the rule of law are Prisoner's Dilemmas.[16] Therefore, it comes as no surprise that Thomas Hobbes's *Leviathan* is the favorite point of departure of rational choice theorists who maintain that all exchange is subject to a Prisoner's Dilemma. The solution to the problem of social order then must be that of government via sanctions to prevent, or punish if necessary, defection in transactions ranging from one-off exchanges to more complex multiparty interactions.[17] Game theorists can then claim that Buchanan's endorsement of a maximal security state was anticipated by Hobbes, who astutely understood the

---

[13] Kenneth J. Arrow's *Social Choice and Individual Values* (Yale University Press, 1951) proved the impossibility of defining public interest or public good; for detailed discussion of his theorem and its implications more broadly, see Amadae, *Rationalizing Capitalist Democracy*, 2003, 82–192.

[14] Neoliberal normativity is not at any point internalized as is the disciplinary gaze Michel Foucault argues is characteristic of modern political economy in his *Discipline and Punish*, 2nd ed. (New York: Vintage, 1995); for a recent news story on workplace monitoring, see www.nytimes.com/2014/06/22/technology/workplace-surveillance-sees-good-and-bad.html.

[15] Michel Foucault, *Discipline and Punish* (New York: Vintage, 1979).

[16] Michael Taylor, *Anarchy and Cooperation* (New York: Wiley, 1976) and his *The Possibility of Cooperation*, rev. ed. (Cambridge: Cambridge University Press, 1987); for a more recent treatment, see Philip Pettit, "*Virtus Normativa*: Rational Choice Perspectives," in his *Rules, Reasons, and Norms* (Oxford: Clarendon Press, 2002), 308–343, especially 319–326.

[17] See the collection of essays edited by Jules L. Coleman and Christopher Morris, *Rational Commitment and Social Justice* (Cambridge: Cambridge University Press, 1998).

impossible Prisoner's Dilemma debacle in which we all find ourselves when we attempt to cooperate with one another (in Chapter 5, "Hobbesian Anarchy"). Buchanan alludes to this in the subtitle of his book: *Between Anarchy and Leviathan.*

The rational choice resolution of the Prisoner's Dilemma with a Leviathan-like state might give one pause.[18] However, if it can be presented as a clear descendant of an earlier form of argumentation already bequeathed to us by Hobbes and amended by Adam Smith as well as by Immanuel Kant, then it might not seem so significant or problematic by itself. However, since even Hobbes spoke of rights, consent, mutual forbearance, and political obligation, his analyses of the problem of social order and that of contemporary neoliberalism are not equivalent.[19] Indeed, as game theorists themselves seem aware, the new mode of governmentality predicated on the Prisoner's Dilemma requires a maximal, rather than minimal, security state.[20] Neoliberal governance simultaneously prescribes the norm of strategic rationality for action and structures institutions to accommodate and reward those actors who evince this behavioral rationale. Social scientists and policy analysts view deviations to be invitations for intervention and do not as a rule recognize a more encompassing set of valid logics for choice.[21] The game theorists Shaun Hargreaves Heap and Yanis Varoufakis are aware that somewhere between the absolutism of Hobbes and latter-day rational choice theory, there was a transition from providing minimal security to policing virtually all transactions. For Hobbes and other modern liberal authors, consent was voluntarily self-motivating and did not rely solely on external enforcement. The authors concur that in a game theoretic world typified by the Prisoner's Dilemma, "the boundaries of the State … will be drawn quite widely."[22]

Buchanan's Prisoner's Dilemma–based analysis of the problem of social order in *Limits of Liberty* demarcates the precise moment when this new logic of governance was articulated and ready to be implemented. It is over this point

---

[18] Brian Skryms, for example, argues that the social contract is better modeled in terms of the Stag Hunt game, but he relies on repeating interactions not relevant to large-scale, anonymous late-modern society; see his *Stag Hunt and the Evolution of Social Structure* (Cambridge: Cambridge University Press, 2003).

[19] For an insightful discussion, see Edward F. McClennen, "The Tragedy of National Sovereignty," *Nuclear Weapons and the Future of Humanity*, ed. by Avner Cohen and Steven Lee (Totowa, NJ: Rowman & Allanheld, 1986), 391–406.

[20] Hargreaves Heap and Varoufakis, *Game Theory*, 2004, 175.

[21] Cass Sunstein and Richard Thaler, *Nudge: Improving Decisions about Health, Wealth, and Happiness* (New York: Penguin, 2009). David G. Rand et al. provide a prominent recent example of how cooperating in a PD-like situation is regarded as a result of "spontaneous intuition," as opposed to a rational course of action, ignoring that actors may instead be exhibiting team reasoning, fair play, or other-regarding behavior or viewing outcomes in terms beyond fungible scarce resources, "Spontaneous Giving and Calculated Greed," *Nature* (2012) 489, 11457, published online, doi:10.1038/nature11467; www.nature.com/nature/journal/v489/n7416/abs/nature11467.html, accessed July 1, 2015.

[22] Hargreaves Heap and Varoufakis, *Game Theory*, 2004, 175.

that Buchanan and public choice, not to mention much rational choice scholar-ship, diverge from John Rawls's approach to justice and good governance.[23] Buchanan proposes that government and law are inseparable from incentives and sanctions. Not only the progressive liberal Rawls, but also the libertarian Nozick holds that mutual consent and mutual forbearance are a mandatory basis for maintaining a civil society.[24] This point of division is not just that between Buchanan and Rawls, but a reflection of a much greater disagreement about the foundation of civil society: the neoliberal practice of government relies on the Prisoner's Dilemma assessment of the problem of social order, which is mutually exclusive with any concept of legitimacy through express consent and voluntary participation or through the tacit recognition that a process of deliberation would ratify the general contours of the rule of law.[25] This breaks sharply with traditional liberal theory, according to which law is bequeathed its legitimacy from due process entailing the actual or implicit consent of its citizens. However, whether the law is deemed to arise from self-evident reflection or impartial judgment, as Adam Smith and Robert Nozick argued, or leaving room for the participatory engagement of those it governs, as recommended by John Lock and John Rawls, neoliberalism marks a significant break by repudiating either the legitimacy of law's content or its binding quality even were all actors unanimously to consent to its particular form.[26] It is not that classic liberals held an idealistic view of human intention and expression, but they did entertain the possibility that people have a choice of acting out of mutual respect in direct contrast to the late-modern neoliberal view espousing that rational actors treat others as complex objects to be manipulated to secure personal gain.[27]

BUCHANAN'S NEOLIBERALISM VS. RAWLS'S CLASSICAL LIBERALISM

Comparing John Rawls's *Theory of Justice* with James Buchanan's *Limits of Liberty* illuminates this significant shift in explaining the nature and emergence of civil society.[28] Even though Rawls wrote his *Theory of Justice* to be wholly consistent with the underlying premises of rational choice theory, he split with the rational choice community in 1985 over the reasonableness of "fair play."

---

[23] John Rawls, *A Theory of Justice* (Cambridge, MA: Belknap, 1971); for discussion, see Amadae, *Rationalizing Capitalist Democracy*, 2003, 262–273.

[24] Robert Nozick, *Anarchy, State, and Utopia* (Oxford: Blackwell, 1975), 26–35.

[25] On the challenges as strengths of deliberative democracy, see Michael A. Neblo, *Deliberative Democracy: Between Theory and Practice* (New York: Cambridge University Press, 2015).

[26] Smith, *Theory of Moral Sentiments*, 1982; for discussion of Smith's system of natural liberty, see Amadae, *Rationalizing Capitalist Democracy*, 2003, 192–219; John Locke, *Two Treatises of Government*, ed. Peter Laslett (Cambridge: Cambridge University Press, 1988).

[27] In Kant's asocial sociability within the context of a realistic approach to human relationships, see Michael W. Doyle, "Kant, Liberal Legacies, and Foreign Affairs," *Philosophy and Public Affairs* (1983) 12: 3, 205–235.

[28] Rawls, *A Theory of Justice*, 1971.

In game theory, it is rational for each individual to cheat whenever the calculable consequences are superior to the costs of compliance.[29] The standard game theoretic position holds that the concept of fair play, or upholding a set of behavioral standards to which one personally consented, defies weighing into a instrumental decision-theoretic calculation because it represents a means and not an end state. Rawls, exhibiting what to some appeared to be a Kantian influence, broke with the rational choice program to advocate a notion of the "reasonable" in contradistinction to the "rational."[30] The difference hinges on whether an agent will voluntarily adhere to law consistent with a liberal state because self-incurred obligation accords with its rationale, or whether the agent solely upholds law via a cost-benefit calculation of instrumentally salient tangible rewards.[31] In the former, obligation acts as a deontic or principled constraint on action that does not augment an actor's preference satisfaction or bottom-line obtainment of resources.[32]

The means by which law may be said to be conditioned by "right," and not solely by "might," has been a central problem of Western political philosophy at least since Hobbes's *Leviathan*, which Buchanan takes to imply "the subjugation of individual men to a sovereign master, with the latter empowered to enforce 'law' as he sees fit."[33] Neoliberalism, in the form advocated by Buchanan, reaches a new adjudication of the problem of social order. Buchanan's understanding of rational egoism excludes a consent-based approach to political legitimacy. As he acknowledges in identifying the current moment of civilization as post-constitutional, in which individual alienation and disaffection from state-prescribed laws is the norm: "once this stage is reached, the individual abides by existing law only because he is personally deterred by the probability of detection and subsequent punishment."[34] Perhaps US law had the semblance of legitimacy in the late eighteenth century, specifically for those responsible for crafting and ratifying the Constitution. However, it is harder for many today to be so sanguine or to recommend a practical return to first-person consent given the two centuries that have passed since the drafting of the Constitution.

---

[29] Gregory Kavka, "The Toxin Puzzle," *Analysis* (1983), 43:1, 33.

[30] John Rawls, "Justice as Fairness: Political Not Metaphysical," *Philosophy and Public Affairs* (1985), 14:1, 223–251.

[31] For the paradigmatic view of this approach to law, see Gary Becker, *Economic Approach to Human Behavior* (Chicago: University of Chicago Press, 1978). For further discussion, see Joseph Heath, *Communicative Action and Rational Choice* (London: MIT Press, 2001), especially 129–172; for complications with consent and political obligation, see A. J. Simmon's *Moral Principles and Political Obligation* (Princeton, NJ: Princeton University Press, 1981).

[32] Nozick endorsed deontological constraint; *Anarchy, State and Utopia*, 1974, 28–35. See how orthodox game theory sheds side constraints from the deliberation of rational agents in Joseph Heath, *Following the Rules* (Oxford: Oxford University Press, 2011), 12–41.

[33] Buchanan, *Limits of Liberty*, 1975, 130.

[34] Ibid., 96.

Buchanan acknowledges that he and Rawls share their concerns with this predicament and that although their analyses are similar, each reaches a different conclusion about the rationale underlying governance.[35] Buchanan believes that the social anarchy of the American 1960s and 1970s was a function of the lack of will to circumscribe citizens' rights through punishment. On the other hand, Rawls concludes that "enforcement may not be possible unless the prevailing distribution meets norms of justice ... notably those summarized in the difference principle."[36] For Rawls, individuals' compliance to social law is not maintained by force, but rather through hypothetical consent to the terms of government, which in his estimation include consideration for society's least well off members. This hypothetical consent, which serves to acknowledge that most of us do not actually agree to the rules that govern us, must at least be conceivable in principle for one's government to have a degree of legitimacy, and thus a display of "right" over mere "might." In the rational choice world, there is no test to differentiate between valid and invalid law because individuals will seek to break the law when it serves their private interest in any and all cases.

This assumption, that all citizens, regardless of their socioeconomic standing, will break the law, eradicates the distinction between wealth generation and distribution, the hallmark of classical liberal theory that correlates to the definition of the perfect duty not to harm others versus the imperfect duties of benevolence and charity.[37] Classic liberals view legitimate law to be self-recruiting because it serves the interests of its members who prefer to live in a peaceful market society rather than in anarchy; they have the burden to show that even members of society without private property will benefit from the status quo property rights system. Even though classical liberals are loath to suggest that individuals without means may need to steal to survive, Locke's proviso that enough be left in common for those lacking subsistence opens the possibility that under extreme circumstances self-appropriation may be consistent with self-preservation.[38] Thus, classical liberals recognized that a system of property

---

[35] Jürgen Habermas also ponders this concern in his *Legitimation Crisis*, trans. Thomas McCarthy (Beacon, 1975).

[36] James M. Buchanan, "A Hobbesian Interpretation of the Rawlsian Difference Principle," *Kyklos* (1976), 29, 23.

[37] This distinction is clearly delineated by Immanuel Kant in his *Groundwork of the Metaphysic of Morals* (Bobbs-Merrill, 1965); Adam Smith also articulates this distinction in his *Theory of Moral Sentiments*, 1982; for discussion, see Amadae, *Rationalizing Capitalist Democracy*, 2003, 205–219. Peter Singer discusses how in a classical liberal property rights system, traditionally its members differentiate between perfect and imperfect duties and recognize that the imperfect duty of charity falls on the shoulders of society's better-off members, "Famine, Affluence, and Morality," *Philosophy and Public Affairs* (1972) 1:1, 229–243.

[38] For a comprehensive discussion of this topic in the early days of classic liberal theory, with an emphasis on Adam Smith's role in strictly maintaining the absolute dominion of property rights, see Istvan Hont and Michael Ignatiev, "Needs and Justice in the Wealth of Nations," in their edited collection *Wealth and Virtue: Shaping the Political Economy of the Scottish Enlightenment* (Cambridge: Cambridge University Press, 1983), 1–44.

rights will especially protect the possessions of the well-endowed from those significantly less materially well off and yet gave a slight nod in the direction of recognizing that theft motivated by dire straits is not equivalent to breach of law prompted by avarice.

By contrast, neoliberalism conflates two recognizable forms of criminality: that of those who are excluded from the fruits of the social contract and that of those who already benefit but are preying on the social contract for further enrichment. Consider the difference between Bernard Madoff and a homeless person who shoplifts out of hunger. In the first case, an individual cheats despite the rule of law required to produce fruit in the first place; in the second case, an individual cheats because there is insufficient fruit to be gained by upholding the rule of law. Early modern political philosophers were keen to draw this distinction. Locke was the clearest and Hobbes not far behind: when an individual's self-preservation is at stake, the rule of nature prevails over the rule of law regardless of whether or not one is in a civil society.[39] Individuals must be able to meet their elementary security needs for the social contract to be viable, attractive, and thus self-recruiting.

The Prisoner's Dilemma model of bargaining and the social contract capture this central paradox of neoliberalism: if free trade is obviously in each individual's interest, then why the need for paramilitary police, enforcement, and sanctions?[40] If a system is so obviously in each person's interest, it should be sufficiently self-enforcing so as not to require microscopic and forceful policing. The paradox arises from the tendency in orthodox game theory to deny the motivational relevance of obligation, commitment, or deontological constraints because they transcend instrumental gain. People are thought to be primarily motivated by tangible rewards reducible to, if not already presented in, monetary terms.[41] These assumptions, which Buchanan and Nash adopt, make it impossible to differentiate between voluntary exchange and extortion because all individuals are presumed to seek asymmetric, unilateral if possible, advantage unless held in check by punitive threats.

Even in the hypothetical case in which one (1) is satisfied with one's original holdings, (2) believes there are no relevant inequalities of bargaining power, and (3) agrees that the terms of the transaction at hand are fair, everyone's first choice is still presumed to be that of suckering the other and stealing all the goods, unless there are prohibitive sanctions. Every bargain necessarily has a PD

---

[39] This is most clearly articulated by Locke, who states, "Men, being once born, have a right to their Preservation, and consequently to Meat and Drink, and such other things, as Nature affords for their Subsistence"; see also Locke's proviso, *Two Treatises of Government*, 1988, 285 (subsection 25) and 288 (subsection 27), and it is also deducible from Hobbes's first two laws of nature, *Leviathan*, ed. Richard Tuck (Cambridge: Cambridge University Press, 1996), 98.

[40] On the increasing militarization of American police forces, see Al Barker, "Where the Police Go Military," *New York Times*, December 4, 2011, SR6.

[41] Even if more complex motives that supersede economic gain are deemed relevant, these are often still treated as seeking the approval of others in line with a cost-benefit calculation; see Pettit, "*Virtus Normativa*," 2002, 340–411.

structure because each agent most prefers to leave the table with all of the goods, and only goes through with the terms of the exchange if a causal process ensures a quid pro quo structure to the transaction or the imposition of penalties for the failure to comply with terms agreed to. The ruthless logic of the Prisoner's Dilemma is familiar. Each can see that "cooperate-cooperate" is a better individual and collective outcome than "defect-defect," but each of us has the ever-present incentive to cheat in pursuit of unilateral, if benighted, success. Therefore, given that we all, as presumably rational actors, seek exclusion for ourselves, the only means by which we can achieve mutual cooperation is via the imposition of sanctions on all transactions.[42] This resolution of the problem of social order is at least as oxymoronic as Jean Jacques Rousseau's observation that we must be "forced to be free": the role of government is to guarantee by force that populations of individuals can achieve the "Pareto optimal" cooperate-cooperate result.[43] Agreement to the terms of trade or governance becomes irrelevant. As long as the terms are better than those achieved by mutual defection, any arrangement of provisions may be enforced.

In *Limits of Liberty*, the Prisoner's Dilemma analysis of social order grows in mythopoeic proportions. In developing his Hobbesian-like narrative, Buchanan deviates from Hobbes and other Enlightenment-era thinkers in holding that humans are naturally unequal and that a status quo inegalitarian distribution will result even prior to the establishment of civil society because some are more capable, talented, and stronger than others.[44] Notwithstanding such observable disparities, trade occurs, supposedly under the traditional economic logic of Pareto optimality: everyone will benefit, regardless of the initial endowment of goods. Buchanan explains, "The gains from trade that are potentially achievable by an agreement on rights are realized by all parties through the disinvestment in socially wasteful effort devoted to both predatory and defense activity. An agreed-on assignment will not normally be stable in one particular sense."[45] Despite the self-evident quality of exchange to achieve mutual gain, "once reached, one or all parties may find it advantageous to renege on or to violate the terms of contract."[46] According to Buchanan, even though it is obvious that mutual cooperation is superior, each agent still has the ever-present incentive to rob the other. To make clear that this "state of nature" problem is that of the Prisoner's Dilemma, he continues,

[42] The concept that each naturally most prefers everyone else to comply with law and to make an exception for one's own person is ubiquitous in rational choice theory; see, e.g., David Lewis, *Convention: A Philosophical Study* (Oxford: Blackwell, 2002), 189–193, and Pettit, "*Virtus Normativa*," 2002, 319.

[43] Jean-Jacques Rousseau, *The Social Contract*, trans. by Maurice Cranston (New York: Penguin, 1968), chap. 7.

[44] Note that the classical liberal authors tended to stress fundamental human equality, e.g., Thomas Hobbes, *Leviathan* (1996), Chapter XIII.

[45] Buchanan, *Limits of Liberty*, 1974, 26.

[46] Ibid.

Within the setting of an agreed-on assignment of rights, the participants in social interaction find themselves in a genuine dilemma, familiarized under the "prisoners' dilemma" rubric in modern game theory. All persons will find their utility increased if all abide by the "law," as established. But for each person, there will be an advantage in breaking the law, in failing to respect the behavioral limits laid down in the contact.[47]

To make headway with Buchanan's formulation of the problem of social order, one must be fully aware of the differences between his analysis and Hobbes's. Recall that for Hobbes, specific rights are a product of government. For Buchanan, as for libertarian philosopher Robert Nozick, rights are prior to civil society, and they therefore provide the point of origin for a "naturally just" initial distribution.[48] The role of the state is only to objectively referee transactions as a matter of protection, and not to adjudicate matters of distribution.[49]

Although this analysis sounds uncannily similar to classic political economy with its night watchman state, it departs from classic political economy in its inability to locate a normative pole external to the central logic of strategic self-interest. It is true that both Adam Smith and Immanuel Kant advocated a minimal security state. However, for both these philosophers, justice constrained rational self-interest through internalized principles of conscience that recognize that respect for others is foundational for civil society as is upholding the rule of law and maintaining agreements personally made. For Smith, the basis of justice resides in non-utilitarian sympathy; for Kant, it resides in transcendental practical reason.[50] Even the libertarian Nozick presumes deontological side-constraints with self-evident validity to structure his classical liberal political economy.[51] Buchanan himself acknowledges that the basic premises on which public choice theory is built make it impossible to locate any source of normativity for so basic a concept as mutual respect, or for treating individuals as ends and not solely as means.[52] From the perspective of rational choice, the suggestion that individuals could be motivated by a sense of duty, mutual forbearance, or commitment to promises made is naïve.[53] Buchanan is clear that his political philosophy embraces a community of devils and does not optimistically assert or require a community of angels or expect even any well-intentioned individuals for that matter.[54]

---

[47] Ibid.

[48] Robert Nozick, *Anarchy, State, and Utopia* (Oxford: Basil Blackwell, 1974), 8–11, 23–26.

[49] Ibid., 95.

[50] For the normative basis of Smith's system of justice, see Amadae, *Rationalizing Capitalist Democracy*, 2003, 193–219; see also Amadae in "Utility, Universality, and Impartiality in Adam Smith's Jurisprudence," *Adam Smith Review* (2008) 4, 238–246; see Kant's *Metaphysical Elements of Justice*, 2nd ed., trans. by John Ladd (Cambridge: Hackett Classics, 1999).

[51] Nozick, *Anarchy, State, Utopia*, 1974, 28–35.

[52] Buchanan, *Limits of* Liberty, 1975, 151–152.

[53] David Gauthier, *Morals by Agreement* (Oxford: Oxford University Press, 1986).

[54] Buchanan's position is already evident in *Calculus of Consent*, 1962, coauthored with Gordon Tullock.

COERCION VS. INCLUSION

The rational choice insistence on not assuming that individuals evince any long-term prudential or other-regarding considerations consistent with Enlightened self-interest in decision making led John Rawls to break with both rational choice theory and his earlier claim in *Theory of Justice* that justice is the most important aspect of a theory of rational choice.[55] Rawls's later idea of "the reasonable" pivots on the concept of fair play: "If the participants in a practice accept its rules as fair, and so have no complaint to lodge against it, there arises a prima facie duty ... of the parties to each other to act in accordance with the practice when it falls upon them to comply."[56] It is thus reasonable to be committed to rules of conduct to which one consents, despite the fact that one's personal payoffs for doing so in each and every instance may not calculably be to one's advantage. The gap between Rawls and Buchanan, and between classic liberalism and neoliberalism, arises because commitment and fair play cannot be readily defended within the narrow confines of rational choice theory. Commitment, required by Hobbes's third law of nature, signifies following through with an agreement independent of externally applied coercive sanctions. Fair play entails abiding by rules to which one hypothetically consents, regardless of whether or not there are occasional opportunities for self-gain by exempting oneself from the laws one hopes govern all other actors.

The assumption that all individuals are naturally seeking exclusion from laws and promises serves a particular function in neoliberal governance.[57] It permits the conflation of forms of crime that prey on the legally sanctioned system and forms of crime that exist despite the system. Public choice theorists, Buchanan included, see human beings as nothing but calculators that make cost-benefit analyses of each decision to determine what is to their own personal advantage, often expressed in dollars. Therefore, it is a first principle in Buchanan's assessment that each individual benefits from participating in a civil society regardless of resource distribution. His dedication to this assumption is evident in the quasi-interpersonally comparable rendition of the Prisoner's Dilemma he places at the heart of the social contract.[58] No matter how small one's share is, mutual cooperation is superior to mutual defection. Similarly – and this is key – *no matter what one's share is, it is in each individual's equal interest to defect.* Buchanan clearly articulates this point: "Once reached, one or all parties may

[55] Whether or not action consistent with Enlightened self-interest (forgoing short term private gain for long term mutual benefit) can be derived from strategic rationality is the subject of ongoing debate, with neoliberal theorists hoping to definitively conclude the affirmative; see discussion of Robert Axelrod's attempted derivation of the Golden Rule from the repeated PD game in Chapter 11, "Tit for Tat"; yet without recourse to extra-instrumental motivations or motives outside the limited parameters of rational choice, the answer is negative without the pressure of incentives.

[56] John Rawls, "Justice as Fairness," in *John Rawls: Collected Papers*, ed. Samuel Freeman (Cambridge, MA: Harvard University Press, 1999), 60.

[57] Martin Hollis, *Trust within Reason* (Cambridge: Cambridge University Press, 1998) analyzes the type of society that must emerge for rational actors.

[58] Buchanan stipulates DC=22,1; CC=19,7; DD=9,2; CD=3,11, *Limits of Liberty*, 1972, 27.

find it advantageous to renege on or to violate the terms of contract. This applies to *any* assignment [of goods] that might be made."[59] Accepting the universal rationality of defection is so important that Buchanan clarifies, "The tendency toward individual violation is not characteristic of only some subset of possible agreements."[60] Neoliberal governance holds that no matter what the distribution of rights, all are alike in their motives to cooperate or to defect from that system of rights. Within this narrative, there is no categorical distinction between the poor individual with few legitimate opportunities and the wealthy banker: each alike has the ever-present incentive to cheat the system; each alike finds mutual cooperation to be superior to mutual defection.

Buchanan continued his discussion of social disruption in his article "A Hobbesian Interpretation of the Rawlsian Difference Principle" (1976). In it, he observes, "Honest assessment of life about us should suggest that there has been an erosion in the structure of legal order, in the acknowledged rights of persons, and that, indeed, modern society has come to be more and more vulnerable to disruption and the threat of disruption."[61] Buchanan finds Rawls's difference principle, which strives to guarantee that all society's institutions that are structured in accordance with inegalitarian distribution of resources must provide some benefit to society's least-advantaged members, to be an attempt to ensure that pockets of disaffected individuals do not arise.[62] He thus, at first, treats difference principle as Hobbesian, because it strategically aids in securing social order by staving off disaffection.

However, ultimately, Buchanan recognizes that Rawls's solution is different from his own, because it assumes that compliance with the system can be maintained through inclusion and through sharing the spoils of mutual cooperation. He is open about this divergence in the conclusions of public choice in contrast to progressive liberalism. According to Buchanan,

Parts of . . . [Rawls'] argument may be read to suggest that individuals *should not* abide by the distribution of rights assigned in the existing legal order unless this distribution conforms to the norms for justice. And persons in the original position *should not* agree on a set of social arrangements that are predicted to place strains on individual norms of adherence and support.[63]

For Rawls, compliance and the duty to comply are wedded to one's sense of commitment to the social system; complicity in society demonstrates allegiance to its overriding principles.[64] Therefore, compliance is linked to tacit or express agreement to the terms of the social contract.

---

59 Ibid., 26.
60 Ibid.
61 Buchanan, "A Hobbesian Interpretation of the Rawlsian Difference Principle," 1976, 21.
62 On Rawls's difference principle, see *Theory of Justice*, 1971, 60–67.
63 Buchanan, *Limits of Liberty*, 1974, 23.
64 Rawls is clear that his view of the social contract has more in common with an Assurance Game in which enforcement is necessary but not sufficient for compliance because actors must also view cooperation as superior to sole defection, although his means of avoiding what

Looking back to the 1970s and Governor Ronald Reagan's showdown with unruly protestors by engaging helicopters with tear gas as a means of crowd control, the impassioned political theoretic debate between Buchanan and Rawls set the stage for the ensuing neoliberal era of community policing via military technologies and mass incarceration for even pedestrian violations of law. In the mid-1970s, Buchanan, reaches the uncompromising conclusion that

[Rawls'] difference principle can be identified as emerging from contractual agreement in the initial position only if the participants make the positive prediction that least-advantaged persons and/or groups will, in fact, withdraw their cooperation in certain situations and that the threat of this withdrawal will be effective.[65]

Buying in potentially disaffected parties via implementation of the difference principle is necessary to Buchanan only when enforcement fails. In his view, the potential gains of mutual cooperation, despite disparate distribution, should be sufficient to secure cooperation when sanctions are used for enforce contracts. He writes,

My own efforts have been directed toward the prospects that general attitudes might be shifted so that all persons and groups come to recognize the mutual advantages to be secured from a renewed consensual agreement on rights and from effective enforcement of these rights. Rawls may be, in one sense, more pessimistic about the prospects for social stability. [For Rawls, e]nforcement may not be possible unless the prevailing distribution meets norms of justice, and notably those summarized in the difference principle.[66]

Buchanan decisively concludes, "Whereas I might look upon the breakdown of legal enforcement institutions in terms of a loss of political will, Rawls might look on the same set of facts as a demonstration that the precepts of a just society are not present."[67] Rawls argues that consent and fair play flow from long-term prudential considerations and are necessary for and consistent with maintaining political stability via an ongoing process of inclusion guaranteed by the difference principle. By contrast, Buchanan looks to force to maintain law and social order without consideration of the distribution of individuals' endowments. His justification of property rights based on existing endowments derived from the power of possession prior to incorporation into a civil society renders the constitutional moment one of leveraging terms through the coercive bargaining tactics of issuing credible threats. Since this forms the founding

---

rational choice theorists consider to be the PD structure of emerging from anarchy and maintaining the social contract was found to be inconsistent with strategic rationality; *Theory of Justice*, 1971, 265–270; for discussion, see Amadae, *Rationalizing Capitalist Democracy*, 2003, 258–273.

[65] Ibid.
[66] Ibid.
[67] Ibid.

moment of organized society under sovereign rule, the socioeconomically empowered are justified and capable of manipulating the threat point of mutual defection to minimize the outcome for underprivileged individuals and to thereby have the wherewithal to negotiate a social contract favoring those with means.[68]

What is missing from the neoliberal view of governance is the crucial role of consent in legitimizing both the foundational practices grounding the exercise of personal autonomy, property rights, and voluntary contract characterizing the free trade and citizens participatory self-rule through democratic government. Although Enlightenment-era liberalism did not encourage the state to play a role in redistributing resources, it did rest on the firm assumption that in any contract setting, individuals' consent to the terms of the agreement provides a rationale for their ensuing compliance with the contract. In neoliberalism, agreement does not play a role in motivating action.[69] Therefore, the content of terms, apart from its relation to the respective bargaining power of the participants, is immaterial. Virtually any terms may be enforced with sufficient force or "political will." Thus, where Hobbes builds up from the voluntary two-person exchange with individuals obliged to one another to obey the sovereign and their own agreements, Buchanan builds up from the Prisoner's Dilemma model of exchange to suggest that strategic actors will pursue personal gain as monads without the dyadic or collective cohesion to self-identify as one nation or public or to organize relationships without inviting external surveillance and enforcement to foster even one's own conformity.

To be charitable to Buchanan's line of thought, we can appreciate the neoliberal hope to study the interactions of self-interested agents, without assuming that they have concern for one another, as contemporary societies largely operate on faceless interactions. The Prisoner's Dilemma construed as a multiparty problem seems well-suited to describe large joint actions in which it appears gratuitous to suggest that each acts to further others' interests, as well as personal goals. However, the stakes in this debate are high. According to one view of the individual and society, each of us acts to maximize expected utility regardless of the effects on others. Law steps in as a public good that realizes prospects for joint gain when myopic self-interest would otherwise leave everyone worse off; law motivates by imposing penalties on would-be cheaters.

---

[68] In a journalistic account of the rise of the phenomenon of cheating in the United States, David Callahan argues that the breakdown of an inclusive social contract is closely associated with the increasing tendency for individuals in all professions and stations to bend rules in their favor; *The Cheating Culture: Why More Americans Are Doing Wrong to Get Ahead* (New York: Mariner Books, 2004).

[69] See, for example, Russell Hardin's, "Contractarianism: Wistful Thinking," *Constitutional Political Economy* (1990) 1:2, 35–52.

CONCLUSION

This chapter makes clear the following points:

1. The Nobel Laureate economist James M. Buchanan demonstrates that the Prisoner's Dilemma model for basic exchange and for the social contract codified in a constitution is generally accepted knowledge.

2. Buchanan's analysis of political economy in effect normalizes the expectation that strategic rationality consistent with noncooperative game theory characterizes rational citizens and consumers.

3. Buchanan's social contract and market exchange leaves no role for consent to terms or self-recruitment to legal compliance predicated on individuals' endorsement of fundamental principles or distributional outcomes. His "post-constitutional" moment in political practice may be equated with neoliberalism.

4. Buchanan's analysis of constitutional order differs from that of both John Rawls and Robert Nozick because the latter adhere to the view that a precondition for social order is individuals' respect of one another's human dignity and right to exist.

5. Neoliberalism cuts across the grain of the well-known debate between classical and progressive liberals who are divided over whether the state should play any distributional role in ensuring that individuals have access to the resources and opportunities required for subsistence and the pursuit of happiness. Buchanan's school of public choice adheres to a strict view of private property rights but relies on individuals to defend and exercise their rights. His view of the constitution and rule of law permits powerful actors to leverage asymmetric advantage in negotiating the content of law and terms of transactions, and either breaking or enforcing those terms in their favor, thus likely resulting in a society with vastly disproportionate access to resources. Whereas Buchanan endorses Nozick's historical evolution of rights, his neglect of the classical liberal no-harm principle necessarily invites deploying coercive threats and enforcement by oppressive punitive measures to sustain a social contract with even more disparate socioeconomic consequences.[70]

---

[70] John Oliver has produced a number of satirical documentary programs examining the impact of neoliberal principles in his program *Last Week Tonight* on stadiums, July 12, 2015, www.youtube.com/watch?v=xcwJt4bcnXs; tobacco, February 15, 2015, www.youtube.com/watch?v=6UsHHOCH4q8; and municipal violations, March 22, 2015, www.youtube.com/watch?v=0UjpmT5noto, all accessed July 15, 2015.

7

Unanimity

Agreement that results from the physics of threat advantage bargaining is not underwritten by genuine consensus. So, even if a Nashian bargain succeeds in producing an agreement to create a property rights scheme, those who are relatively disadvantaged by the agreement will seek to destabilize it by engaging in post-contractual rent-seeking behavior. As the discussion of the properties of property rights schemes illustrates, there are ample opportunities for rent-seeking. In Nashian bargaining, hold-out behavior may make agreement impossible, or if successful, inefficient. On the other hand, since Nashian bargaining is not based on genuine consensus, even if it is successful the agreement may not endure. Moreover, whatever agreement is reached, those who feel they are exploited will seek to destabilize the agreement, and there is ample opportunity for them to do so.

Jules Coleman, 2002[1]

Free market liberalism and democratic government have traditionally accepted that unanimous agreement is the highest form of validation for establishing terms of trade and law. The previous chapter discussed how the Prisoner's Dilemma view of the social contract disregards any motivationally binding quality of the exact terms of an agreement. Yet, at the same time, rational choice theorists still present unanimous agreement as the central means of securing legitimacy for trade and democracy. This chapter questions whether the shell of agreement left can serve to ground the unanimity principle. In voting procedures, unanimity signifies that all members are on board with a resulting outcome and are therefore presumably inherently motivated to endorse it with their actions. Similarly, in market exchange, unanimity implies that every participating in a transaction finds the terms of exchange sufficiently agreeable to uphold them.

[1] Jules Coleman, *Markets, Morality and the Law* (Cambridge: Cambridge University Press, 2002), 274.

UNANIMITY VS. UNANIMOUS AGREEMENT TO TERMS

The idea that agreement to terms is motivationally irrelevant is so radical for liberalism that even Buchanan only suggests it in a mixed and inconsistent fashion. Following his work with Gordon Tullock in *Calculus of Consent*, Buchanan defends a contractarian approach to liberalism reminiscent of Thomas Hobbes's idea that the fundamental role of government should be to fulfill each individual's direct interest in procuring security.[2] Even though he rejects the idea that agreement to terms mobilizes compliance, he still hopes to identify principles of government that would achieve the virtually unanimous assent of all citizens. Given Buchanan's repudiation of the significance of agreement to terms, and his staunch conclusion that the sword will be necessary to maintain order even after an accord is reached, it is mystifying that there could be any vital role left for agreement to play in his social order.

Buchanan's crucial though residual reliance on agreement typifies a deep tension within neoliberal political theory. Even if we accept that morals and norms have been reduced in status to private preferences and expected utility rankings, achieving unanimous agreement on a policy still seems a secure basis on which to organize collective action. Buchanan looks to agreement in the form of unanimity to provide a rationale for his Leviathan. Many theorists who attempt to come to terms with the corrosive Prisoner's Dilemma appeal to the unanimous agreement of agents that mutual cooperation is better than mutual defection.[3] Here, agreement signifies aligned preferences: all prefer mutual cooperation to mutual defection, or CC>DD for every actor. The unanimity principle therefore suggests that government is warranted in applying the threat of sanctions on potential defectors.

The Prisoner's Dilemma approach to rationalizing governance delineated by Buchanan ingeniously separates the achievement of unanimous agreement from the precise terms that actors acquiesce to. This strategy allows him to isolate two distinct premises: (1) that unanimity is well defined and serves as a meaningful basis for political coordination and (2) that it is possible to categorically distinguish between agreement that exchange is mutually beneficial on the one hand, and agreement over the exact terms of exchange on the other. Unanimity is promoted as a principle legitimating governance while agreement over the precise terms of distribution is deemed superfluous to motivating individual or collective action. It is not evident that the two premises can be simultaneously intelligible. This tension becomes evident when we recall that part of the

---

[2] James M. Buchanan and Gordon Tullock, *The Calculus of Consent* (Ann Arbor: University of Michigan Press, 1965).

[3] See also Russell Hardin, "Hobbesian Political Order," *Political Theory* (1991), 19:2, 156–180 and "Does Might Make Right," in *Authority Revisited*, ed. J. Roland Pennock and John William Chapman (New York: New York University Press, 1987).

original value of the market as an institution derived from the fact that it evolved through the unanimous assent of its participants. This idea is explicitly rendered in Thomas Hobbes's third law of nature, discussed in Chapter 5, "Hobbesian Anarchy." This assent is inseparable from the agreement to specific, voluntarily accepted terms of exchange. Although it seems absurd to isolate the motivational force of unanimous agreement from the content of the agreement itself so as to conclude that any agreement is favorable to none, this is exactly what Buchanan does. Any enforceable agreement negotiated by credible threats may be characterized as consensual. Therefore, and more importantly, such reasoning provides justification for levying sanctions against any defector from an exchange regardless of whether or not the terms of exchange themselves were voluntarily agreed to.

Buchanan justifies the constitutional state on the grounds that everyone prefers some government to the state of no government typified by Hobbes's unenviable state of nature. As in the ultimatum game wherein one actor chooses a distribution of a reward, offering the remainder to the other person, and if the second decision maker accepts the offer, then any division of the total resource bundle is unanimously assented to. His dovetailing of general agreement with realism about human nature only comes together at the price of viewing property rights as de facto as opposed to normatively regulative. This is because many divisions of spoils could result depending on the disagreement threat point of what outcome obtains in the case that no settlement is reached by the agents. Instead of a normative default point characterized by mutual respect and recognition of each person's right to exist without being harmed, each actor is free to exercise credible threats predicated on harm to obtain a more favorable personal outcome. Buchanan hopes to secure private property by stipulating that only unanimous consent can serve as a condition for relinquishing it. However, if agreement to specific terms of a transaction has no motivational content, and therefore no intrinsic import, then it follows that the practice of property rights is only secure insofar as coercion is employed to defend that system. The Achilles' heel of Buchanan's defense of property rights, which otherwise seems attractive for ostensibly relying on the unanimity principle, is that it ultimately creates a situation wherein the individual with the greatest force will be able to set the terms of exchange and interaction.[4] The less desperate actors are, and the more resources they have to fall back on in case a mutually agreeable outcome remains out of reach, the better the likelihood that they will be able to offer meager terms to their interaction partners. These actors can hold out longer for an agreement and have a better chance to extract

---

[4] Herbert Gintis offers a similar view of the origin of property rights that specifies a payoff matrix in terms of a Chicken (Hawk Dove) game instead of a PD under the assumption that agents will likely avoid physical confrontation because it is too costly, and that holders of property in the status quo will fight harder to maintain it because of prospect theory showing that owners value property in hand over identical property not yet acquired; *Bounds of Reason* (Oxford: Oxford University Press, 2009).

terms to their advantage, although occasionally actors who under these circumstances have no guaranteed right to subsist could threaten extreme violent behavior as a tactic to improve their distribution.

UNANIMITY SCRUTINIZED

Jules Coleman's essay on unanimity demonstrates how philosophically complex this concept is.[5] On the surface, the principle of unanimity appears self-evident and free from controversy. It seems to be roundly compelling and constructive. It is profoundly democratic but also obviates the difficulties of the tyranny of the majority because there is no minority that could be oppressed. It captures the idea of market efficiency, or the Pareto condition that individuals only accept states of the world that improve their conditions, increase their well-being, or promote their interests. It also seems to convey the sense of voluntary participation because each individual has personal veto power to reject any outcome under consideration that she or he deems personally inferior. In the case that a collective decision requires unanimous consent to be enacted, then any single individual has veto power to say "no." Hence, if Buchanan's insistence on unanimity as the basis of his constitutional order can deliver on all of these promises to ground mutually attractive principles governing a constitutional order and to identify the attractive feature of the free market through individuals' voluntary and unanimous participation, then it would seem to be unassailable from all angles of attack.

Throughout his career, Buchanan has claimed to follow his intellectual predecessor's, Knut Wicksell's, use of unanimity to provide an argument for government. Wicksell directly attacks the heart of the matter in considering what level of public goods production versus taxation would achieve unanimous consent from citizens.[6] As has been repeated ad nauseam in the Prisoner's Dilemma rendering of the public goods problem, it is demonstrably in each individual's interest to partake in public goods, even if each seeks to free ride on others' efforts.[7] The uncontroversial role of government is to levy punishment on defectors, thereby achieving a state of affairs on the Pareto frontier whereby all individuals gains more than they pay through the collective effort. However, I argue in Part II, Chapter 9, that the rational choice application of the Prisoner's Dilemma model of exchange to analyze social order deviates considerably from Wicksell's, which holds that as long as individuals gain more than they receive

---

[5] Coleman, *Markets, Morality, and the Law*, 2002, 277–289.

[6] Knut Wicksell, "A New Principle of Just Taxation," in *Classics in the Theory of Public Finance*, ed. Richard A. Musgrave and Alan T. Peacock (London: Macmillan, 1958), 72–119.

[7] For Coleman's commentary on the Prisoner's Dilemma model of social order, see his *Markets, Morality and Law*, 2002, 243, and 388, fn 6; see also Russell Hardin, *Collective Action* (RFF Press, 1982); Dennis Mueller, *Public Choice III* (Cambridge: Cambridge University Press, 2006).

from a collective venture, or government, then voluntarily contributing is rational.[8]

Coleman explains Wicksell's basic premise: "Because the provision of public goods is capable of making each person better off, there should be a way of providing and distributing their benefits and costs that can secure each person's agreement." He continues, "There should exist, in other words, a public good-tax package tailored to each person which will secure his consent. In the aggregate, there is a relationship between the optimal provision of public goods and unanimity."[9] Here unanimity refers to all individuals ratifying a comprehensive tax package based on its formula being responsive to each individual's costs versus benefits from collective action. Wicksell's tax scheme was formally developed by his student Erik Lindahl. The Wicksell-Lindahl system determines each individual's tax package by comparing the marginal cost of a public good to the marginal benefit each individual receives from it and then builds this up to a plan that can acquire each individual's consent. It seems uncontroversial that individuals would sign up for a tax scheme in which the benefits they receive outweigh the costs of their personal contributions. The concept of efficiency lies behind such an application of the unanimity principle. Wicksell, moreover, assumes that efficiency has an objective benchmark: the output must be more than the input given market prices and an individual's willingness to bear the costs. For example, if I contributed $400 per year of my income and received $600 per year of services I value, then I would find this plan agreeable.[10]

A close observation of Wicksell's application of the unanimity principle reveals two significant deviations from Buchanan's. First, Wicksell specifies the precise terms of each citizen's acceptance of a tax package whereas Buchanan insists that specific terms are superfluous because any government is better than no government. He does implicitly suggest that individuals would not consent to a state in which they were worse off than before the agreement, and yet because his agreements are susceptible to manipulation through coercive threats in the case of a failure to settle, he is unable to guarantee that the terms individuals accept are attractive. This departure is surprising given its radical nature. Buchanan's subjective stance on value leads him to reject Wicksell's use of shadow prices, evident from public records of transaction histories, to devise tax packages favorable for each individual. Although he looks to Wicksell's use of unanimity to inform his rationale for government, he rejects its central premise that the agreement that matters is the precise terms

---

[8] Wicksell, "A New Principle of Just Taxation," 1958, 72–119; Mancur Olson cites Wicksell in his *Logic of Collective Action* (Cambridge, MA: Harvard University Press, 1965), 89 and makes clear that prior to game theory a collective venture in which all individuals' gain exceeds their costs of contribution was deemed individual and collectively rational.

[9] Coleman, *Markets, Morality and the Law*, 2002, 278, and 278–279.

[10] In marginalist analysis, the point would be to identify the plan that equalizes the marginal payment against the marginal cost of production and encompasses all members in a cost-effective manner.

favorable to each individual. For Wicksell, terms that citizens would find attractive can be objectively specified by knowing, for example, what the costs of a public service are, how much individuals require, and what those individuals' costs could be without public access.

The second marked departure that Buchanan takes from Wicksell is at first more difficult to identify but is no less radical than the first. Whereas Wicksell's use of unanimity tracks market-style efficiency in maintaining that all individuals are better off by their own estimation, Buchanan assumes that the existence of unanimous agreement itself indicates the achievement of efficiency. Coleman pinpoints the crucial difference:

These two arguments for the unanimity rule are based on two very different conceptions of efficiency and of the relationship between it and unanimity. Though Buchanan thinks of himself as above all else a Wicksellian, the differences between them may be more impressive than are the similarities. In Buchanan's argument for the unanimity rule, we infer efficiency from unanimity; in the Wicksell-Lindahl scheme, we establish efficiency and then infer unanimity from it.[11]

In essence, Wicksell's concept of efficiency is based on an objective, or at least quasi-public, test of whether every individual's net gain from paying a tax outweighs that individual's net cost. Wicksell assumes that this can be decided as a precondition and proxy for agreement; the point at which an individual's marginal tax rate exceeds that person's marginal gain from public services is not up to imagination or random judgment. Having sewage and garbage removed has a cost. If this cost is worth it to me, given the alternative of living with removing my own waste, then the public good is efficient for me. If it can be provided to everyone in accordance with this analysis, then the public good is generally efficient.

By contrast, Buchanan's use of the unanimity principle is purely semantic, or analytic: "In the semantic sense, to say that a state of the world is efficient is just to say that it has secured unanimous agreement."[12] In this alternative view, nothing underlies agreement; it does not track any objectively evident property in the world. Buchanan thus deviates from neoclassical economics in his insistence that there is no external marker of the achievement of market efficiency besides the fact that individuals choose to make the trades they do. He is a fan of neither general equilibrium theory nor the idea that it is possible to discern an individual's well-being independently from each agent's subjective account of personal satisfaction.

Buchanan's principle of unanimity, being more minimalist and less ambitious than Wicksell's, might seem to be a more certain foundation for a constitutional order. Indeed, given agents' generally accepted propensity to free ride, a concern at the forefront for Buchanan, Wicksell's tax scheme might seem to suffer from the fact that individuals will fail to report their honest preferences

[11] Coleman, *Markets, Morality and the Law*, 2002, 283
[12] Ibid., 281.

to sweeten their benefits and lessen their burdens.[13] Buchanan anticipates this practical infeasibility of the Lindahl-Wicksell tax plan: citizens will report their preferences, underestimating the value of public goods, and their own marginal rate of return of services in turn for paying taxes. Buchanan's minimal constitutional order thus seems to acknowledge that individuals are not motivated to uphold bargains because each will continually press his or her advantage. Holding out and renegotiating, or exhibiting preferences that diminish their actual experience of gain characterizes citizens' relationships to one another and government. This position coincides with Buchanan's understanding that the social contract, indeed any contract, necessarily resembles a Prisoner's Dilemma in which each most prefers unilateral gain if at all possible, and the eventual outcome of an exchange will be determined by the disagreement point set by credible harmful threats. Yet to provide a rationale for his constitutional order, Buchanan appeals to the seemingly obvious fact that all individuals still unanimously prefer mutual cooperation to mutual defection in every PD scenario.

We thus return to the observation that for Buchanan's constitutional order to be attractive, not only must the unanimity principle be meaningful, but it must also be possible to differentiate between an agreement that every government is better than no government and an agreement on an actual tax scheme endorsed by any particular government. Buchanan tries to uphold the power of unanimity in the form that everyone prefers Leviathan to anarchy consistent with neoliberal political theory, but he dismisses the idea that terms deemed favorable to an individual will lead to voluntary adherence to an agreement. What, then, remains intact of the significance of unanimity if it does not serve to indicate that the state to which individuals agree is actually more favorable than other alternatives and results in voluntary participation? Most obviously, it is apparent that unanimous agreement is not, in the lexicon of game theory, strategy proof. Agents may veto arrangements that, though actually favorable, may be leveraged to greater personal advantage through a "hold-out" strategy. One could threaten to defect unless one were permitted to gain more and contribute less, threatening to otherwise ruin a collective venture, specifically if unanimous agreement is a condition for that venture to go forward. As Coleman explains, even though Buchanan hopes to circumvent the problem of agreement to precise terms in isolating the universal and obvious point that all prefer cooperation to anarchy, the actual practice of agreement is inseparable from strategic maneuvering over the conditions of distribution. Unlike in Wicksell's tax scheme, the Prisoner's Dilemma floor of mutual defection itself is subject to coercive threats. This moving floor is the basis for establishing whether individuals are better off in civil society or without, and that particular distributional arrangements may be legitimately coercively enforced.

To make the point dramatically, either the benchmark of unanimous agreement is so weak as to suggest that any time individuals join in a collective

---

[13] For further elaboration, see Coleman, *Markets, Morality, and the Law*, 2002, 283–284.

action, they have achieved a mutually favorable basis for social order, or it is so strong that it requires an externally valid indication to check that universal agreement is not merely mass psychosis. For example, if individuals were to dig their own graves, the mere fact of their action serves to indicate that it comports with their preferences; alternatively, if actors jointly jumped off a cliff, then if they had falsely believed in a resulting reward of a heaven on earth, their actions would need to be scrutinized by a shared public benchmark of well-being. Wicksell relies on the latter criterion. However, Buchanan avers that there cannot be any independent means to check that agents are actually efficiently served by the state to which they have agreed. Once the independent test of merit of assent has been rejected, unanimity has no more clout than the suggestion that the existence of a school of fish indicates a universal agreement to swim to a single place at a single time. As we shall see in looking at Posner's law and economics in Chapter 8, in contemporary expected utility and game theoretic political economy, consent implies no more than showing up. Being at a certain place at a certain time signifies that given one's options, taking the action to be here now seemed the best choice at the time of decision making.

We may recall that a strength of Buchanan's position was supposed to be its minimalism that accepts moral skepticism yet searches for an Archimedean point upon which to erect government as a means to secure gains from trade that would otherwise be lost as each seeks to dominate the other. This Archimedean point is the purported obviousness that we are all trapped in a multiparty Prisoner's Dilemma defined by the mutually and individually preferred hope for joint gain, even given each individual's not-so-secret hope to sucker all others in the bid to achieve personal success. Buchanan insists on the objectivity of the favorability of mutual cooperation (CC) at the same time that he denies that any particular terms will be behaviorally motivating. Therefore, he approves of the threat of sanctions for any defection, rendering specific terms superfluous. The specific terms that will be enforced are those negotiated on the basis of bargaining strength.

COERCIVE BARGAINING AND UNANIMITY

One is left with the impression that terms will be generated through coercive bargaining, and that sanctions will be threatened in accordance with terms that reflect agents' strength, as demonstrated through their threat advantages. In effect, there are two steps to Buchanan's argument here. First, he asserts that that property held prior to the social contract was gained through de facto possession and will be protected by the unanimity rule. Coleman explains,

In free markets, no individual is required to give up that to which he already has a legitimate claim. The unanimity rule provides each person with a veto over any policy that imposes net costs on him or anyone else. This feature of the unanimity rule is usually discussed favorably in the context of bargaining over the terms of the constitutional contract specifying the conditions of the move from anarchy to polity.[14]

Thus, according to Buchanan, possession precedes the social contract as a function of actors' power to acquire and hold, which sets the basis for incorporating that original state of access into a property rights regime. Second, negotiations take place according to the de facto power each individual has based on personal holdings.[15] By putting together the idea that each individual has a set of pre-constitutional property holdings defended forcibly, and that these holdings provide each agent with a specific bargaining strength in negotiating the terms of the constitution, legislation, and individual one-off exchanges, the full implications of Buchanan's position are apparent. His delineation of the Archimedean agreement that some form of government is jointly beneficial regardless of the specific terms enshrines pre-constitutional property rights gained by de facto power into a constitutional order that is sustained by applying coercive sanctions to maintain distributional differences negotiated by credible threats.

Buchanan is not overly concerned with the precise topology of differential allocations because he specifically privileges property rights in accordance with the status quo distribution. He tacitly assumes the legitimacy, or at least inevitability, of coercive bargaining or threat advantage at the same time that he argues that any distribution defended by the force of coercive government is universally accepted as superior to the alternative of a chaotic and violent state of nature.

One cannot help admiring the elegance of Buchanan's defense of a constitutional order that automatically privileges the status quo and locates a universal point of agreement amid a sea of moral relativism and social conflict over distributional claims. Buchanan combines a surly pragmatism with a minimalist, uncontroversial normative pivot point in unanimous consent to achieve that which neither Robert Nozick nor John Rawls was able to. Nozick, although not too far from Buchanan in rejecting any specific claims that free markets actually deliver general utility or well-being, was still unable to account for the origins of his deontological property rights system by frustrating even undergraduates with his unapologetic claim that "individuals have rights."[16] Rawls, whose original position many readers found to be too hypothetical and remote, ultimately moved beyond rational self-interest in grounding his constitutional order.[17] Buchanan seems to strike the right balance between pragmatically accepting that agents come to civil society with property previously acquired through forceful conquest, and maintaining, in accord with generally accepted

---

[14] Ibid., 286
[15] Ibid.
[16] Robert Nozick, *Anarchy, State, and Utopia* (Oxford: Basil Blackwell, 1974), 182, xix.
[17] Discussed in S. M. Amadae, *Rationalizing Capitalist Democracy* (Chicago: University of Chicago Press, 2003), chap. 8. Some see Rawls as differentiating himself from liberalism further in his *Political Liberalism* (New York: Columbia Classics, 2005).

wisdom, that civil society is virtually universally esteemed over either pre-market society or society rife with corruption. Who would join a market society if the price of membership entailed losing resources previously acquired? Yet, again, for his constitutional order to be intelligible, it is necessary to both grant the unanimity principle some Archimedean purchase and permit that the unanimous CC>DD agreement can stand independently of any inherent principle of distribution.

The salient, though consistently overlooked, point is that in this derivation of the unity in preferring mutual cooperation to mutual defection, the neoliberal application of Pareto efficiency is itself a function of threat advantage. The achievable part of the curve versus the portion that relies on one person being made worse off is determined by the disagreement point, which can be defined as the outcome if both sides exercise credible threats on the other. Determining the minimum that the least–well-off individual must gain to achieve an improvement over mutual defection in the Prisoner's Dilemma game is the product of the physics of coercive threats (see Figure 2 in Chapter 2).

Ultimately Coleman worries that Buchanan's dual hope to eliminate distributional conflict from the constitutional moment and to privilege status quo distributions through the acceptance of bargaining via threat advantage will run into two devastating difficulties. On the one hand, it is not clear that distributional conflict can ever be surmounted. Not only will distribution be a constant source of antagonism, but also any government that does arise will only provide the occasion for rent seeking by which government officials support institutions and policies that regularly pay them or their associates, regardless of whether they add value. Thus, the unanimity that is supposed to supply the glue for the constitutional order is both indicative of strategic positioning in accordance to bargaining strength and devoid of motivational content because agents will continually seek to augment their terms vis-à-vis others. Buchanan's unanimity that is put forward to balance between accepting de facto property rights derived from might and providing a rationale for government proves unable to provide the consensual cement that is necessary to prevent individuals from endlessly squabbling over the outcome of every transaction and the personal implications of every law, not to mention actors' perpetual incentive to break any rule when it pays. Even the introduction of force appears prone to the same problem, namely that its application will be motivated by hope for gain rather than for justice.

These considerations bring us back full circle to the debate between Buchanan and Rawls over what the most convincing constitutional principles are for a liberal political order, and over the character of agreement and unanimous consent. Whereas Rawls assumes that agreement to terms has motivational impact, Buchanan and other theorists upholding rational decision theory assume that agreement over social norms and laws is only indicative of aligned preferences. A genuine consensus entails each individual agreeing to the normative validity of general rules and signifies compliance secured by internalizing the rationale informing rules as a guide to personal action. By contrast, Buchanan's unanimity principle implies no more than that given a pair-wise choice over outcomes, a body of agents exhibits the same preference. The difficulty is that the universe of social outcomes is much more replete than a choice between mutual cooperation and mutual defection. This superficial expression of similar interest is thus readily shattered as soon as each individual seeks to press personal advantage either through rent seeking or negligent compliance.

CONCLUSION

This chapter has made the following points:

1. Unanimous agreement to a particular set of constitutional principles, or to the terms of a transaction, is believed to offer a foolproof foundation for a constitutional order or market exchange because every person sanctions its coming to pass.[18]

2. There are two ways to consider the power of unanimity. On the one hand, if individuals signal that they agree to an outcome by voting or accepting terms of exchange, then their ratification demonstrates their endorsement of the collectively shared outcome. On the other hand, unanimous agreement can be understood to signify that if actors' preference rankings over outcomes are known from their stated preferences or behavior, then a policy maker can design laws and institutions to best accommodate their interests using this information.

3. Even though James M. Buchanan follows Knut Wicksell's school of economic thought, he deviates by adopting an understanding of unanimity that reflects an actor's behavior at the moment of choice instead of objectively discerning an analysis of the costs and benefits of producing a collective product tailor-made to actors' stated interests or observed choices.

4. Even though minimalist and therefore attractive, Buchanan's approach ends up treating the achievement of unanimous agreement as fleeting and dependent on particular circumstances. Even if original actors agreed that one mutually cooperative outcome is better than mutual defection, an individual, in seeing a more attractive third choice, may cancel his or her prior agreement as a holdout move to achieve a better outcome. Thus, actors may initially agree that one point on the Pareto frontier of mutually beneficial outcomes is superior to mutual defection, yet another outcome could be introduced that may lead an individual to veto the previous pairwise choice to achieve a better outcome.

5. Furthermore, as is the standard difficulty with the Prisoner's Dilemma resulting from the fact that each individual is better off defecting from an

---

[18] On the traditional power of unanimous agreement to secure a constitutional order see Georg Simmel, *The Philosophy of Money*, (London: Routledge, 1978), 444.

agreement, regardless of what the other does, no agreement motivates compliance.

6. Thus, whereas for classic liberals, unanimous agreement over elementary principles (such as guiding action in accordance with the Pareto principle of making at least one person better off and no one worse off) compels action, and the status quo disagreement outcome is defined in accordance to the no-harm principle, in a neoliberal regime unanimity over either principles or outcomes lacks any motivational force, and actors will strategize their votes and bargaining terms in ways that either manipulate the best-worst outcome for others or veto otherwise agreeable outcomes to achieve an even more attractive outcome by risking mutual defection.

7. In all circumstances, actors only comply if they face punitive sanctions for defecting.

8

Consent

> Ultimately what sets apart the classical liberal and [law and] economic concep-
> tions of rights and liability rules is the case in which liability rules are thought
> sufficient to justify a transfer.... For it can never be any part of the classical liberal
> account that by compensating someone for taking what is his without his consent,
> an injurer respects the victim's rights; whereas the core of [rational choice] eco-
> nomic analysis is the possibility that by compensating a victim, an injurer (at least
> sometimes) gives his victim all that he is entitled to, thereby legitimating the
> taking.
>
> Jules Coleman, 1988[1]

Richard Posner's law and economics, on examination, is even more distribu-
tionally biased in favor of wealthier individuals than Buchanan's minimal
constitutionalism. Buchanan's use of the unanimity principle to capture the
importance of voluntary consent is at least designed in principle to enable
individuals to veto conditions that would erode their current status, even if it
fails to live up to this promise given the inevitability of coercive bargaining and
de facto possession. Posner's law and economics, on the other hand, overtly
privileges wealthy individuals by asserting that resources should be in the hands
of those who value them most, as measured by agents' willingness *and ability* to
pay for them. In effect, Posner's scheme puts resources up for auction to the
highest bidder.[2] Rights themselves are subject to dispersion based on considerations of power. The concept of consent that Posner uses to ground his law and
economics method is attenuated at least to the extent of Buchanan's unanimous
agreement, if not more so.

Posner is one of the most prolific living public intellectuals in the United
States, having published at least forty books, and is also a presiding judge on the

---

[1] Jules Coleman, *Markets, Morality and the Law* (Cambridge: Cambridge University Press, 1988),
62–63.
[2] Discussed in ibid., 87–91, especially 90–91.

US Court of Appeals for the 7th Circuit.[3] The legal program he tirelessly promotes can be summarized by the mantra "justice is wealth maximization." According to the law and economics school, the role of law is to ensure that resources are granted to those who can best use them, with use evaluated in terms of the ability to generate wealth. According to Posner, law only has the positive function of enforcing rules. It does not reflect privileged moral values. It is not inherently normative.[4]

In defending his view of law, Posner attempts to retain the classic liberal claim that rights reflect individuals' freedom to enter into voluntary exchange at the same time that he upholds the position that markets are superior to any other form of economic organization because they result in the most efficient production of value. Liberal political economy has typically embraced the unity of its commitment to individual liberty and its belief that such a system of liberty will spontaneously yield optimal outcomes measured in preference satisfaction or social welfare.[5] This linkage between freedom and social efficiency was consistently challenged in late twentieth-century political economic theory because either the principle of individual liberty or the principle of market efficiency is prioritized at the expense of the other.[6] Buchanan and public choice theorists unequivocally privilege individual liberty and downplay that efficiency has its own metric independent of the exercise of individual choice. Posner and law and economics scholars emphasize the achievement of efficiency measured in terms of wealth maximization to the exclusion of a robust concept of individuals' freedom of choice.

Ahead I explore how Posner works to maintain that his slender defense of consent is meaningful, even if it is ultimately more vacuous than Buchanan's unanimity, which retains agents' *ex ante* veto power over outcomes. The first section examines Posner's concept of consent, and the following section analyzes his use of the Kaldor-Hick's compensation principle, which permits policies that create welfare improvements to some at a cost to others. These sections lay the groundwork for explaining Posner's neoliberal replacement of *ex ante* consent with *ex post facto* consent as the basis of consumer's exercise of choice.

---

[3] On the questionable rigor of Posner's arguments concerning consent, see Coleman, *Markets, Morality, and the Law*, 1988, 118–121.

[4] For a good overview of law and economics, see Lewis Kornhauser, "Economic Analysis of the Law," *Stanford Encyclopedia of Philosophy*, published November 26, 2001, revised August 12, 2011, available online, http://plato.stanford.edu/entries/legal-econanalysis/, accessed July 15, 2015.

[5] Istvan Hont and Michael Ignatiev, "Needs and Justice in the *Wealth of Nations*," in their edited collection *The Shaping of Political Economy in the Scottish Enlightenment* (Cambridge: Cambridge University Press, 1983), 1–44.

[6] Amartya Sen has formalized this challenge in "The Impossibility of a Paretian Liberal," in his *Choice, Welfare and Measurement* (Cambridge, MA: Harvard University Press, 1982), 285–290.

CONSENT IN POSNER'S SYSTEM OF JUSTICE AS WEALTH
MAXIMIZATION

Since Adam Smith, political economists have claimed that individual freedom is
the surest means to generate social wealth. However, by the mid-twentieth
century, the difficulty of proving that individual freedom – anchored in basic
rights to personhood, property, and contracts – produces an optimal level of
social welfare became increasingly evident. Buchanan and even Nozick, for
instance, avoid claiming that markets produce socially optimal or efficient states
indicated by any measure apart from consent.[7] The libertarian upholds the
sanctity of rights independent of any extraneous consideration that rights should
or will serve to achieve optimal social outcomes or efficient distributions.[8]

Posner, on the other hand, seeks to uphold rights to consensual transactions while actually privileging the enterprise of maximizing social wealth. He
does this by building on the work of Nobel Prize–winning economist Ronald
Coase, who argues that efficiency can be achieved as a separate consideration
from the assignment of rights.[9] Rights are no longer intrinsically meaningful
or defensible; instead, they exist and are justified as means to effectively
generate wealth. Posner takes this prioritization of efficiency over rights a
step further: he endorses the notorious Kaldor-Hicks compensation principle,
which argues that social welfare can be improved by policies producing both
winners and losers as long as the winners could hypothetically compensate the
losers for their loss and still retain a surplus. The principle is infamous in
welfare economics because it suggests that social welfare may be increased
even if those negatively affected by policy do not share in the surplus generated
from their loss of entitlement. Although originally advanced as a means to
limit a monopoly's powers without compensation for its loss, the Kaldor-
Hicks principle can also be wielded to argue for the reallocation of resources
from the poor to the rich.[10] According to Kaldor-Hicks logic, traffic projects
through poorer neighborhoods may be justified if commuters gain more time
and money from the new road construction, and residents are forcibly
removed at "shadow market" prices via eminent domain (even if they never

[7] This is clear in Robert Nozick, *Anarchy, State, Utopia* (Oxford: Basil Blackwell, 1974),
149–177.
[8] The main point of Sen's "Impossibility of a Paretian Liberal," 1982, is to prove that both rights
and the condition of market efficiency cannot simultaneously be served.
[9] Ronald H. Coase, "The Problem of Social Cost," *Journal of Law and Economics* (1960), 3:1,
1–44.
[10] One reading of Lawrence Summers's infamous memorandum during his World Bank presidency
relies on the Kaldor-Hicks compensation principle to argue that wealth may be generated
through toxic dumping in the developing world without compensating the recipients of the
waste; discussed in Daniel Hausman and Michael S. McPherson, *Economic Analysis, Moral
Philosophy, and Public Policy* (Cambridge: Cambridge University Press, 2006), 12. See their
discussion of Kaldor-Hick's efficiency, 144–147. For a criticism of the law and economics use of
eminent domain, see Richard Epstein, *Takings: Private Property and the Power of Eminent
Domain* (Cambridge, MA: Harvard University Press, 1985).

voluntarily agreed to them).[11] The losers neither share in surplus gain nor receive compensation for the personal loss of having their property seized.

Posner's application of the Kaldor-Hicks principle has three intricate moving theoretical parts, each requiring a distinct exposition. First, it is important to understand the intellectual legacy of the Coase theorem and its implications for a theory of rights, as well as its conceptual basis in the neoclassical economics of diminishing marginal productivity. Second, it is important to see how Coase's theorem essentially views economic value as objective, or at least externally provided to the judge or policy maker. Third, it is necessary to understand the Kaldor-Hicks compensation principle and how it compares to the Coasean idea of assigning rights to achieve market efficiency. In direct contrast to Buchanan's libertarian priority of rights over externally defined criteria of efficiency through the exercise of veto power, Posner's utilitarian position privileges social wealth maximization by stripping meaningful content from the exercise of consent.

Coase's theorem incorporates numerous idealizing assumptions into its analytic structure specifying that perfect competition must obtain with all actors having perfect knowledge of relevant information, markets entailing no transactions costs, and no buyers or sellers sufficiently large to impact prices. The theorem holds that any precise legal establishment of entitlements, that is, well-defined property rights, will yield a state in which resources are put to their most productive use. After rights are assigned, resource use will depend on the prices for goods and the rates of marginal productivity for competing products. Even though the resulting distribution of wealth is dependent on the initial allocation of rights, Coase argues that resources will be put to their most productive use.[12] Coleman provides a useful synopsis of Coase's theorem:

> Allocative efficiency, or the maximum productive use of resources, does not depend on the initial assignment of entitlements. The initial assignment is only the starting point for negotiations. The point at which negotiations cease represents the efficient allocation of resources. The initial assignment of entitlements, however, does affect the relative wealth of the competing parties simply because the assignment determines which party has to do the purchasing (or what economists misleadingly call bribing).[13]

Rights are a tool to facilitate free market exchange, not a reflection of conditions protecting individuals' freedom or a representation of the entitlement individuals had prior to their entry into the market.

---

[11] For legal reasoning consistent with this, see John Attanasio, "Aggregate Autonomy, the Difference Principle, and the Calabresian Approach to Products Liability," in David G. Owen, ed., *Philosophical Foundations of Tort Law* (Oxford: Clarendon Press, 1995), 229–320.

[12] See Coleman, *Markets, Morality and the Law*, 1988, 74, for discussion; the claim is controversial but can be defended purely in accordance with the analytic structure of the theorem; it is a matter of internal consistency and remains unclear if the theorem can map onto the actual world to be put into practice intact.

[13] Ibid., 71

Coase's theorem is based on a stringent set of assumptions.[14] First, rational individuals are not only acting to maximize subjective utility but must be maximizing monetary gain directly, without additional considerations. Second, the economy is perfectly competitive, and thus the prices of goods are independently established and will not be affected by any specific assignment of rights. Third, the bargains achieved once rights are determined are not subject to income effects, so neither party would make a different choice regardless of how rich or poor each were. His breakthrough in legal theory was expressed in terms consistent with neoclassical economics that regarded the conditions for efficiency as externally given criteria of leveraging productive resources to their greatest profit-maximizing use as a function of diminishing marginal productivity.[15] The idea is that if a property could offer a greater profit yield through an alternative use, then the actor able to achieve that outcome would purchase the property from its current owner.

Coase's paper "The Problem of Social Cost" was a response to welfare economist Arthur Pigou's tax scheme.[16] Coase puts forward a proposal to achieve optimal outcomes given the tendency of competing land uses, such as farming and cattle grazing versus railways, to generate conflict. For example, trains damage farmland, as they produce sparks causing wildfires on adjacent ground. Coase argues against Pigou in suggesting that society will be better off if energy were spent ensuring efficiency rather than on assessing blame for harm. Coase claims that the resulting efficient outcome will arise regardless of the initial assignment of rights, in accordance with the assumptions that prices for goods are designated outside the decision problem and that individuals are only concerned with maximizing profit. He further charges Pigou's invocation of taxes with inefficiency, arguing that the entrepreneurs themselves will be the best judge of profit-making decisions and will be likely to enter into private bargains even after Pigou's legal remedy has been imposed.

Coase thus makes the question of which party should have the right against infringement secondary to the fact that any entitlement allocation will yield an efficient deployment of resources. Once rights are assigned, the individual who could make better economic use of a resource will purchase it from the current owner if that prospective owner has the financial wherewithal to do so. Coase argues for his method of adjudication with the full knowledge that the optimal wealth distribution will ultimately depend on the allocation of resources. From its onset then, law and economics gives priority to optimal productivity over considerations of distributive fairness. Rights are fully a secondary consideration to economic growth and wealth maximization. Moreover, in relying on principles from neoclassical economics that render diminishing marginal

---

[14] See Robert Cooter's entry on the "Coase Theorem" in *The New Palgrave Dictionary of Economics*, vol. 1 (London: Palgrave MacMillan, 1987).

[15] Note that here Coase's reasoning reflects that of Knut Wicksell more than that of James M. Buchanan, as discussed in Chapter 7, "Unanimity."

[16] Coase, "The Problem of Social Cost," 1960.

productivity objective in the sense that all producers face the same constraints, Coase's theorem obviates the consideration of coercive bargaining by insisting that no single individual can alter any prices.

Posner and law and economics move beyond Coase in adopting a game theoretic (rather than neoclassical) standard of economic value and embracing the Kaldor-Hicks compensation principle, neither of which is endorsed by Coase.[17] Indeed, a charitable reading of Coase could recognize his contribution as an additional consideration on which a judge may reflect when deciding a case. A judge could consider both the merits of rights claims based on moral reasoning and historical circumstances, and the potentially worthwhile goal of promoting social efficiency. However, law and economics proposes that wealth maximization should be the judge's primary, if not only, point of reflection in casting judgments. Coase lauded efficiency for leveraging resources to their greatest productive power given publicly shared standards of diminishing marginal utility and profit maximization functions.[18] Posner instead seeks to place resources in the hands of the individual who most values them given an opaque test of willingness and ability to pay expressed in hypothetical or real auctions over entitlements.[19]

In accordance with his refrain "justice equals wealth maximization," Posner requires a concrete measure of wealth. Instead of subjecting profit maximization to industry-wide constraints, his view, consistent with John von Neumann and Oskar Morgenstern's game theoretic definition of price, is that wealth is increased through exchange whenever the buyer's subjective utility for a good (reflected in that individual's willingness and ability to pay for a good) is greater than the seller's (reflected in that individual's willingness to part with the good). The purchaser gains surplus value equal to the difference between his maximum willingness to pay for a good and what he actually pays, while the seller gains value equal to the dollar amount gained minus her subjective utility for the good. If we consider a car valued at $5,000 for the seller (her minimum price to part with it) and valued at $10,000 for the buyer (his maximum price to acquire it), the net gain in economic worth will be $5,000, even if the actual exchange price is $8,000, giving the buyer a surplus value of $2,000 and the seller a surplus value of $3,000.[20] Law and economics departs from neoclassical economics' objective, market-wide aggregate establishment for prices as it considers individuals' subjective utilities in isolated cases, which can rarely be empirically

---

[17] An excellent source on the development of law and economics and Gary Becker's and Richard Posner's contributions is Nicholas Mecuro and Steven G. Medema, *Economics and the Law*, 2nd ed. (Princeton, NJ: Princeton University Press, 2006), 94–155.

[18] Coleman, *Markets, Morality, and the Law*, 2002, 74.

[19] Coleman alludes to this opacity, ibid., 74–75.

[20] This is articulated by John von Neumann and Oskar Morgenstern, *Theory of Games and Economic Behavior* (Princeton, NJ: Princeton University Press, 1944/1953/2004), 556–557, all editions have the same pagination.

tested, unless through watching how consumers reveal their preferences through actual choices.[21]

Posner's test of wealth creation, as well as his unrelenting view that resources are socially best allocated when held by the highest bidder, who by definition puts them to their most efficient use, has no external test and consistently determines that individuals who have more financial wherewithal are inherently more justified in owning property. Posner asserts that the justification of owning property is the ability to generate wealth. He implicitly suggests that wealthier individuals are more deserving of resources as a matter of definition: an individual who is able to pay more for a resource demonstrates that she values the resource more highly than the seller and hence maximizes wealth via the mere acquisition of the resource. To make good on this assertion, Posner provides his own conception of consent and relies on the Kaldor-Hicks compensation principle to argue for the reallocation of entitlements premised on potential gains made by winners who could compensate losers and still have a surplus left over.[22] Even though the winning party who acquires the resource and consumes its surplus value does not compensate the loser, overall wealth is generated. This follows from Posner's reasoning that the differential between the buyer's and seller's willingness to pay for and part from the good is realized in the exchange, no matter which party acquires the surplus. Therefore, a reassignment of property rights can be as effective as an actual market exchange in realizing this surplus value. Coleman astutely crystallizes Posner's conclusion: "where markets are too expensive or otherwise unworkable why not simply assign the entitlement to the high bidder straightaway?"[23] Thus, justice as wealth maximization argues that individuals who have the wherewithal to acquire property or to generate greater income streams from it than current owners have priority over acquiring rights to that resource.

The controversial US Supreme Court case *Kelo vs. City of New London* embodies the principles of law and economics and provides an example of how Kaldor-Hicks reasoning can be used to reallocate resources by coercing individuals to part from private property through legal redress. This case is controversial because of how clearly it seems to violate the Fifth Amendment, that is, that private property is inalienable insofar as one private party is not entitled to take another's private property for private use.[24] The five-to-four majority sided with the law and economics reasoning that the current property owners were not able to put the land to its full economic use and the land should thus be given

---

[21] See, e.g., Amiram Gafni, "Willingness-to-Pay as a Measure of Benefits: Relevant Questions in the Context of Public Decisionmaking about Health Care Programs," *Medical Care* (1991) 29:12, 1246–1252.

[22] Coleman, *Markets, Morality, and the Law*, 2002, 84.

[23] Ibid., 88.

[24] See US Supreme Court decision, *Kelo vs. City of New London*, 545 US 469, 2005, full text available on Cornell University's Law School Legal Information Institute website, www.law.cornell.edu/supct/html/04-108.ZS.html, accessed July 15, 2015.

to a private developer who promised to build condominiums and other lucrative projects that would increase the city's tax base. The private developer in question was permitted to pay the property owners a shadow market price (the average value of property prior to the development initiative) without their consent. By the Kaldor-Hicks test, the new owners do not have to pay those who lost entitlement any of the surplus they hope to gain by leveraging the full value of the property, nor do they have to pay compensation for forcibly removing private residents from their homes without consent.[25]

Two elements of law and economics reasoning are starkly clear. First, individuals' rights are not a sacred standard-bearer for personal freedom of choice or autonomous use of property. Second, with its application of the Kaldor-Hicks compensation standard, law and economics not only permits the law to create winners and losers, but it also indiscriminately permits wealthier individuals to exploit poorer individuals without sharing the surplus gained from such exploitation.[26] To Coleman, whereas the Coase theorem may be defensible when embedded in a robust theory of rights and moral desert, the Kaldor-Hicks compensation principle can only be legitimate if used to protect society from monopoly power.[27] Coleman worries, "The remarkably bold claim Posner advances is not just that wealth maximization is justified because it is consented to; it is that wealth maximization of the Kaldor-Hicks variety ... is justified because it is consented to."[28] This point, that Posner defends the mobilization of Kaldor-Hicks according to the view that the losers would consent to their treatment, is so crucial that it is valuable to read it in Posner's own words: "I want to defend the Kaldor-Hicks or wealth-maximization approach not by reference to Pareto superiority as such or its utilitarian premise, but by reference to the idea of *consent* that I have said provides an alternative basis to utilitarianism for the Pareto criterion."[29] The task ahead is to understand Posner's use of "consent," given both the configuration of the legal suit *Kelo vs. City of New London* and the practical import of the Kaldor-Hicks compensation principle legitimating one party's gain in wealth through non-consensual means, thus imposing costs on the losing party.

KALDOR-HICKS EFFICIENCY: NO NEED TO COMPENSATE LOSERS

One way to make sense of Posner's routine judgment that Kaldor-Hicks efficiency is an effective means to maximize wealth is to step back to gain

[25] For the unsatisfactory elements of this means of condemning property, see Richard Epstein, *Takings: Private Property and the Power of Eminent Domain* (Cambridge, MA: Harvard University Press, 1985).
[26] There is also the issue of the lack of transparency of the test of Posner's rule of entitlement in accordance with the highest bidder – for its inherent obscurity, see Coleman, *Markets, Morality, and the Law*, 2002, 91.
[27] Ibid., 92.
[28] Ibid., 117.
[29] Ibid.

perspective on how rational decision theory per se has an overarching impact on the meaning of consent. Coleman notices that in rational choice theory, arguments for self-interest and those for consent are misleadingly merged under the idea that it is self-evident that the promotion of self-interest is indistinguishable from consent. Individuals inherently consent to that state or action that makes them better off. Coleman explains, "The source of this confusion is the axiom of rational choice that rational, fully informed individuals naturally agree to pursue goals, promote policies and support institutions that maximize their expected utility."[30] By this interpretation, consent is analytically built into the premise that purposive agents maximize expected utility. It is assumed to be consistent with the axioms of rational choice theory that "one consents to what is in one's interest; what is in one's interest, one consents to."[31] In fact, both Posner and Buchanan, as devotees of rational decision theory, fall into the same confusion, although with distinct implications: "Both the ... [rational choice] economist and the constitutionalist blur this distinction in ways that rob their respective enterprises of normative grounding. The ... [rational choice] economist does so by building an individual's consent into an ordering of his preferences. The constitutionalist does so by building efficiency into an individual's expression of consent."[32] For Buchanan, if an individual acts, then this conduct is indicative of having pursued an action consistent with personal preferences and must by definition exhibit efficiency. For Posner, actors' preferences may be surmised from patterns of their choices, or what the existing market price is for their property; actors are supposed to consent to any outcome demonstrating efficient resource allocation consistent with their preferences.

Conflating consent with expected utility maximization is doubly misleading. First, it assumes that consent is incorporated into individuals' expected utility rankings over possible world states. Second, the analytic equation of consent and self-interest makes it virtually impossible to appreciate that individuals are able to accede to states of affairs that may in fact go against their own evaluations of personal satisfaction or self-interest.

Let us examine the latter point first, as it is more readily communicated. Coleman explains,

In order to make good the distinction we recognize between autonomy and utility, we must admit that it does not follow logically that people would consent to what they prefer. This sounds odd at first, but it need not. The root idea, I think, is that efficiency or utility is a property of social states themselves. Autonomy speaks to the concern for the process by which one moves from one social state to another. Put another way, an ordering of one's preferences over social states is path independent, whereas which social states one consents to or agrees to is path dependent.[33]

[30] Ibid., 135.
[31] Ibid., 137.
[32] Ibid., 136.
[33] Ibid., 137.

This characteristic inability to address the quality of process versus the utility of ends challenges virtually all rational choice scholarship mobilized to draw political theoretic conclusions.[34] To demonstrate the point more explicitly, Coleman draws attention to the now standard supposition, either in the Nash equilibrium or the Prisoner's Dilemma account of the social contract, that agents' desire for goods simultaneously implies that each would rather acquire the other's goods via theft than mutual exchange. This proposition seems sufficiently debatable that it is a wonder that it has become a mainstay of theories of collective action, free riding, and even simple exchange.

Drawing attention to the very scenario implied by the Prisoner's Dilemma rendering of the exchange problem, Coleman observes,

> My having the gold and you not is thus an incomplete description of a social state [even if one that is typically used to illustrate that PD actors prefer the state in which they have all the goods at the other's expense]. More complete ones are: my having the gold and your not in virtue of, say, my [means of] earning it or my purchasing it from you; and my having the gold and your not in virtue of my defrauding or robbing you of it. I prefer having the gold to not having it when I secure it legitimately, but not when I come to it through illegitimate means.[35]

This quote is important because it gets to the heart of the Prisoner's Dilemma rendering of the social contract and exchange. For the reason that preferences are purely over outcomes independently of how they are attained, actors are automatically presumed to prefer suckering others rather than obtaining goods through legitimate and bilateral exchange. Rational choice stipulates that expected utility functions contain all the information relevant to purposive agents, so Coleman's path dependence could be incorporated into the individuals' utility functions.[36] However, in practice, this nuanced interpretation of utility is not permitted in the expected utility theory required in game theory, which can only guarantee solutions under the assumption that actors base strategies for action on their evaluation of outcomes and may permit die role to govern their choice of actions.

The first point alluded to earlier, that consent can be analytically contained within individuals' expected utility functions, also bears further scrutiny because it is the foundation for Posner's insistence that consent is consistent with Kaldor-Hicks efficiency and his argument that justice should be restricted to wealth maximization. Posner's position that consent is inseparable from individuals' preferences over outcomes penetrates down to the roots of the law and economics perspective on property rights. The analytic incorporation of consent into preference satisfaction permits Posner to insist both that wealth

---

[34] See Amartya Sen's discussion, "Liberty, Unanimity and Rights," in his *Choice, Welfare and Measurement* (Cambridge, MA: MIT Press, 1982), 291–317.

[35] Coleman, 2002, 137–138.

[36] This is what Donald C. Hubin suggests in his "Groundless Normativity of Instrumental Rationality," *Journal of Philosophy* (2001), 98, 445–465.

maximization upholds autonomous agency via consensual agreement and that wealth maximization is not only coherent and legitimate but also is the key function of justice.

Posner's argument rests on two pillars. One is the idea that individuals' expected utility functions permit setting an economic price for all eventualities. The second is the idea that property rights are best viewed as reflecting a pricing system for different eventualities. This first point is intelligible if we imagine a state of affairs, not so different from the manner in which Nash sets up his bargaining problem, in which I not only order various potential states of the world from most to least preferred but also have a clear sense of intensity of my preference satisfaction or loss for every conceivable outcome. For example, I would really like to go out to dinner tonight with friends, but I could also stay home and get some work done.[37] Perhaps I am indifferent between these two alternatives such that an additional bonus or attraction could sway me toward one alternative or the other. However, say there is the possibility that on my way to dinner I might have a bicycle accident, after which I would neither eat dinner with my friends nor work. According to Posner's logic, this newly introduced less attractive possibility can be measured on a metric of monetary value, so that even if I am only slightly bruised and my bike a bit damaged, receiving a $500 cash payment could make me indifferent between the first two options and the bicycle accident. In law and economics, all outcomes are measurable on a monetary scale that reflects individuals' utility functions, based on how each individual measures worth against the backdrop of actual resources and actual income. Thus, given my income, the inconvenience of a bike accident to me may be evaluated as less costly than for a stockbroker. For this reason, law and economics theorists recommend that more talented or wealthy individuals get more compensation for losses in accident claims.[38]

Posner's use of a monetary metric that covers all contingent states of affairs and permits individuals to rank outcomes on a finely graded scale also depends on the existence of monetary prices as independent quasi-Platonic entities external to all decision problems. This idea of monetary value that stands apart as a metric all can use to differentiate between various states of affairs is foundational for Posner's defense of justice as wealth maximization. Ultimately, wealth production is evident in the world by individuals' abilities to generate income streams; the more income generated from given resources, the more wealth has been maximized and justice is satisfied. Likewise, the greater the positive gradient between an acquirer's willingness and ability to pay for and

---

[37] Recall that to be useful in game theory, my evaluation of world states must be amenable to judgments of the sort comparing what probability must hold for my most and least preferred outcome for a lottery to be equivalent to obtaining the middle outcome for certain; von Neumann and Morgenstern, *Theory of Games*, 1944/53/2004, 20–31.

[38] This is now standard reasoning in tort law that was applied in the federally mandated September 11, 2001, Victim Compensation Fund; see E. Kolbert, "The Calculator: How Kenneth Feinberg Determines the Value of Three Thousand Lives," *New Yorker*, November 25, 2002.

the yielder's willingness to part from a resource, the more wealth has been generated in a transaction.

Posner's view that wealth maximization is the leading mission of justice turns Adam Smith's commutative system upside down, yet it similarly implies that any growth of the economic pie will be beneficial to all agents. Adam Smith gave priority to individuals' rights and argued that inalienable rights generate prosperity. He provided an objective standard for this outcome: that the cost of living will go down, thereby rendering even less affluent individuals better off.[39] Posner, by contrast, argues that rights have no intrinsic validity but only serve the function of maximizing wealth. Posner does not attempt to provide a detailed argument for how less well-off individuals' cost of subsistence will necessarily decrease in his system, or how justice as wealth maximization will offer inclusive and improving access to resources for all members of society. Nevertheless, he still insists that economic growth, achieved by his two standards of rising income streams and rewards for higher willingness and ability to pay for resources, is the highest goal of a social order.[40]

NEOLIBERALISM'S EQUATION OF EX ANTE AND EX POST CONSENT

To further argue his case, Posner must still somehow incorporate consent into his system of justice as wealth maximization. His utter dependence on the analytic containment of consent within preference satisfaction measured against a backdrop of monetary value is readily evident in his commitment to law and economics liability law. Here, the contrast between Buchanan's constitutionalism and Posner's utilitarianism reveals itself again. Buchanan and Posner must confront the question of what level of threatened punishment is appropriate or necessary to satisfactorily defend a property rights regime consistent with their positions. Each gravitates toward a distinct rationalization of punishment consistent with his respective concept of consent: Buchanan emphasizes unanimity and veto power; Posner turns to consent implicitly defined by agents' utility functions. Both theorists' systems work to enforce status quo property rights with a bias toward augmenting the status of the more well-off to the detriment of the less well-off. Although law and economics is not inherently a conservative or elitist system of jurisprudence, Posner's version results in social outcomes that are even more regressive than Buchanan's constitutionalism.

Coleman's discussion of the contrast between property rules and liability rules, which delineate when one party harms another party, is invaluable for seeing how law and economics equates rights with prices agents are willing to pay to avoid harm as expressed in their expected utility scales or revealed

---

[39] For discussion see S. M. Amadae, *Rationalizing Capitalist Democracy* (Chicago: University of Chicago Press, 2003), 205–219, and Hont and Ignatiev, "Needs and Justice," 1983.

[40] See his essay, Richard A. Posner, "Wealth Maximization and Judicial Decision-Making," *International Review of Law and Economics* (1984) 4, 131–135; and "Wealth Maximization Revisited," *Notre Dame Journal of Law, Ethics, and Public Policy* (1985), 2, 85–106.

preferences evident in actual choice. Buchanan's constitutionalism, on the other hand, views rights as an inalienable quality protected by individuals' veto power over actions against them. The decisive difference between property rules and liability rules is that the former requires as a condition of transfer of rights the possessor's *ex ante* agreement; the latter dispenses with the need for *ex ante* agreement and instead places the full responsibility of legal recourse on *ex post* compensation for a price determined by law.

Property rights, characteristic of Buchanan's constitutionalism and Nozick's libertarianism, hold the following: "If the content of B's entitlement is specified by a *property rule only*, then he has a legitimate claim against A that any transfer of his resources from B to A must proceed according to terms established by *ex ante* agreement. Agreement is necessary and sufficient for legitimate transfer."[41] Property rules require as conditions of exchange that individuals agree to terms; otherwise, the transaction is invalid, and the role of justice is to seek redress from party A who exacted the property of B. Property rules may well add additional penalties on the harm caused by A and may criminalize actions as well. Liability rules have a different character: "If the content of B's entitlement is specified by a *liability rule only*, then B has two claims: One is to the liberty to seek a transfer through *ex ante* agreement with A; the other is to recompense in the event A forgoes negotiations and imposes a transfer upon him."[42] The crucial distinction between the protection of rights via liability rules and via property rights is that the former allows A to take B's property without prior agreement so long as A pays the damage at a legally approved (shadow market) price. According to Posnerian law and economics, as long as this price is consistent with an agent's personal utility function over outcomes, then *ex post* compensation in accordance with prices the markets demonstrate in every-day transactions meets the criterion of consent even without that agent's active participation.

To provide an example of how this reasoning works, consider the difference between property and liability rights over one's body. Let us consider two cases of their illegality upheld by either *ex ante* property rules or by *ex post* liability rules. We know that in developing countries, especially Brazil, there is a functioning market in live organs, and it is the established practice in some nations to use live organs as collateral for loans.[43] Therefore, there is a market reflecting the value of live organ donation that actually sets a price of about $3,000 on the sale of a kidney, what typical agents in less fortunate circumstances are voluntarily willing

---

[41] Coleman, *Markets, Morality, and the Law*, 2002, 40.

[42] Ibid., 40.

[43] Nancy Scheper-Hughes, "The Global Traffic in Human Organs," *Current Anthropology* (2000), 41:2, 191–224. Gary Becker and Richard Posner address the topic of live organ sales on their blog post "Should the Purchase and Sale of Organs for Transplant Surgery be Permitted?" Their blog is called The Becker Posner Blog, available at www.becker-posner-blog.com/2006/01/should-the -purchase-and-sale-of-organs-for-transplant-surgery-be-permitted-becker.html, posted January 1, 2006, accessed July 15, 2015.

to accept to part with an organ. If bodies were only protected by liability rules, then it would be possible to steal a kidney while a patient is undergoing another surgery and then to subsequently pay the damages of $3,000 if a claim is pressed. On the strict liability reading of entitlements, the acceptance of *ex post* compensation is equivalent to consent under the reasoning that it is already established that individuals voluntarily undergo such transactions for products for a like price. Of course, under a property rights regime, *ex ante* agreement is legally necessary to validate the transfer. If this condition is violated, then market compensation is insufficient to right the wrong that has been perpetrated.

Indeed, it is in equating consent on an *ex post* compensation basis in accordance with prices set by external rules that Posner is able to suggest that Kaldor-Hicks efficiency functions in accordance with consent. The former residents of New London did not *ex ante* agree to the transfer of their property rights. They were stripped of decisional autonomy and did not share in the surplus that provided the rationale for the transfer in the first place; they were merely offered *ex post* compensation at an earlier shadow market price. By Posner's scheme, this structure of rights transfer exemplifies consent when the victim accepts compensation.

Posner approaches the structure of rights in direct opposition to Buchanan because he views rights only as instrumentally useful for maximizing wealth. He seeks to embrace consent to propose that law and economics respects individual freedom of choice, yet he eviscerates the concept so that it permits nonconsensual property transfers in accordance with a principle of *ex post* compensation, the price of which is settled outside of mutual agreement. He awards his concept of efficiency the highest position and seeks to provide it with residual justification by claiming it is consistent with consent. Once Posner has rationalized *ex post* compensation as a legitimate form of consent, he has precisely what he needs to endorse the Kaldor-Hicks compensation principle as a favorable means of wealth maximization.

If Posner's liability rules were used to defend persons' bodies, then accepting compensation after an accident or theft would be equivalent to prior consent to the terms of the transaction. Buchanan's property rules, on the other hand, do not validate *ex post* agreement; compensation does not transform a property violation into a permissible action. Posnerian law and economics has the unsalutary effect of putting every resource up for sale for market prices determined independently of actual consent.

By analytically building consent into individuals' expected utility functions, and by referring to "willingness to pay" as a coded phrase implying "ability to pay," Posner sets the stage for justifying two radical ideas. First, an individual with greater financial wherewithal is positioned to extract resources from less well-off individuals purely given the former's greater ability to command resources. Second, this involuntary extraction of resources from the less affluent to the more affluent definitionally satisfies the criterion of wealth maximization. Posner's legal position, with its eviscerated concept of consent, ends up justifying the forcible expropriation and transfer of resources to those who most value them measured by their greater financial wherewithal.

As an example, consider the notorious memorandum written by the World Bank's chief economist Lawrence Summers that explicitly endorses dumping toxic waste on less developed countries.[44] This public policy can be justified using law and economics reasoning in the following manner. Land use may be evaluated according to the stream of income it generates, or its sale value. The Somali national coastline neither generates an income stream nor has much sales value as real estate. In fact, considering its accessibility for dumping waste, it may be some of the lowest-valued available land. According to Summers's analysis, if it costs more to dump waste in developed countries, it is economically viable for wealthy countries to select to dump on the Somali coast instead. At this point, Buchanan could make an argument for property rights based on constitutionalism, under which it is dubious that the country of Somalia would agree to dumping on its beaches; if locals had the wherewithal to do so, they have the right to exercise veto power. However, Summers and Posner would adopt a different argument strategy. By recourse to the concept of virtual consent underlying liability law, Posner would recommend that wealthier countries can choose to dump on poorer countries and can validate this "transaction" by paying a fee that would compensate for the loss of rent or property value otherwise commanded by the property: no *ex ante* consent is necessary to validate this *ex post facto* consensual exchange.

To carry this example further, residents of less developed countries personally have less economic value as they generate less value viewed objectively as lifetime earnings potential.[45] If we extrapolate outward to consider the impact of the dumping on the Somali coast to acknowledge that some of this pollution in fact may be highly toxic waste linked to serious health injuries, then we could still present an economic argument that the lives of Somalis are worth less in terms of the total dollar amounts they are able to earn in their lifetimes. Therefore, even considering the health damages, it may still be economically efficient according to Posner and Summers to dump waste on poor countries.

Although both Buchanan and Posner endow wealthier individuals with greater advantages, Buchanan's constitutionalism seeks to prevent interactions from emerging that would disadvantage agents because each should ideally have veto power over actions that would produce such a consequence.[46] However, Posner's law and economics gives agents no say over preventing their resources from being taken by agents with greater purchasing power. Posner could argue that his

---

[44] Summers's memorandum is in Hausman and McPherson, *Economic Analysis, Moral Philosophy, and Public Policy*, 2006, 12–13.

[45] Also in Summers's memorandum; see ibid., 11–13.

[46] For Buchanan, individuals' wherewithal to leverage rights will come down to their threat advantage; Ken Binmore suggests that less well-off individuals can maintain superior bargaining power because they have much less to lose than better-endowed individuals; Ken Binmore, *Natural Justice* (Oxford: Oxford University Press, 2005), 152–157.

entitlement scheme in accordance with liability rules and *ex post* compensation is more efficient because it more accurately reflects the prices agents would and could pay to personally defend their rights, whereas, as we have seen, Buchanan requires the force of law to threaten sanctions to incur punishment beyond that of mere compensation. Regardless, however, Buchanan's property rights regime will still be more costly to enforce because the apparatus needs to be in place to try and penalize those agents whom Posner's justice as wealth maximization permits to pay compensation at a market price without additional penalty for causing harm.

CONCLUSION

Public choice and law and economics represent two schools of neoliberal governance. Classical liberalism derives from natural law theory to ground either the self-evidence of rights to personhood and property or the obligation to comply with agreements according to the principle of consent and voluntary cooperation.[47] Liberalism secured a basis for voluntary market exchanges within the context of a night watchman state by arguing that upholding justice is the best means to generate general prosperity, which emerges from individuals' security of their persons and property, and the freedom to exchange without interference in accordance with their own interests. Neoliberalism remains unable to identify a purchase for justice and common norms of conduct outside of the individual's maximization of expected utility.[48] Its proponents argue that expected utility can encompass all categories of action. Yet in practice, expected utility theory, specifically as it is applied to generate a rational choice approach to political economy, renders intractable the incorporation of moral side constraints, such as the classic liberal duty of not injuring others, into its metric of evaluation, and equates preferences with the total range of concerns and logics for action characterizing individuals' decisions.[49]

Two familiar challenges arise. First, it is unclear how the boundaries of personhood and property are determined, if not through coercive bargaining underlying the formation of a legal regime and the perpetual contestation of legal rights. The second acutely critical point is that, even if we were to accept that agents are satisfied with their original holdings, believe bargaining power is equal, and agree to terms of exchange, according to the bargaining theory championed by John Nash, every agent would still foremost prefer to steal from other members of society unless constrained by the application of force.

---

[47] Adam Smith's *Theory of Moral Sentiments* is suggestive of the first possibility; Thomas Hobbes's and John Locke's liberalism reflects the second set of considerations. For a brief overview of classical liberalism, see Gerald Gaus, "Liberalism," *Stanford Encyclopedia of Philosophy*, first published November 28, 1996, revised December 22, 2014, http://plato.stanford.edu/entries/liberalism/, accessed July 15, 2015, sections 1.1 and 2.1.

[48] Philip Pettit, "*Virtus Normativa*," *Rules, Reasons, and Norms* (Oxford: Clarendon Press, 2002).

[49] Joseph Heath, *Following the Rules* (Oxford: Oxford University Press, 2011), 1–41.

Consent to terms of rule, whether in two-party contracts or in an *n*-member society, provides the fulcrum of classic liberalism. It is possible that Nash, along with noncooperative game theory more generally, provides a deep analytic insight into the nature of exchange that free trade must be encapsulated in a coercive system of constraints to be rendered feasible.[50] This necessarily follows from first claiming that every consideration of value to agents is contained in their preferences over outcomes and then assuming that the only thing that actors value is possession of fungible goods which have value independent from social institutions and relations.[51] Expected utility functions, often calibrated in dollar terms, cannot easily incorporate an evaluation of how much respecting another's right of ownership is worth: it would require that I estimate the value of your property to your equivalent of its value to me so that I am not tempted to steal it. Noncooperative game theory and expected utility theory are thus poor tools for capturing all behavior considerations relevant to agents, such as long-term prudential concerns that cannot be measured on a case-by-case basis; mutual respect, or the voluntary limitation of one's action to avoid injuring others despite no direct increase in personal satisfaction; and commitment, or the wherewithal to constrain one's future actions in conjunction with others to achieve mutual goals.

Decision theory by itself, without game theory, is complex and may be modified to encompass moral considerations such as side constraints.[52] However, the noncooperative game theoretic position that underwrites the Prisoner's Dilemma view of the social contract and bargaining cannot readily admit long-sighted prudence, mutual forbearance, or commitment. The iron law of this body of theoretical work holds that any agreement must necessarily devolve into coercive bargaining unless encompassed by the threat of force to maintain compliance. Buchanan's political philosophy rests on this assumption by focusing on the significance of coercive sanctions over and beyond the terms of agreement to suggest that any distributive pattern of a social contract of private agreements is enforceable. Posner ups the ante by arguing that agents do not have personal veto power to avoid an exchange. His view taps into the

---

[50] The best discussion of this is Ken Binmore's introduction to John Forbes Nash Jr.'s *Essays on Game Theory* (London: Edward Elgar, 1997).

[51] This treatment of expected utility structures Nash's original presentation; see John Nash, "The Bargaining Problem," originally in *Econometrica* (1950) 18, 155–162, republished in John F. Nash Jr., *Essays on Game Theory* (Cheltenham, UK: Edward Elgar, 1996), 155–162.

[52] See Amartya Sen, "Maximization and the Act of Choice," and "Goals, Commitment, and Identity," in his *Rationality and Freedom* (Cambridge, MA: Belknap, 2004), 158–205 and 206–224; note that Sen's treatment of decision theory would result in contractions with its standard formulation; see Amadae's review, "Amartya Sen's Rationality and Freedom," *Economics and Philosophy* (2004), 20:2, 381–389. Note that Donald Davidson accepted decision theory but asserts that linguistic capacity must precede preferences in inverse order to the rational choice priority of preferences over meaning and truth conditions; "Truth and Meaning," in his *Inquiry into Truth and Interpretation* (Oxford: Clarendon Press, 1984) 7–36; for discussion, see Joseph Heath, *Communicative Action and Rational Choice* (Cambridge, MA: MIT Press, 2001), 22.

standard assumption that rights are not voluntarily respected but must be enforced at a cost. He views his system of justice as wealth maximization as enforcing a system of rights that reflects individuals' ability to pay to have their rights enforced.

Both public choice and law and economics have drifted far from classical liberalism, which endorses types of action not reducible to momentary cost-benefit analysis of outcomes. Individuals exposed to such teachings are prone to becoming skeptical about the rationality of types of conduct that resist ready reduction to expected utility maximization. Research shows that a small population of noncooperative actors can influence the larger community, leading to a preponderance of noncooperative outcomes.[53] The principles of mutual respect of rights and commitment are fundamental to classic Western liberalism and therefore remain well worth contemplating and enacting even though they are not readily assimilated into the language of expected utility and tangible rewards.

By way of conclusion, the following points are prominent:

1. Although James M. Buchanan and Richard Posner propose differing views on consent, both are divorced from classical liberalism. For the classical liberal agent, agreeing to terms is motivationally binding and is built on the respect of personhood and recognition of the others' right to exist, which are the elemental commitments grounding the rule of law and free market exchange.

2. Both public choice and law and economics are retrogressive in the sense of tacitly endorsing ever widening access to resources and demonstrations of power, even more so than Robert Nozick's libertarianism. Buchanan accepts that actors' de facto power, along with its juxtaposition to punitive sanctions derived from the exercise of power, sets the terms defining both the rule of law and private contracts. Posner warrants property transfer without *ex ante* consent if monetary compensation at a shadow market price covers loss, and therefore renders all property accessible to those with a greater financial wherewithal to pay for it.

3. However, like within the family of modern liberal political theories, neoliberalism is not limited in its expression to non-egalitarian outcomes, although it offers little purchasing power to resist this implication given its rejection of the no-harm principle as the ground from which to build relations and institutions. Cass Sunstein's school of libertarian paternalism, for example, also shares with law and economics the tendency to identify actors with their preferences and opportunities, therefore finding it permissible for the state to intercede in their decision making to correct

---

[53] In this research, cooperative actors acted on considerations of altruism and norm following; the opposite is also possible, that a smaller number of cooperators may have a disproportionate effect on a population. See C. F. Camerer, "When Does 'Economic Man' Dominate Social Behavior," *Science* (2006), 311, 47–52.

cognitive deficiencies and shortfalls of rationality, nudging citizens to make the right choice.[54]

4. Neoliberalism's treatment of persons as though they were interchangeable with nonhuman actors, or even as strategy profiles derived from expected utility functions standing as programmable algorithms determining choice, reveals its shared roots in nuclearized sovereignty. A major US Cold War objective was to maintain sovereignty through issuing credible threats of calibrated destruction ranging from conventional warfare through limited nuclear options to full-scale thermonuclear exchange. Command and control for this eventuality demanded that the United States could execute a chain of command and set of instructions over war fighting that could persist regardless of whether officers of state were deceased or missing. Rational choice removes the need for a conscious or present decision maker at any given moment of choice. Similarly, for law and economics and libertarian paternalism, if actors' preferences are known along with all eventualities and market prices, then systems of exchange and institutions can be programmed and automated to achieve optimal equilibria.

5. For law and economics and public choice, rational agency is confined to strategic competition. Both view freedom experienced in terms of determining the significance of one's choices or choosing from a menu of alternative rationales for action to be naïve at best and inviting failure at worst. Yet the two schools differ in their views of which social order best serves individuals' interests. Although neither has a normative purchase in the sense of recognizing any course of action for its ethical merit other than satisfying individuals' preferences, Buchanan hopes that government will be limited to protecting those property rights that individuals bring into the social contract and continually exert their power to renegotiate and augment. Posner holds that the entire purpose of property rights in the first place is to render possible economic gain; hence, he requires that everything be for sale to maximize wealth, even, in effect, the entitlement to rights themselves.

---

[54] Richard H. Thaler and Cass R. Sunstein, *Nudge: Improving Decisions about Health, Wealth and Happiness* (New York: Penguin, 2009).

9

Collective Action

> The facts that there is a lot of collective action even in many large-number contexts ... and that, therefore, many people are not free riding in relevant contexts suggest [that] ... the actors in the seemingly successful collective actions fail to understand their own interests.

> Despite the fact that people regularly grasp the incentive to free ride on the efforts of others in many contexts, it is also true that the logic of collective action is hard to grasp in the abstract. The cursory history... [of collective action] suggests just how hard it was to come to a general understanding of the problem. Today, there are thousands of social scientists and philosophers who do understand it and maybe far more who still do not. But in the general population, few people grasp it. Those who teach these issues regularly discover that some students insist that the logic is wrong, that it is, for example, in the interest of workers to pay dues voluntarily to unions or that it is in one's interest to vote .... It would be extremely difficult to assess how large is the role of misunderstanding in the reasons for action in general because those who do not understand the issues cannot usefully be asked whether they do understand. But the evidence of misunderstanding and ignorance is extensive.

> Russell Hardin, 2003[1]

The failure of collective action in large-scale interdependent efforts, this chapter shows, is the consequence of causal negligibility rather than the intention to free ride on others' efforts. Rational choice theory puts forward a different explanation: that collective enterprises are doomed to failure because of the overriding logic of the Prisoner's Dilemma game, extended from two individuals to countless individuals. This deep-seated worry over individually enacted globally destructive action, which resonates with contemporary concerns over climate change, overfished seas, and boycotts, has

---

[1] Russell Hardin, "The Free Rider Problem," *Stanford Encyclopedia of Philosophy*, 2003, accessed January 30, 2015, http://plato.stanford.edu/entries/free-rider/.

been labeled "the tragedy of the commons," and the free rider problem.[2] Just as in a two-person PD game, individual self-interest recommends defecting and thus leads to mutual impoverishment, so, too, in a situation with multiple persons, such logic will necessarily lead to the failure of collective action, whether in the case of an environmental commons or the provision of public goods such as infrastructure.

This chapter examines within historical context the neoliberal common lore that it is rational to free ride on collective enterprises and that large-scale ventures must fail accordingly, and it shows how such common lore contradicts both conventional wisdom and recent theoretical investigations of the topic. First, the chapter briefly discusses the application of game theory to climate change and highlights how the Prisoner's Dilemma and strategic rationality continue to dominate social scientific approaches to developing public policies to solve social dilemmas. Second, it introduces the logic behind the Prisoner's Dilemma assessment of collective action via the extension of the two-person game to multiple individuals. Third, it discusses Mancur Olson's 1965 *Logic of Collective Action* to show that even quite recently, prominent theorists concluded that individuals could cooperate without the introduction of sanctioning devices in small and mid-sized groups.

The chapter's conclusion dovetails with Elinor Ostrom's 1990 work *Governing the Commons: The Evolution of Institutions for Collective Action*. Whereas Ostrom provides empirical evidence to demonstrate that actors are capable of achieving mutually beneficial outcomes in variously sized ventures, this chapter discredits the theoretical structure underpinning the thesis that working together in small, medium, and large groups epitomizes Prisoner's Dilemma logic. It demonstrates that the challenge confronting multi-actor undertakings is the failure of purposive agency due to everyone's inability to make an appreciable difference, rather than anyone's inclination to exploit others' efforts. It will again show how strategic rationality is insufficient to serve as humankind's sole meaningful rationale for action. The chapter concludes by contrasting the classical liberal view of perfect competition as a public good and guarantor of mutual prosperity with the neoliberal inability to exit the Prisoner's Dilemma in interactions of any size and attendant reliance on enforcement mechanisms to offset mutual ruin.

A PLANETARY PRISONER'S DILEMMA: GAME THEORY MEETS GLOBAL WARMING

In 2005, the British government commissioned Sir Nicholas Stern, an economist at the London School of Economics, to lead a study on the economic challenges

---

[2] Daniel T. Rogers reports how these phrases have entered common parlance, *Age of Fracture* (Cambridge, MA: Belknap, 2012).

posed by climate change.[3] The *Stern Review: The Economics of Climate Change* (2006) illustrates the incorporation of rational choice theory into the analysis of the world's most far-reaching and existential problem. It demonstrates how the neoliberal orientation to individual choice and action has become accepted as conventional wisdom: "the problem of collective action in social contexts ... is the Prisoner's Dilemma writ large."[4] The report's main argument is that the costs that will be incurred if we avoid taking measures to protect the atmosphere from the unbounded release of carbon dioxide will far outweigh the costs required to make the changes necessary to rein in greenhouse gas emissions.

Part VI, Chapter 21, of the report, "Framework for Understanding International Collective Action for Climate Change," hones in on the game theoretic approach. Chapter 21 moves quickly from introducing game theory and the Prisoner's Dilemma to specifying their relevance for studying collective action, public goods, and the tragedy of the commons. This keystone chapter of the *Stern Review* reflects the academic consensus regarding the implications of rational choice for studying collective action in its varying expressions.

Climate change presents a colossal challenge of collective action. Whereas everyone depends on the atmosphere to hospitably regulate climate, no appreciable causal impact results from any single individual's dispersal of carbon dioxide into the atmosphere. The *Stern Review* first provides an overview of the state-of the-art economic analysis of collective action:

Economists seek to understand the incentives relevant to situations that require collective action, and have studied the institutional arrangements that can facilitate co-operation. The study of collective action is concerned with understanding how to overcome the market failures that lead to the under-provision of public goods where individuals or countries face an incentive to free ride on the actions of others.[5]

Economists view the failure of markets to provide solutions to social dilemmas in the form of public goods as "market failures." An inhabitable climate is considered to be an under-provisioned public good because each individual contributes to releasing carbon dioxide into the atmosphere to promote personal goals while other actors absorb the cost. A market failure means that an allocation of resources could be altered so that at least one individual is better off and no individual is worse off. Noncooperative game theory explains how even in a mutual-best-reply equilibrium achieved through individuals' rational choices, a suboptimal and hence inefficient outcome may arise, specifically when there are scarce resources that all actors alike seek as in the PD model of market exchange.

[3] N.H. Stern, *Stern Review: The Economics of Climate Change* (London: Great Britain Treasury, 2006), available online: http://webarchive.nationalarchives.gov.uk/20100407172811/http://www.hm-treasury.gov.uk/stern_review_report.htm, accessed July 15, 2015, see 143–167.
[4] Russell Hardin, *Collective Action* (Johns Hopkins University Press, 1982), xiii.
[5] *Stern Review*, 2006, 451.

The *Stern Review* definitively recommends game theory as the leading approach to studying collective action of all kinds, including global warming. It resoundingly approves of Thomas Schelling's acceptance of the centrality of strategic rationality and Prisoner's Dilemma logic as the foundation of neoliberal political economy:

Game theory has been used to explore the underlying structure of some common problems. The Prisoner's Dilemma Game has been used to explore a wide range of situations in which individuals act rationally in the light of their own situation and yet find themselves faced with an outcome that leaves them worse off than if they were able to co-operate.[6]

The report outlines five core findings from game theoretic Prisoner's Dilemma analysis of how to achieve cooperation: (1) incentives can be changed, (2) reciprocity can be invoked in repeating interactions, (3) transparency and monitoring can be introduced to enhance compliance mechanisms, (4) persistent pressure to renegotiate agreements will be applied if they seem out of line with an actor's coercive bargaining power, and (5) reputation may establish credible commitment.[7]

The authors of the review come to an apt conclusion about the prospects for successful cooperation obtained by following the dictates of rational choice. They emphasize that standard rational choice assumes narrow self-interest and rules out procedural rationality and other-regarding normativity: "Though extremely useful as a starting point for analysing international collective action, most of these theories tend to focus only on *self-interest very narrowly defined*, and so leave out perspectives on responsibility and ethical standards – for example, the views on what constitutes human decency that are expressed by the public."[8] However, rather than adopt a more encompassing framework, the authors recommend policies that are strictly in line with the standard operationalization of strategic rationality. Game theory typically works from the assumption of narrowly construed individualistic maximization and, by design, forbids normative consideration of the processes by which ends are realized. Ethical considerations are thus dismissed as irrelevant. Moreover, joint maximization consistent with team reasoning is also rendered impracticable. The *Stern Review* is brilliant for recognizing the limitations of game theory but nonetheless accepts that the rational choice policy prescriptions "are always imperative to implement correctly."[9] As is standard throughout the mainstream game theoretic approach, although theorists acknowledge strategic rationality's restricted vision of human conduct, they still tend to tacitly endorse its

---

[6] Ibid.; it cites R. Gibbons, *Game Theory for Applied Economists* (Princeton, NJ: Princeton University Press, 1992) with respect to the common technical understanding of the PD game to study collective action.

[7] *Stern Review*, 2006, 452.

[8] Ibid., 453; emphasis added.

[9] Ibid.

simplifying assumptions and address its limitations in an ad hoc, provisional fashion.[10] In the end, the *Stern Review* exemplifies and perpetuates the characteristic neoliberal wisdom and centerpiece of the rational choice assessment that the necessary failure of collective action is due to Prisoner's Dilemma logic.[11]

THE STRATEGICALLY RATIONAL FAILURES
OF COLLECTIVE ACTION

The provision of public goods and the tragedy of the commons tend to be grouped together for the same treatment; however, they are theoretically distinct. In a public goods exercise, individuals are asked to contribute a certain amount in exchange for receiving a benefit that exceeds the cost of participation. Individuals will tend to contribute less than the amount that would be most efficient for every individual to share in the highest possible net benefit versus net cost. In the tragedy of the commons, each actor incurs a cost on the environment distributed across all actors to receive a personal benefit. The cost is amplified as each actor makes the same ultimate choice and, as in the Prisoner's Dilemma, ends up achieving suboptimal net individual gain because of it. In each case, failure results from the overall inability of individuals to achieve an efficient collective outcome because each has the ever-present interest to contribute less or not at all.

Garrett Hardin provided the initial rational choice treatment of the tragedy of the commons in his 1968 article "The Tragedy of the Commons" in *Science*.[12] Although Hardin scarcely touches on game theory, he still introduces

---

[10] See a subsequent discussion in ibid., 461. Note that at the same time that it is recognized that "the game theory that underpins analyses of international co-operation for global public goods tends to take as its starting point a narrow perspective of self-interest as the only motivation for action, distinguishing it from ethical approaches," the idea is put forward that ethical and strategically rational approaches can be combined. *Prisoners of Reason* argues that game theory cannot be combined with ethics because of its foundational assumptions. The *Stern Review* makes clear that the ethical perspective it views as consistent with game theory is that of "reciprocity and reputation," which have a very particular connotation in game theory in tit for tat punishment and credible threats, 461. Note also that the *Stern Review*'s citation for brokering game theory and ethics is David Gauthier, "Morality and Advantage," *Philosophical Review* (1967), 76, 460–475, but as discussed in *Prisoners of Reason*, Chapters 5 and 6, Gauthier's approach was found to be unsuccessful (see, e.g., Martin Hollis, *Trust within Reason* [Cambridge: Cambridge University Press, 1998]).

[11] Ostrom addressed this point in no uncertain terms in *Governing the Commons* (New York: Cambridge University Press, 1990), 46.

[12] Garrett Hardin, "The Tragedy of the Commons," *Science* (December 13, 1968), 1243–1248; Hardin cites von Neumann and Morgenstern; he categorizes the tragedy of the commons as a "technical problem," like tic-tac-toe, with no solution. Thus, Hardin gestures toward but does not mention the PD, he acknowledges the need to use single criterion valuation, and he denies Adam Smith's invisible hand. The best discussion of this early work on population ecology and game theory is in Paul Erickson's *The World the Game Theorists Made* (Chicago: University of Chicago Press, 2015).

its significance for analyzing collective actions problems. His main focus is the problem of overpopulation and the Earth's carrying capacity for human civilization. He concludes that the only solution that can adequately address the burgeoning human population is to introduce laws with sanctions for overbreeding. Hardin's solution is "mutual coercion mutually agreed upon," anticipating the neoliberal belief in the necessity of coercive force to achieve cooperation.[13] Hardin likewise specifies the necessity of a single metric for value and defers to the interpersonally transferable treatment of utility consistent with game theory.[14]

Researchers following Garrett Hardin interpreted the tragedy of the commons with respect to overgrazing, overpopulation, and the over discharge of greenhouse gases into the atmosphere as a multi-actor Prisoner's Dilemma game.[15] According to Hardin's original paper,

> As a rational being, each herdsman seeks to maximize his gain. Explicitly or implicitly, more or less consciously, he asks, "What is the utility to me of adding one more animal to my herd?" This utility has one negative and one positive component.
>
> (1)   The positive component is a function of the increment of one animal. Since the herdsman receives all the proceeds from the sale of the additional animal, the positive utility is nearly +1.
>
> (2)   The negative component is a function of the additional overgrazing created by one more animal. Since, however, the effects of overgrazing are shared by all the herdsmen, the negative utility for any particular decision-making herdsman is only a fraction of −1.[16]

Rational choice theorists regard this situation as a standard Prisoner's Dilemma with many actors.[17] The ostensibly self-evident reliance on tangible, interpersonally transferable rewards automatically results in the applicability of the PD model, which excludes the possibility that actors have resources to work together to achieve common goals.

---

[13] Quote from Garrett Hardin, "Tragedy of the Commons," 1968, 1246. Russell Hardin concurs with this view of collective action and the social contract, finding the challenge to be a Prisoner's Dilemma and not a Stag Hunt; *Collective Action*, 1982, 168–169.

[14] Hardin, "Tragedy of the Commons," 1968, 1244.

[15] R. M. Dawes, "The Commons Dilemma Game: An N-Person Mixed-Motive Game with a Dominating Strategy for Defection," *ORI Research Bulletin* (1973), 13, 1–12. For discussion, see Ostrom, *Governing the Commons*, 1990, 3.

[16] Hardin, "Tragedy of the Commons," 1968, 1244.

[17] For a sampling of literature making this point, see Thomas Schelling, "Hockey Helmets, Concealed Weapons, and Daylight Saving: A Study of Binary Choices with Externalities," *Journal of Conflict Resolution* (1973) 17:3, 381–428; Michael Taylor, "The Provision of Public Goods," in *Anarchy and Cooperation* (New York: John Wiley, 1976); Philip Pettit, "Free Riding and Foul Dealing," *Journal of Philosophy* (1986), 83:7, 361–379; and Stephen Gardiner, "The Real Tragedy of the Commons," *Philosophy and Public Affairs* (2001), 30:4, 387–416. For the contrasting perspective that I discuss in the following section, see Richard Tuck, "Prisoner's Dilemma," in *Free Riding* (Cambridge, MA: Harvard University Press, 2008), 19–29.

TABLE 9. *Tragedy of the Commons Modeled as Prisoner's Dilemma*

	Individual B	
	Status Quo	Increase Herd
Individual A		
Status Quo	3, 3	1, 4
Increase Herd	4, 1	2, 2

In Hardin's scenario, each individual faces a choice of whether to add an animal to his herd or to maintain the status quo. Each individual has a direct benefit from adding an animal. Each individual bears only a fraction of the cost of increasing his herd size. Therefore, regardless of what the other agent or agents decide to do, each actor is always better off seeking gain. The accompanying PD matrix captures this theoretical structure, with numbers designating ratio of output yield against cost.[18]

Of course, in the two-person example, the commons is occupied by two parties who make independent choices that result in a suboptimal outcome, assuming that every individual is strategically rational.[19] Such a situation satisfies the two definitive features of the PD. First, "it is *collectively rational* to cooperate: each agent prefers the outcome produced by everyone cooperating over the outcome produced by no one cooperating." Second, "it is *individually rational* not to cooperate: when each individual has the power to decide whether or not she will cooperate, each person (rationally) prefers not to cooperate, whatever the others do."[20] Thus, again, the Prisoner's Dilemma game lies at the foundation of neoliberal political economy and provides the rationale for the governance that must be introduced to ensure mutually beneficial conduct. Modelers' reliance on interpersonally transferable utility and game theory's restriction to outcomes and standard assertion of individualistic maximization lock individuals, whether in two-person or multi-actor contexts, into the unrelenting Prisoner's Dilemma bind of mutual indigence: one model fits all sizes of interdependent action situations.

The tragedy of the commons is therefore seen as an extension of a two-person Prisoner's Dilemma in which each agent has the incentive to overload a

[18] Gardiner, "The Real Tragedy of the Commons," 2001.

[19] The standard definition of "rational" applies: maximizing expected utility of a commonly shared criterion for success without recourse to norms, side constraints, or commitment; solidarity and team reasoning; or other-regarding preferences. These assumptions are specified, for example, in Pettit, "Free Riding and Foul Dealing," 1986, 361–379. On value, see especially p. 373; see also Thomas Schelling, "Hockey Helmets, Daylight Saving and Other Binary Choices," reprinted in *Micromotives and Macrobehaviour* (New York: W. W. Norton, 1978).

[20] See Gardiner's explanation of why the tragedy of the commons is best modeled as PD, "The Real Tragedy of the Commons," 2001, 392.

common resource for personal gain. A common-pool resource is naturally available for exploitation, so it is difficult to restrict individuals from contributing to the overuse that results in the globally experienced suboptimal yield. In a public goods problem, on the other hand, agents must be coerced to contribute to a joint effort for a common source of value to be created. Nevertheless, the exercise of providing a public good can be compared to the tragedy of the commons scenario by focusing on the costs rather than the benefits to an individual for contributing to the collective enterprise.

Thomas Schelling discusses how a two-person Prisoner's Dilemma can be used as the basis for an extended analysis of how numerous individuals may confront an interdependent reward scheme. Everyone would benefit if all cooperated. However, each individual has the incentive to defect in the hope of benefiting either from others' restraint from overusing a common-pool resource or from others' contributions to a public good. As a result, all individuals achieve a suboptimal outcome compared to what they would have achieved if all had cooperated. According to Schelling,

> The influence of one individual's choice on the other's payoff we can call the *externality*. Then the effect of his own choice on his own payoff can in parallel be called the *internality*. We then describe prisoner's dilemma as the situation in which each person has a uniform (dominant) internality and a uniform (dominant) externality, the internality and the externality are opposed rather than coincident, and the externality outweighs the internality.[21]

Schelling uses the term "internality" to refer to an actor's personal gain from acting and "externality" to refer to an actor's cost to other actors as a consequence of her or his action.[22] In the collective action problem under discussion, the uniformity that Schelling identifies is in the benefit to the individual from defecting and the cost incurred by others by an actor's defection. From the perspective of strategic rationality, each actor's impact on others, or causal effluvium, is superfluous to individual choice. Every individual has the incentive to defect because the direct benefit outweighs the personally experienced cost of failing to cooperate. Costs and benefits are treated as interpersonally transferable utility. No additional considerations weigh on individuals' choices; reflections of the merit of processes, meta-preferences over whether an individual would prefer to live in a mutually cooperative society or in a society of individual maximizers, and reasoning as a member of a team are ruled out without examination.

---

[21] Thomas C. Schelling, "Hockey Helmets, Concealed Weapons and Daylight Saving: A Study of Binary Choices with Externalities," *Journal of Conflict Resolution* (1973) 17:3, 381–428, at 385–386, this analysis directly compares the internality and externality and depends on interpersonally transferable utility; for discussion, see Tuck, *Free Riding*, 2008, 26–27.

[22] Traditionally, in economics, the term "externality" has been used to refer to impacts on third parties outside of transactions, but the game theoretic usage consistent with Schelling's analysis changes the language, using "externality" to imply any cost that impacts actors outside of an actor's personal expected utility function.

Using this analysis of costs and benefits, which in public goods problems presumes "the same binary choice and same payoffs" for each individual, the provision of public goods and tragedy of the commons can both be modeled using the PD game.[23] In the two-person case, the direct benefit to the individual from defecting is the net gain from free riding on others' efforts (i.e., the difference between cooperating and defecting, holding all other actors' choices the same) plus the gain of avoiding the cost of contributing. In the collective action problem encompassing two or more actors, every individual will bear the costs of his defection, but this cost will be distributed among participants. The gain from action is individual; the cost of the action is shared by all members of the group. The tragedy of the commons then provides the opportunity to test a central thesis of *Prisoners of Reason*: either both the tragedy of the commons and the chronic under-provision of public goods are unique instances of neo-liberal subjectivity manifested through enacting Prisoner's Dilemma logic, or these problems have been characteristic of human society throughout its history and are now receiving rigorous investigation through the value-neutral tools of rational choice.

THE IMPERCEPTIBILITY OF MANCUR OLSON'S LOGIC OF COLLECTIVE ACTION

Although game theorists have often misread it to show otherwise, Mancur Olson's 1965 *Logic of Collective Action* does not assume or reinforce the Prisoner's Dilemma analysis of small-sized, medium-sized, or large-scale groups. According to Olson, the size of a group matters; according to game theory, the same PD logic applies independent of group size.[24] Olson's analysis hinges on the causal negligibility of actors' impact on realizing collective goals, rather than the clear external cost imposed on others in the two-person and small-n PD game. By contrast, the public goods and tragedy of the commons argument presented by rational choice theorists concludes that a large-scale failure of collective action is the limiting case of a Prisoner's Dilemma extended from two individuals to a very large number of people.[25] According to this view, the two-person dilemma that rational self-interest necessarily leads to a mutually impoverishing Pareto suboptimal result will only be multiplied when each additional individual is added. The

---

[23] Schelling's model, as in uniform multi-person PD games used for "resource dilemmas," assumes that payoffs reflect salient objective and interpersonally transferable resources, "Hockey Helmets," 1973, 386. Note that the PD model is similarly applied to the case of two or more actors in the development of a free market from duopoly and oligopoly; for discussion, see Tuck, *Free Riding*, 2008, 169–171.

[24] This section takes up an insight developed by Richard Tuck in contrasting Olson's *Logic of Collective Action* (Cambridge, MA: Harvard University Press, 1965) with the rational choice–based Prisoner's Dilemma model routinely applied to public goods and the tragedy of the commons. See Tuck, *Free Riding*, 2008.

[25] See Tuck for discussion, *Free Riding*, 2008, 28.

multi-person, large-scale collective action must therefore necessarily fail in accordance with this same logic. However, this is not Olson's argument.

The stakes of explaining how collective action fails, either because of the Prisoner's Dilemma logic of predatory gain extended from two individuals to countless individuals on the one hand or because of the breakdown of instrumental agency in very large groups on the other, are high. The former reinforces the irrationality of cooperating at all levels of social organization. The latter argues that the solutions to global resource dilemmas may build on the tactics small and medium-sized groups have developed to collaborate effectively. If we accept that the two-person PD model accurately reflects two-person interactions in bargaining, exchange, as well as a state of nature, then it seems prima facie evident that it must extend to social relations with numerous individuals who make a binary choice over whether to contribute to or defect from an interaction with a tangible rewards PD payoff structure.

Reading Olson closely confirms that the Prisoner's Dilemma analysis is not only largely irrelevant but also contradictory to his overarching thesis and that Olson himself acknowledges that he does not employ game theory.[26] Olson argues that small and medium-sized groups are able to cooperate successfully, but that even those actors who have tools transcending conventional instrumental rationality, including moral regard for others and solidarity, will be stymied by the collective action problem once their size crosses a threshold that renders any single individual's actions causally inappreciable. Olson's work fits comfortably with the conventional experience that small groups (such as cartels) and medium groups (such as trade unions) are sufficiently capable of voluntarily cooperating and must be curbed by legal restraint.[27]

Olson presented a mathematically formalized tripartite categorization of collective action problems: small, medium, and large. Small groups may have a privileged ability to cooperate if their rate of profit exceeds the number of members. For these small groups, each individual is rewarded by a profit margin that multiples his contribution to such an extent that no individual is tempted to withhold his contribution.[28] Consider the following example of a "privileged group," using Olson's terminology, which is readily able to cooperate. Olson's

---

[26] Olson, *Collective Action*, 1965, 43.

[27] This point is discussed by Tuck, *Free Riding*, 2008. Around the globe, freedom of association is actively thwarted to prevent unionization, rather than the inverse predicted by the conventional PD approach to collective action; *Human Rights Watch Report*, "Indivisible Human Rights: The Relationship of Political and Civil Rights to Survival, Subsistence and Poverty," published September 1992, available online at www.hrw.org/sites/default/files/reports/general929.pdf, accessed July 15, 2015, ISBN 1-56432-084-7.

[28] There may be some questions about the exact mechanism of guaranteeing the rate of return. For example, if a group has four members, and if each contributes $100, and each gains a return of $400, it is unclear whether one individual could fail to contribute her $100 in the hopes of still recouping $300 by free riding on other members. The approach of hegemonic stability solves this puzzle by suggesting that even in considering Olson's privileged group, no group of more than one member can cooperate. For discussion, see Duncan Snidal, "The Limits of Hegemonic Stability Theory," *International Organization* (1985) 39:4, 579–614, specifically 599.

analysis incorporates the impact each individual's contribution may have on the total resource supply and its yield. Suppose that it has ten members, each of whom contributes $1,000, and as a function of the total collective input of $10,000, each receives a share of $10,000 as a result of a tenfold yield. If a single defection, leaving only $9,000, were to produce only an eightfold yield, so that every individual would receive $7,200, then it is easy to see that the defector would have been better off by contributing. Had she contributed, her total profit would have been $1,800 higher. Such a payout structure clearly does not reflect the Prisoner's Dilemma model, and thus Olson's privileged group falls outside of the standard game theoretic treatment of the failure of cooperation.

For mid-sized groups, the Prisoner's Dilemma model seems apt, particularly if agents are only concerned with material rewards, not the processes by which rewards are generated. In this case, even though all members gain more from contributing than the cost of their contribution, each has a greater material incentive to free ride. Olson speaks to this reasoning: "Though all of the members of the group therefore have a common interest in obtaining this collective benefit, they have no common interest in paying the cost of providing that collective good. Each would prefer that the others pay the entire cost, and ordinarily would get any benefit provided whether he had born part of the cost or not."[29] However, he neither acknowledges nor promotes this logic. Instead, Olson assumes that mid-sized groups can achieve a cooperative outcome as a result of other-regarding preferences or normative commitments.[30] In a medium-sized group, an individual's failure to cooperate is noticed by everyone because the total public good to be shared is appreciably diminished, as in the two-person PD. The individual who contemplates defecting then may reason that "he would be worse off when the collective good is not provided than when it was provided and he met part of the cost," hence preferring mutual cooperation to lone defection.[31] The Prisoner's Dilemma therefore does not pertain to Olson's analysis of the failure of mid-sized groups. He doubts that this rationalization of free riding is sufficient because he understands that medium-sized collectives are able to effectively achieve joint objectives. He proposes that the ability of groups to collaborate successfully is a function of their size, and he focuses his inquiry on how the dynamics of participation change in relation to the ability of actors to influence the group's collective outcome.[32]

Olson homes in on how group size interacts with the relationship between the marginal cost of purchasing additional units of a public good and the marginal rate of production for that good. From a collective standpoint, it is easy to identify when each individual's cost for a good, when collectively amalgamated, is less

[29] Olson, *Logic of Collective Action*, 1965, 21.

[30] Olson admits both social goods and non-narrowly self-interested action can lead intermediate groups to cooperate; ibid., 50, 60–61.

[31] Ibid., 43.

[32] Olson goes on to make these two points explicitly, ibid., 21; on p. 43, Olson suggests that strategic rationality might counsel particular action.

than the group's ability to supply the good, in which case obtaining the good together is cost-effective from everyone's perspective. Group size is important because the ratio of individual contribution to the total amount of public good provides each individual with a ratio between her own contributions and how much she receives in return as a function of this payment. For example, if an individual contributes $100 – leaving all others' contributions and receipts at the same level – and the individual in question receives $110, or a 10 percent return, then it pays to contribute to the collective venture. Olson highlights the conditions under which participation is directly in each individual's interest:

> The marginal cost of additional units of the collective good must be shared in exactly the same proportion as the additional benefits. Only if this is done will each member find that his own marginal costs and benefits are equal at the same time that the total marginal cost equals the total or aggregate marginal benefit.[33]

Although an individual may contemplate or actually choose to free ride, this is not the point of Olson's theoretical exploration. An individual may or may not contribute, and here neoclassical economics is agnostic as to whether or not an actor acts in accordance with narrow self-interest or other-regarding considerations that may encompass acting in accordance with shared norms. Olson is only concerned with determining at which point group size renders contributing to a joint venture economically unsound because individuals are no longer able to perceive any discernable impact from contributing. In fact, in referring to a Prisoner's Dilemma-like situation, in which one agent may harm another by failing to contribute, Olson observes that a group's ability to cooperate "depends upon whether any two or more members of the group have a perceptible interdependence, that is, on whether the contribution or lack of contribution of any one individual in the group will have a perceptible effect on the burden or benefit of any other individual or individuals in the group."[34] The key attribute of the Prisoner's Dilemma in all of its applications is that all individuals demonstrably affect others' outcomes by their choice to cooperate or defect, hence their fear of being suckered. Not only is Olson's treatment distinct from the free rider problem central to rational choice theory, but also his central thesis is that the failure of collective action arises from the eventual insignificance of any single actor's ability to make a difference as the group's size grows increasingly large. This is the condition that obtains under perfect market competition in which no buyer or seller has sufficient presence to alter the trading price of a good.

Olson's *Logic of Collective Action* is widely misunderstood to be suggesting that groups face obstacles to achieving cooperative outcomes because of the Prisoner's Dilemma logic.[35] A close reading of his argument, however, reveals

---

[33] Ibid., 30–31.
[34] Ibid., 45.
[35] Hardin, *Collective Action*, 1982, 7; see Tuck's *Free Riding*, 2008, for a deeper analysis of the contemporary confusion in reading Olson through the lens of game theory, and specifically the PD game.

that he did not doubt the power of mid-sized groups to collaborate as a function of their ability to act on resources beyond narrow self-interest.[36] Olson's central point is that very large groups will fail to achieve collective goals because their size renders the causal efficacy of any single individual's contributions impotent when evaluated from the perspective of any member of the group.

Three points loom large. First, it is astonishing that the Prisoner's Dilemma bandwagon effect was so great that few theorists were aware of the sea change in thinking about the rationality of cooperation that attended this new fascination.[37] Second, not long ago, as Richard Tuck evidences in *Free Riding*, both empirical practice and elite theory accepted the rationality of collaboration, which has the structure of joint maximization.[38] Third, once pried free from the inauspicious Prisoner's Dilemma logic, Olson's analysis points to a unique challenge confronting large-scale global concerns, such as climate change, deforestation, and overfishing. Any attentive study of these arresting contemporary problems will face a very different puzzle if it turns out to be true that individuals are paralyzed by the sheer scale of the predicament. Olson beseeches us to be aware that "in a large group in which no single individual's contribution makes a perceptible difference to the group as a whole or to the burden or benefit of any single member of the group, it is certain that a collective good will *not* be provided."[39] The failure of large-scale collective action is more a function of the apathy or despair resulting from the insufficiency of instrumental agency than of strategic calculations for achieving self-gain despite others.

THE MALEVOLENT BACKHAND OF NEOLIBERAL GOVERNANCE[40]

The rational choice two-person Prisoner's Dilemma model applies to any case in which individuals exist in a state of, as Mancur Olson puts it, "perceptible interdependence," including basic market exchange and bargaining, the social contract, and even the state of nature.[41] The *Stern Review* documents that since the mid-1970s, it has been standard practice for analysts to extrapolate from the two-person scenario to circumstances composed of any number of individuals, that each person's pursuit of preference satisfaction despite others represents a

[36] Olson admits in his discussion of labor unions that in the late nineteenth-century labor movement, when the number of American workers was less than half a million, collective action was achieved; but after the turn of the century, when the number grew to more than 2 million, large-scale collective action became increasingly difficult; *Logic of Collective Action*, 1965, at 77, discussion is between 76 and 91.
[37] I am grateful to Tuck's *Free Riding*, 2008, for making this clear in the case of how Olson is largely misread to anticipate the PD analysis of collective action.
[38] Michael Bacharach, *Beyond Individual Choice* (Princeton, NJ: Princeton University Press, 2006).
[39] Olson, *Logic of Collective Action*, 1965, 44.
[40] Subtitle drawn from Hardin, *Collective Action*, 1982, 6.
[41] Olson, *Logic of Collective Action*, 1965, 45.

normal and rational manner of conduct. Major theorists – Thomas Schelling, Michael Taylor, Russell Hardin, Philip Pettit, and Dennis Mueller, to name a few – have found this analysis compelling.[42] Such accounts of the tragedy of the commons, free riding, and the failure of collective action accomplish two goals. First, they normalize the rationality of pursuing self-interest in opposition to others' interests under all circumstances. Second, they subsume earlier theoretical findings under this generalized multi-person Prisoner's Dilemma game. This approach is easy to teach and easy to digest. The PD logic looks unassailable in all contexts: regardless of what other individuals do, I am consistently better off defecting. Once this rationale has been thoroughly vindicated in a two-person situation, it must extend through the levels of small, medium, and very large-sized groups with the same outcome: self-interest leads to mutual ruin.

Russell Hardin helped lead the way in promoting this cynical, as he refers to it, view of individuals' inability to collaborate in assemblies of all sizes.[43] In *Collective Action*, Hardin observes,

> The problem of the back of the invisible hand has become known as the problem of collective action, or the Prisoner's Dilemma, or the free rider problem, or the condition of common fate, depending on the context or discipline in which it is used. This multiplicity of terms reveals a failure to generalize the nature of the problem. It has been generalized only recently, most notably ... in the game-theoretic Prisoner's Dilemma ... and in Mancur Olson's *The Logic of Collective Action*.[44]

Hardin confirms that he and other mainstream rational choice theorists see the Prisoner's Dilemma as a general case with myriad applications. He goes on to envelop Olson's diagnosis of the failure of collective action and to catalogue the wider variety of circumstances to which the PD can be applied. However, the fact that the Prisoner's Dilemma model does not even pertain to Olson's theory of collective action signifies that this perverse game gained a momentum of its own and provided the opportunity to reinterpret the significance of social actions that even in the mid-twentieth century were permeated with much different meanings.

Recasting the tragedy of the global commons in light of any single individual's despair over changing anything, for better or worse, rather than hope to free ride, profoundly alters the diagnosis and potential remedy of this problem. Olson directly addresses this point, noting that his theory of large-scale collective action "does *not* necessarily assume the selfish, profit-maximizing behavior that economists usually find in the marketplace." He clarifies, "The concept of the large or

---

[42] Philip Pettit has contributed many articles about the Prisoner's Dilemma including in his "Preserving the Prisoner's Dilemma", *Synthese* (1986) 86, 181–184; "*Virtus Normativa*: Rational Choice Perspectives," *Rules, Reasons, and Norms*, ed. by Philip Pettit (Oxford: Clarendon Press, 2002); "The Prisoner's Dilemma and Social Theory: An Overview of Some Issues," *Australian Journal of Political Science* (1985) 20:1, 1–11; and "Free Riding and Foul Dealing," *Journal of Philosophy* (1986) 83:7, 361–379.

[43] Hardin, *Collective Action*, 1982, xv.

[44] Ibid., 7.

latent group offered here holds true whether behavior is selfish or unselfish, so long as it is strictly speaking 'rational.'"[45] Olson's entire thesis revolves around the breakdown of instrumental rationality in enormous groups. The Prisoner's Dilemma logic represents the opposite because "each criminal's actions have a real and appreciable effect on the other criminal's utility."[46] Especially in our contemporary era of looming collective tragedies, it behooves us to grasp the import of Olson's message and to reappraise our seven-decade infatuation with strategic rationality and the Prisoner's Dilemma impasse. Not only is the paralysis of being unable to causally impact the outcome in large groups wholly distinct from the PD prescription to advance by negatively impacting others, but Olson's investigation also suggests that PD logic is not necessarily relevant in any setting. Olson discusses how groups whose members can impact one another have various means to successfully collaborate, including moral scruples and solidarity.[47] These two acute points puncture the hot air balloon of the generalized Prisoner's Dilemma approach to the tragedy of the commons, free riding, and collective action.

Elinor Ostrom's *Governing the Commons* (1990) critiqued the Prisoner's Dilemma industry by providing numerous case studies in which agents overcame the challenge of cooperating, and contributed to her receipt of the Nobel Memorial Prize in economic sciences in 2009.[48] The task at hand is to move beyond examining individualized empirical examples and to challenge the entire apparatus of strategic rationality as an exhaustive account of coherent action and the Prisoner's Dilemma as a relevant model for tackling cooperation. An equally important task is to determine why few researchers have noticed the wide berth between the game theoretic neoliberal understanding of political economy and the modern liberal approach preceding it. There are of course prominent spokespersons who offer criticisms of aspects of the strategic rationality paradigm.[49] Still, not only does game theory, as a well-established pedagogy, hold its place in graduate and professional courses, but its findings, from the tragedy of the commons and rationality of free riding to the irrationality of voting, also continue to be undergraduate staples and important tools in

---

[45] Olson, *Logic of Collective Action*, 1965, 64.

[46] Tuck, *Free Riding*, 2008, 24.

[47] We might break down Olson's suggestions scattered throughout *Collective Action*, 1965, as commitment, other-regarding action, and solidarity (e.g., team reasoning, see Michael Bacharach, Natalie Gold, and Robert Sugden, *Beyond Individual Choice: Teams and Frames in Game Theory* (Princeton, NJ: Princeton University Press, 2006), and mechanisms consistent with strategic rationality, such as reputation.

[48] See especially Ostrom, *Governing the Commons*, 1990, 1–15.

[49] Tuck, *Free Riding*, 2008; Jon Elster, *Ulysses and the Sirens: Studies in Rationality and Irrationality* (Cambridge: Cambridge University Press, 1998); Amartya Sen, "Rational Fools: A Critique of the Behavioral Foundations of Economic Theory," *Philosophy and Public Affairs* (1977) 6:4, 317–344; Michael Taylor, *Rationality and the Ideology of Disconnection* (Cambridge: Cambridge University Press, 2006).

economists' toolkit.[50] Why has this single approach to rendering purposive agency intelligible, with its exclusive focus on ends rather than processes, single criterion mensuration, individualistic maximization, and narrow self-interest, prevailed?

To slightly rephrase the question, where did the "backside" of the invisible hand come from? Either Adam Smith failed to notice it in Enlightenment Europe or rational choice theorists invented it during the Cold War. Rational choice theorists tend to refer to any application of coercive threats to incentivize agents to conform to the law to solve the ubiquitous Prisoner's Dilemma as the backside of the invisible hand.[51] However, whereas Smith counted on a mutually beneficial social contract based on the impartial spectator, self-evident property rights, and individuals' incentives to profit through exchange rather than theft, the neoliberal society relies on heavy-handed enforcement strategies that enable some individuals to disproportionately appropriate the contributions and endowments of others as their means of success.[52]

Olson's *Logic of Collective Action*, which relies on standard neoclassical economic tools and not the strategic rationality of game theory, provides a helpful vantage point from which to survey the intimate details of the stark inversion from mutually beneficial classical liberalism to mutually debilitating neoliberalism. The rational choice extension of the Prisoner's Dilemma from two individuals to 10,000, 1 million, or 7 billion provides the key to understanding this transposed appraisal of meaningful action. Where the two-person exchange is modeled as a PD because each individual prefers to leave with all the goods rather than settle on an amicable exchange, and this single logic of action is amplified to any size collection of individuals, coercive force must always be introduced to produce social order. Indeed, force can be used to change the default of bargaining so that individuals will settle for less than they would were normative bargaining via the no harm default, consent and voluntary compliance practiced instead.

Because Adam Smith also relies on police, specifically to protect those with property from those with less, rational choice does not appear to deviate from classical liberalism at first glance. However, Smith did not advocate leveraging force as a means to dispossess less well-off individuals. Nor did he propose the "externality" approach to wealth production. The neoliberal way to generate profit is to externalize costs and internalize gains. According to Russell Hardin and the standard rational choice analysis, the backside of the invisible hand requires the introduction of coercive sanctions to get individuals to cooperate, and thereby avoid the Prisoner's Dilemma suboptimal outcome.

---

[50] On this latter topic, see Tuck, *Free Riding*, 2008, 30–62; see also Geoffrey Brennan and Philip Pettit, for whom sanctioning can be accomplished by by-standers; "Hands Invisible and Intangible," *Synthese* (1993), 94, 191–225; *Stern Review*, 2006, chap. 21.

[51] This is Hardin's interpretation, *Collective Action*, 1982, 6–9.

[52] Russell Hardin, "Does Might Make Right," *NOMOS XXIX* (New York: New York University Press, 1987), 201–217.

A clear difference between modern liberalism and postmodern neoliberalism is the former's recognition that the "competitive market – the epitome of private institutions – is itself a public good."[53] In the free market, as with any case of collective action, no single actor can causally affect another actor because of the shared fact of negligibility. Under perfect competition, there can be no price setters, whether producers or consumers. In opposition to the standard PD logic, the free market must be protected from firms' desire to form cartels, therefore jointly controlling the price of goods to achieve a higher profit rate than that available to consumers through perfect competition.[54] Just as with trade unions, a legal structure must be imposed to regulate collaboration: force must be used to protect free market competition from actors' predilection to collude.[55] Reading between Olson's lines, the moral of his study is that just as firms must face perfect competition, so, too, must laborers, because only in this way will the free market be preserved. The Prisoner's Dilemma analysis of market competition also leads us to conclude that no government is necessary to prevent firms from colluding to artificially maintain high prices. Where neoliberal PD analyses identify little role for governmental intervention to maintain free competition, for Smith, the state must maintain the appropriate conditions to uphold the system of natural liberty. Popular opinion seems to applaud firms, such as Walmart and Amazon, with sufficient market share to set prices in entire sectors of products, with no concern to insist that government maintain the conditions of perfect competition.

The rational choice normalization of the rationality of free riding further demonstrates the rupture between classical liberalism and postmodern liberalism. In the Prisoner's Dilemma, each actor most prefers for the other to cooperate while personally defecting, thereby reaping the reward of the other's contribution and leaving that other individual worse off. The straightforward Prisoner's Dilemma model of two firms that each seek to be the sole provider of a good at a higher price also does not accurately capture the choice problem. This situation is more accurately reflected as a sequenced decision problem in which one or several firms make a product for a price that balances supply and demand.[56] A new firm then has the choice of competing at that price level or seeking an advantage by selling units at a lower price, which then forces the other firms to do the same. The concern with collusion is that if there are only a few companies in the industry, they may choose to keep the price fixed at a higher level by confining production. In free competition, however, there will be a sufficient number of actors such that no single producer (or buyer, for that matter) can alter the market price.

---

[53] Ostrom, *Governing the Commons*, 1990, 15.

[54] Tuck, *Free Riding*, 2008, 172–204.

[55] Ibid., 160–162; Tuck's point is that collaboration can be considered rational, and there certainly was no consensus that it was irrational prior to the mid to late twentieth century.

[56] For discussion, see Tuck, *Free Riding*, 2008, 168.

This analysis of the free market, which stresses the need to legally enforce firms' individual maximization when there are lower numbers of suppliers, does not entertain the possibility of either free riding or gaining leeway by imposing costs on other actors. The private property regime is built to facilitate perfect competition, which serves the function of increasing productivity over all sectors until, over the long run, profit rates equalize across industries. This equalization of profit margins serves the function of lowering the cost of subsistence and providing goods and services in areas where there is unmet demand, and thus higher profitability.[57] In Smith's *Wealth of Nations*, private property, a minimal state, and legally enforced perfect competition lead to general prosperity. Actors reap rewards according to their prudence and hard work. Productivity is compensated by profit, and the fact that individuals profit from their own efforts is the key to mutual prosperity. Individuals will seek opportunities for gain and thus will be drawn to economic activities that yield supply where demand permits a higher profit margin. The Prisoner's Dilemma model of basic exchange is therefore irrelevant to modern capitalism because, in principle, individuals profit from their productivity, and not from imposing costs on other actors through malfeasance, shoddy products, or exporting costs.

Neoliberalism dismisses self-evident property rights and determines that even if property rights were unanimously agreed upon, everyone would still seek first and foremost to cheat others. Adam Smith's theory of wealth production tethers the individuals' initiative to better their condition to the overall profit rate available in the economy. Neoliberalism offers individuals three means to accumulate wealth at a rate above average: (1) maintaining monopolistic control over a product, application, or algorithm; (2) manipulating risk; and (3) imposing costs on others.[58] None of these was the sanctioned engine for classical liberal growth. However, the widespread, if reluctant, endorsement of neoliberalism may resonate with a worry that overall profit rates of the sort Adam Smith envisioned may be insufficient for all individuals to achieve their economic aspirations. Lester Thurow's *Zero Sum Society* (1980) articulates the worry that auspicious rates of economic growth permitting positive-sum gain are no longer feasible, and that profiting from undermining others offers the only currently available means to profit.[59] Game theory's formation to reflect the laws of thermodynamics supports this hypothesis, and it describes action consistent with relentless competition strictly limited by scarcity.

---

[57] This analysis depends on all actors having the economic wherewithal to register demand.

[58] The 2007 financial crisis resulted from the displacement of risk from profitable firms to their clients; Lawrence Summers's World Bank memorandum of 1993 articulates clear arguments for displacing the costs of developed countries' advanced living standards onto inhabitants of the developing world; memorandum is in Daniel Hausman and Michael McPherson, *Economic Analysis, Moral Philosophy and Public Policy* (Cambridge: Cambridge University Press, 2006), 11–13.

[59] Lester Thurow, *Zero-Sum Society: Distribution and the Possibilities for Economic Change* (New York: Basic Books, 1980).

242   *Government*

CONCLUSION

This chapter compared Mancur Olson's *Logic of Collective Action* to the Prisoner's Dilemma analysis of collective action failure to argue that the game theoretic approach is both unprecedented and counterintuitive. Throughout modernity, debate among economists and political theorists has been over the proper role of the state: Should it be restricted to sustaining property rights through a system of commutative justice backed by police enforcement, or should it play a redistributive role in ensuring citizens' economic security as a necessary component of securing their self-preservation? Throughout modernity, theorists accepted that if the government could offer public goods in the form of security and infrastructure that did not depend on redistribution any more than providing police and justice did, then this role for government was not in and of itself problematic.[60] Olson weighed into this debate acknowledging the rationality of cooperating with small or medium ventures in which what one contributed in costs was outmatched by what one gained in distributions. However, he argued that for very large groups approximating perfect market competition, individuals could not be instrumentally motivated to contribute because, by definition, no one's single contribution could make any appreciable difference on either the collective outcome or individual shares.

Olson argued that this lack of causal impact in large-scale ventures necessarily impedes all actors, those motivated by moral or other-regarding considerations as well as those motivated by narrow self-interest. Game theorists have assumed that two-person cooperation involving scarce resources is a Prisoner's Dilemma game and can be extended from two individuals to a large number of individuals, each of whom most prefers for every other agent to cooperate while making an exception for himself. Yet this PD analysis of collective action and free riding misses the crux of Olson's argument. For Olson, group size matters. According to his analysis, the reason very large groups are bound to have a difficult time securing cooperation, even if those actors are motivated by side constraints or concerns for others' well-being, is individuals' perception that their efforts make no causal impact on whether the collective undertaking succeeds or fails. Hence, Olson's *Logic of Collective Action* serves to draw attention to collective action failure as a result of the logic of negligibility rather than everyone's first preference to free ride. Olson does not suggest that the Prisoner's Dilemma model of collective action is necessarily inaccurate because if all actors did most prefer to defect from contributing yet collected gain from others' efforts, then collective action must fail. His particular genius though lies in showing us how the fact of causal negligibility in large group settings will by itself provide a significant challenge to successful participation.

[60] For example, Knut Wicksell, "A New Principle of Just Taxation," in *Classics in the Theory of Public Finance*, ed. by Richard A. Musgrave and Alan T. Peacock (London: Macmillan, 1958), 72–119. Robert Nozick permits supply of the public good of justice independent from individuals' ability to pay, *Anarchy, State and Utopia*, 1974, 26–28.

We can surmise the following points:

1. If the Prisoner's Dilemma model, and the assumptions on which it relies to resist a satisfactory mutually beneficial solution (only outcomes with salient fungible properties matter to agents, gratuitous altruism contradicts instrumental rationality, and joint maximization and commitment are incoherent) accurately characterize two-person exchange and common resource dilemmas, then it would seem accurate to conclude by implication that individuals with these preferences and predicaments will end up in a large-scale collective action referred to as the tragedy of the commons.

2. However, whether or not the PD is relevant to two-person interactions, its causal logic holding that individuals acquire gain by offsetting costs onto others, as is the case if one individual suckers the other, is not the key to understanding large-scale collective action failure of the sort analyzed by Mancur Olson.

3. Drawing attention to the long-standing field-wide centrality of the Prisoner's Dilemma is not meant to diminish its ability to serve as a useful analytic model for understanding some dimensions of shared resource dilemmas. Analyzing economic problems in detail in terms of their instrumental impact is a worthwhile exercise, and noncooperative game theory may play a constructive role. However, it is possible that the idealized assumptions necessary for operationalizing the models become interpreted as strategic imperatives on par with the Kantian means-ends hypothetical imperative, and yet exceed its restricted status by being treated as the sole standard for rational action. In this case either the demands of modeling have overstepped the boundary of playing a descriptive role to serving as a normative guideline for individual decision making and a prescriptive role for policy analysis, or theorists are projecting their own attitudes onto those being modeled.[61]

4. Therefore, in teaching both the Prisoner's Dilemma model of two or small numbers of actors and the large-scale collective action problem, thoroughness requires clarifying the assumptions structuring strategic rationality, and their potential role in fostering neoliberal subjectivity and designing neoliberal institutions. Large-scale collective action and public goods, like perfect competition in the free market, fall outside the PD model because they are characterized by every individual's causally negligible role in making any appreciative difference to himself or anyone else by contributing. There is no instrumental reason to contribute to a collective action, whether or not an individual has the disposition consistent with strategic rationality to free ride.

---

[61] On Kant's hypothetical imperative, which exists alongside his categorical imperative, see Immanuel Kant, *Groundwork of the Metaphysic of Morals*, ed. and trans, by John Ladd (New York: Harper and Row, 1964).

5. Distinguishing between the motivational complex and challenges leading to a suboptimal outcome in a Prisoner's Dilemma and in a vast collective venture is crucial for satisfactorily addressing this predicament and tragedy. The neoliberal action consistent with noncooperative game theory may become increasingly normalized if the tragedy of the commons is viewed as the outcome of a population of hopeful free riders each of whom most prefer to be the lone defector. Moreover possibly the tragedy of the commons may at least in part be an outcome of resignation and frustration that not even the usual avenues of joint effort, commitment, and other-regarding action are sufficient to make a noticeable difference in the globally realized outcome.

PART III

EVOLUTION

Introduction

> Game theory is a general lexicon that applies to all life forms. Strategic interaction neatly separates living from nonliving entities and defines life itself. Strategic interaction is the sole concept commonly used in the analysis of living systems that has no counterpart in physics or chemistry.
>
> Game theory provides the conceptual and procedural tools for studying social interaction, including the characteristics of the players, the rules of the game, the informational structure, and the payoffs associated with particular strategic interactions. The various behavioral disciplines (economics, psychology, sociology, politics, anthropology, and biology) are currently based on distinct principles and rely on distinct types of data. However, game theory fosters a unified analytic framework available to *all* the behavioral disciplines. This facilitates cross-disciplinary information exchange that may eventually culminate in a degree of unity within the behavioral sciences now enjoyed only by the natural sciences . . . . Moreover, because behavioral game-theoretic predictions can be systematically tested, the results can be replicated by different laboratories. This turns social science into true science.
>
> Herbert Gintis, 2009[1]

Prisoners of Reason argues that neoliberal capitalism and governance reflect a new theoretical rationale that was initially developed and applied to prosecute the Cold War through exercising credible military threats, many of unimaginably devastating nuclear destruction. Policy analysts and social scientists came to accept strategic rationality as the epitome of human reasoning, thus rendering obsolete formerly accepted perspectives on intelligible action. Most prominently, in its orthodox operationalized form, strategic rationality only permits agents to pursue outcomes in competition with others over scarce, instrumentally salient resources. This book's primary goal is to render clearly how this

---

[1] Herbert Gintis, *The Bounds of Reason* (Oxford University Press, 2009), 45.

new understanding of rational action sustains neoliberal markets and government in direct contradiction to classical liberalism.

Noncooperative game theory adopts a stance toward action that accepts actors' preferences as given; models action choices descriptively, normatively, or prescriptively depending on the context; and seeks solutions or equilibrium outcomes that will obtain assuming that actors maximize expected utility with varying attitudes toward risk. In John von Neumann and Oskar Morgenstern's original articulation, this science of choice claims to be an exhaustive treatment of rationality in all decisions involving more than one individual. Therefore, it offers to agents the imperative to comply with its principles of choice, tacitly linked to survival and success, or to suffer the fate of irrationality and loss. Once this behavioral protocol is accepted, then the alternative logics of action, including principled commitment, shared intention, and joint maximization, as well as gratuitous other-regarding acts of beneficence become unintelligible and are lumped together with irrational conduct such as playing blackjack as though there were ten aces in a standard deck of cards.

Classical liberalism, which refers to the modern-era writings of Thomas Hobbes, John Locke, Adam Smith, and Immanuel Kant, and extends forward in time to include the more recent texts of Robert Nozick and John Rawls, reflects profound thinking about how to achieve a mutually agreeable and beneficial social order notwithstanding the fact that the social world evinces dissensus and conflict.[2] Yet these theorists converged on holding that individuals recognize their own and others' right to corporeal existence, to the property extending their capacity to act in the world, and to the goods and services owed to them through contractually bound exchanges. The no-harm principle, which unites the right to noninterference with the negative virtue or perfect duty to refrain from interfering in others' right to exist, is the self-approved Archimedean reference point that enables classical liberal society to emerge from anarchy. The minimal state remains small because it reinforces people's commitment to noninterference in private affairs with its power of enforcement, and it remains limited because it builds up from individuals' voluntary respect of others' right to not be harmed. Living in a peaceful society depends on actors' not injuring other people, not stealing property, and complying with commitments voluntarily incurred.

Neoliberal political theory invalidates the logics of action-sustaining classical liberalism. The modern classical liberal system of prudential responsibility and allegiance to the duly constituted rule of law acknowledges realms of value above and beyond fungible resources, security of personhood and possession being the most obvious, and the wherewithal to choose manners of conduct that are consistent with living in a civil society, rather than in a perpetual state of

---

[2] In identifying and addressing a constellation of ideas, I have pointed to specific theorists, which is not to underestimate or ignore the value of recognizing "the dizzying variety of ways" in which the term "liberalism" has been used in political thought and social science; see Duncan Bell, "What Is Liberalism?," *Political Theory* (2014) 42:6, 682–715, quote from abstract, 682.

nature. Neoliberalism, by contrast, offers as imperative the single logic of strategic action that condones unbounded utility maximization that reflects salient instrumentally obvious fungible rewards, and might produce order if actors manage to converge on a mutual-best-reply equilibrium that easily could be suboptimal. No actor concedes to any other actor his or her fundamental right to exist. Rather, as in offensive neorealism, actors assume that personal survival will probably be at cross-purposes with others', and that self-defense through the augmentation of one's resources is a natural imperative. In this world of finite resources that all actors alike seek in competition with one another, the Prisoner's Dilemma is prominent because every actor prefers to have all the goods, leaving everyone else with none. Unlike in classical liberalism, actors fail to respect others' corporeal persons and their private belongings as an prerogative of personhood and agency. Instead, the neoliberal actor has the foremost preference of securing the highest payoffs possible irrespective of the impact on others and, moreover, views any encounter that could result in exchange as an opportunity to issue credible threats so that in any eventual unstable and precarious cooperative outcome, the other will settle for the least amount possible. Neoliberalism builds order atop what game theorists view to be the fundamental building blocks of order: actors' unbounded pursuit of expected utility that tracks ontologically present value as a prior condition for establishing patterns of conduct.

Part I examined how game theory was developed to model strategic combat over finite resources and threats of harm with relevance to rational deterrence. The nuclear security dilemma resulted in the logical vindication of the officially declared policy of the United States to prepare to fight and win a nuclear war among superpowers, notwithstanding the fact that mutual assured destruction (MAD) is and has been an inescapable existential condition. Whereas, at first, the weakness of minimum deterrence through MAD had been the incredulity of issuing immoral promises to destroy innocent civilians after deterrence had failed, political theorists' acceptance of game theory beyond the realm of military strategy, already prefigured in von Neumann and Morgenstern's *Theory of Games and Economic Behavior*, impugned the reasonableness of the institution of moral promising across the board for being inconsistent with strategic rationality.[3]

Part II explored the contours of neoliberal political theory that differ categorically from classical liberalism in viewing the sovereign's power of enforcement as the necessary and sufficient condition to govern, accepting that the social contract is built up out of coercive instead of normative bargaining, rejecting fair play and commitment in favor of incentives and threats, eviscerating unanimity and consent of motivational significance, and understanding the failure of collective action to be the limiting case of the two-person Prisoner's

---

[3] John von Neumann and Oskar Morgenstern, *Theory of Games and Economic Behavior* (Princeton, NJ: Princeton University Press, 1944).

Dilemma in which each actor foremost seeks to sucker the other. Neoliberal theory identifies de facto power and possession as individuals' means to action in the world, excluding express or tacit consent to abide by agreements and rules of conduct they themselves endorse. It accepts that the opening point for negotiation in either dyadic exchange or the social contract is the application of credible threats. It views no agreement to terms as sacrosanct because such agreements are perpetually subject to renegotiation for better terms and will be reneged on whenever it is possible to do so by suckering others with impunity.

It is not astonishing that some agents might view their range of options and the meaning of their actions in these terms. Doubtless throughout history, pure strategic rationality without regard for others has reflected coherent plans of action for some people like Thomas Hobbes's Foole and David Hume's knave. However, what is new in neoliberalism and its attendant certification of the gold standard of rational agency codified in orthodox noncooperative game theory is the view that actors have no choice but to embody such rules of engagement as a condition of survival. This view is supported by the branch of research discussed in Part III, "Evolution": evolutionary game theory.

Part III discusses the role that game theory has played in analyzing evolution. Evolutionary games and replicator dynamics represent a field of game theoretic analysis. Yet it is little known that Richard Dawkins's bold argument in his "selfish gene" theory was derived from the application of noncooperative game theory to the study of biological evolution. Dawkins himself acknowledges this point. He writes, "I have a hunch that we may come to look back on the invention of the ESS [evolutionary stability strategy] concept as one of the most important advances in evolutionary theory since Darwin."[4] The ESS concept is defined using game theory.[5] Dawkins goes on to observe that "the ESS concept ... is applicable wherever we find conflict of interest, and that means almost everywhere."[6] In essence, the assumption of unremitting individualism built into noncooperative game theory resulted in the deduction that selfishness is the necessary character of all living forms. Evolutionary game theorists have not been shy in applying their conclusions to social theory.[7]

The two chapters in this part honor the burgeoning scholarly effort to determine whether the forces of individually targeted evolutionary natural selection permit the development of other-regarding or altruistic behavior.[8]

---

[4] Richard Dawkins, *The Selfish Gene* (Oxford: Oxford University Press, 2009), 84.

[5] See Shaun Hargreaves Heap and Yanis Varoufakis, *Game Theory*, 2nd ed. (New York: Routledge, 2004), 219.

[6] Dawkins, *Selfish Gene*, 2009, 84.

[7] Robert Trivers's introductory statement to the first edition of Dawkins's *Selfish Gene* in 1976 directly refers to "Darwinian Social Theory," xx.

[8] This body of research attempts to determine whether evolution can sustain modes of action that defy narrow individualistic maximization; for an engagement in this discussion, see S. M. Amadae and Daniel Lempert, "Long-Term Viability of Team Reasoning," *Journal of Economic Methodology* (published online May 11, 2015), doi: 10.1080/1350178X.2015.1024880.

Prisoners of Reason concludes by demonstrating how Richard Dawkins and Robert Axelrod have applied the laws of noncooperative game theory to suggest that coherent purposive action on all levels, from subhuman life forms, to individuals, and also to firms and nations, must comply with its dictates or fail to be successful as determined by an external preexisting metric. This latter body of research relies on orthodox game theory because it accepts single criterion tangible rewards and the logic of consequences.

Game theory is celebrated as a means to unify the social and behavioral sciences, and it provides a single methodology to distinct levels of organization.[9] Dawkins, who built analytic models using noncooperative game theory, subsequently applies the results of these models for making claims about human nature and offering public policy recommendations. Axelrod, who played a significant role in establishing the school of neoliberal international relations theory, applies his argument across the board from evolutionary biology to individuals in society, to nation states navigating anarchy. Both Dawkins and Axelrod place great credence on repeating games to demonstrate that the semblance of cooperation in nature and society can be explained by reciprocal altruism. Their central idea is that agents will cooperate when it is directly in their interest along the Tit for Tat principle of "you scratch my back, and I'll scratch yours."

---

[9] Gintis, *Bounds of Reason*, 2009; Ken Binmore, *Game Theory and the Social Contract*, 2 vols., *Just Playing* (Cambridge, MA: MIT Press, 1994) and *Playing Fair* (Cambridge, MA: MIT Press, 1998); see also Brian Skyrms, *Evolution of the Social Contract*, 2nd ed. (Cambridge: Cambridge University Press, 2014).

10

Selfish Gene

> Genes are competing directly with their alleles for survival, since the alleles in the gene pool are rivals for their slot on the chromosomes of future generations. Any gene that behaves in such a way as to increase its own survival chances in the gene pool at the expense of its alleles will, by definition, tautologously, tend to survive. The gene is the basic unit of selfishness.
>
> Richard Dawkins, 2009[1]

Richard Dawkins's stated mission is to reject that any organism could have evolved to be altruistic, or that group selection is a viable evolutionary mechanism.[2] In *The Selfish Gene*, originally published in 1976, the British ethologist and evolutionary biologist argues that the process of biological evolution ensures that all organisms, on an individual basis, are inherently selfish. Dawkins fully realizes that his research could have far-reaching implications for our understanding of life. Moreover, he does not shy away from articulating the full ramifications of his vision for human beings, culture, and society. He is open about the repercussions of his argument:

Like successful Chicago gangsters, our genes have survived, in some cases for millions of years, in a highly competitive world. This entitles us to expect certain qualities in our genes. I shall argue that a predominant quality to be expected in a successful gene is ruthless selfishness. This gene selfishness will usually give rise to selfishness in individual behavior.[3]

Although quick to disclaim that he is not "advocating selfishness as a principle by which we should live," Dawkins still conveys a consistent message that no biological agent is naturally selfless.[4] He boldly states that "much as we might

---

[1] Richard Dawkins, *The Selfish Gene* (Oxford: Oxford University Press, [1989]/2009), 33, 36.
[2] See ibid., 1–11.
[3] Ibid., [1976]/2009, 2.
[4] Ibid., 267, added in final edition.

wish to believe otherwise, universal love and the welfare of the species as a whole are concepts that simply do not make evolutionary sense."[5] Aware he was putting forth a vision of the nature of life itself, perhaps equivalent to that of epic proportions, he recalls, working alongside the British theoretical evolutionary biologist John Maynard Smith and the American sociobiologist Robert Trivers, that 1975 "was one of those mysterious periods in which new ideas are hovering in the air."[6] Although not directly acknowledged, the new ideas were those of game theory.[7] Dawkins views his work as part of a movement that offers "a change of vision . . . [that can] achieve something loftier than a theory." Specifically, "It can usher in a whole climate of thinking . . . [a]nd a new way of seeing."[8] Dawkins notes that he "wrote *The Selfish Gene* in something resembling a fever of excitement."[9]

It becomes clear in assessing Dawkins's expression of evolutionary game theory that its rendition of "Darwinian social theory" has had the effect of seamlessly linking human society with primordial evolution.[10] Dawkins himself reports that some of his readers rue the day they read *The Selfish Gene* because it led them to adopt nihilistic pessimism.[11] He does not flinch from the "tendency to shoot the messenger . . . displayed by other critics who have objected to what they see as the disagreeable social, political or economic implications of *The Selfish Gene*."[12] Dawkins's derivation of the self-sustaining property of narrow self-interest is a consequence of his reliance on the game theoretic Nash equilibrium of mutual best reply. The goal of this chapter is to understand the change in vision Dawkins introduced, and its close relationship to noncooperative game theory and its central logical problematic, the Prisoner's Dilemma. It is straightforward to observe that Dawkins's biological agents seem comparable to the individuals deemed to populate capitalism: "Each gene is seen as pursuing its own self-interested agenda against the background of the other genes in the

[5] Ibid., 2, first edition, 1976.
[6] Ibid., xvii, preface to second edition, 1989.
[7] John Maynard Smith was one of the pioneers of evolutionary game theory; see his "Evolution and the Theory of Games: In Situations Characterized by Conflict of Interest, the Best Strategy to Adopt Depends on What Others Are Doing," *American Scientist* (1976) 64:1, 41–45; "The Theory of Games and the Evolution of Animal Conflicts," *Journal of Theoretical Biology* (1974) 47:1, 209–221; "Optimization Theory in Evolution," *Annual Review of Ecology and Systematics* (1978) 9, 31–56; *Evolution and the Theory of Games* (New York: Cambridge University Press, 1982). See also Robert Trivers, "The Evolution of Reciprocal Altruism," *Quarterly Review of Biology* (1971) 46, 35–37.
[8] Dawkins, *The Selfish Gene*, preface to second edition, 1989, xvi.
[9] Ibid., xvii.
[10] Quote from Robert L. Trivers, foreword to the first edition of *Selfish Gene*, 1976, in 2009 version, xx.
[11] Dawkins reports this in the "Introduction to the 30th Anniversary Edition," *The Selfish Gene*, 2009, xiii.
[12] Ibid.; note that evolutionary game theory is widely agreed to have implications for social theory; see Shaun Hargreaves Heap and Yanis Varoufakis, "Evolutionary Games: Evolution, Games, and Social Theory," *Game Theory: A Critical Text*, 2nd ed. (New York: Routledge, 2004), 211–266.

gene pool."[13] As in noncooperative game theory, every individual maximizes gain of a scarce fungible resource in competition with other like actors without constraint.

DAWKINS'S SELFISH GENE THEORY

At first glance, it would not seem that evolutionary biology would have much to say to economists because in the former, organisms are proposed to optimize an objective fitness function while in the latter, agents are believed to optimize subjective utility functions. This alternative stress on the objective versus subjective nature of value is a tension between these two fields that now exist under the same umbrella of game theory, although I have argued throughout *Prisoners of Reason* that value is treated as though it has primordial ontological status throughout most orthodox game theory. Evolutionary game theory is predicated on the view that the payoffs to games reflect objective conditions that must be met for individuals to survive and propagate their genetic programming into the successive generation of organisms.[14] What follows from this is that nature acts as a consistent scorekeeper that selects for the rational behavior of individuals who must maximize objective fitness value as a condition of their survival.

Dawkins presents a purely deductive model supported by circumstantial evidence providing anecdotes of organisms' competitions for survival. He seeks to prove that natural selection does not permit agents to be selfless or to be selected on the basis of groups that could be composed of selfless individuals.[15] The deductive model is based on identifying individual units, or actors, that optimize their acquirement of a value in competition with others. By Dawkins's description, "We are survival machines – robot vehicles blindly programmed to preserve the selfish molecules known as genes."[16] If we accept that evolutionary natural selection targets individuals to determine comparative fitness in the competition for survival, then Dawkins concludes, "it seems to follow that anything that has evolved by natural selection should be selfish."[17]

---

[13] Dawkins, "Introduction to the 30th Anniversary Edition," 2009, ix. Note that the science of evolution, often referred to as Darwinism, has dovetailed with theories of social progress or development from the time of Thomas R. Malthus, *An Essay on the Principle of Population*, 1798. See also Joseph Schumpeter, who coined the now popular neoliberal phrase "creative destruction" to refer to capitalist production; "Capitalism," *Encyclopaedia Britannica* (London: Encyclopaedia Britannica, 1946).

[14] Philip Pettit provides sophisticated discussion of the analogous structure of explanation operating at the level of genes, physical bodies, individuals, and social groups; see "Functional Explanation and Virtual Selection," in his *Rules, Reasons, and Norms* (Oxford: Clarendon Press, 2002), 245–256.

[15] For discussion of the pressures of individualistic selection toward narrow self-interest, and the attempt to investigate the possibility of group selection, see Hargreaves Heap and Varoufakis, *Game Theory*, 2004, 248–249.

[16] Dawkins, "Preface to the First Edition," 1976/2009, xxi.

[17] Dawkins, *Selfish Gene*, 1976/2009, 4.

A profound question follows: What exactly is the identity of the individual who evolved through natural section to be selfish? In other words, given that the application of game theory requires identifying a player who optimizes a function, what entity or agent plays this role in evolutionary biology? We might consider an entire species, or kinship groups in a species, or individual members of a species, or perhaps an individual's entire chromosomal DNA code, or perhaps a single chromosome, or a bundle of genes, or a single gene. Dawkins answers the question directly:

We saw that selfishness is to be expected in any entity that deserves the title of a basic unit of natural selection. We saw that some people regard the species as the unit of natural selection, others the population or group within the species, and yet others the individual. I said that I preferred to think of the gene as the fundamental unit of natural selection, and therefore the fundamental unit of self-interest.[18]

Admitting his analytic derivation, the evolutionary theorist observes, "What I have now done is to *define* the gene in such a way that I cannot really help being right!"[19] In this passage, Dawkins alludes to the correlation between identifying the level at which the selection process works and simultaneously then specifying the unitary actor on which selection occurs. Among the many options available, Dawkins concludes, from analytic argumentation, that evolutionary selection operates on genes. Dawkins's gene must behave as an entity that can be precisely duplicated so that it can exist across time in multiple generations, perhaps lasting for hundreds of millions of years.[20] It is most important for Dawkins that genes, or replicators, that play the game of life according to the rules of noncooperative game theory "are the game's pure strategies which are assumed to be copied (from parent to child or from originator to imitator) without error."[21] Genes function as strategies embodied in biological material, and they therefore have a connotation of immortality. Even though evolution is perceived to be about adaptation and transformation in time, Dawkins's genes are identical to themselves and do not evolve or adapt.

Dawkins's individuals, then, are bits of genetic code that propagate through time. It is difficult to understand what "selfishness" can mean if not proposed at the level of individual humans or animals.[22] Genes are decidedly not conscious. They group together, apparently in cooperation, to comprise bodies: "In the fierce competition for scarce resources, in the relentless struggle to eat other survival machines, and to avoid being eaten, there must have been a premium on

[18] Ibid., 33.
[19] Ibid.
[20] Ibid., 15.
[21] Hargreaves Heap and Varoufakis, note Dawkins's coinage of the term "replicator," and its conceptual meaning defined in terms of game theory's pure strategies specifying acts in games (as opposed to mixed strategies that essentially roll dice to choose an action), *Game Theory*, 2004, 221.
[22] For commentary, see Hargreaves Heap and Varoufakis, *Game Theory*, 2004, 249.

central coordination rather than anarchy within the communal body."[23] Somewhere between the submicroscopic level of the gene and the macroscopic individual body, purposive agency seems to have been achieved. In Dawkins's words, "One of the most striking properties of survival-machine behavior is its apparent purposiveness." He does not care to distinguish between conscious action versus apparently purposive action "because it is easy to talk about machines that behave *as if* motivated by a purpose, and to leave open the question of whether they are actually conscious."[24] Throughout the applications of game theoretic models, they are applied "as if" those agents being modeled exemplify the logic of strategic rationality.

The apparent purposiveness that Dawkins alludes to is a quality of a machine with a measuring device installed that can direct subsequent action to achieve a specified state. For example, perhaps a machine built to run at 100 degrees Fahrenheit can control its temperature by burning more or less fuel accordingly, such as a system with a thermostat. This mechanical system would act as though it had the purpose of maintaining its temperature at 100 degrees. Dawkins supplies the example of a guided missile that flies as though it has the intention to strike its target.[25]

There are the gene, the gene's biological container constructed with the cooperation of other genes, and the gene's apparent purposive action to achieve or maintain a state.[26] Dawkins wishes to preserve the gene as the primary unit of evolutionary selection at the same time that he denies that groups of individual animals could be such a unit. Similarly, he argues that no gene could have evolved to be purely altruistic, that is, to serve the lives of other animals or other genes.[27] The gene is the ultimate master of the survival machine.[28] A bit cryptically, Dawkins observes that "genes are master programmers, and they are programming for their lives."[29] It appears that he means that genes control acts, much like a game theoretic strategy specifies what act to take in every conceivable situation. However, there is a bit of ambiguity in Dawkins's self-programming genes. Whereas the implication seems to be that a self-directing program could modify itself at will to achieve a new outcome, Dawkins instead seems to mean that the gene tells the body what to do in every type of circumstance. Hence, genes do not modify themselves to reflect different behavioral traits over time to optimize an animal's survival. Instead, a gene programs

---

[23] Dawkins, *Selfish Gene*, 1976/2009, 47.

[24] Ibid., 50.

[25] Discussion is on Dawkins, *Selfish Gene*, 1976/2009, 50–51; Dawkins emphasizes that even though the machines we know are built by humans to serve a purpose they intend, this is not the view of evolutionary biology he puts forward.

[26] Ibid., 62; but it is key for Dawkins that the genes can exist independently of the body, and not vice versa.

[27] Ibid., 60.

[28] Ibid., 59; Dawkins suggests that consciousness may break free from its genetically controlled bondage.

[29] Ibid., 62.

behavior that, if it aids the survival of the animal, will be reproduced in the next generation. The narrative Dawkins provides is that of characteristics (or predispositions for specific behavior) that are imprinted in genes and are perpetuated if they aid in an animal's propagation and reproduction. The inference of selfishness is implied by the presumed competitiveness of biological existence: given the struggle for scarce resources, the genes that will be perpetuated are necessarily those that act efficiently to propagate themselves by accumulating more resources than others. In other words, it is presupposed from the outset that animals are successful, meaning that they have more copies in the next generation than others, by optimizing an objective fitness function in direct competition with others. The following section discusses how this view of evolutionary biology is directly deduced from noncooperative game theory.

SELFISH GENE THEORY AND GAME THEORY

Understanding selfish gene theory requires understanding how game theory, specifically the Nash equilibrium of mutual-best-reply, is proposed to be the deductive model explaining the mechanism by which evolution functions.[30] This section discusses the basic model of game theoretic evolution, which is projected back onto anecdotal studies of particular niches of animal behavior. To apply game theory, a unit of action performing optimization must be identified. This of course is Dawkins's gene. This unitary actor is a parcel characterized by a complete set of strategies defining action choices in all conceivable circumstances.[31] Strategies are rules for action. Dawkins explains, "Animals have to be given by their genes a simple rule for action, a rule that does not involve all-wise cognition of the ultimate purpose of the action, but a rule that works nevertheless, at least in average conditions."[32] In essence, Dawkins's gene is a unit of action that has a strategy set determining what to do in the normal circumstances it routinely encounters. Successful action is that defined as the continuation and replication of existence such that the genetic material a gene is programmed into is represented in the next generation of actors. Dawkins elucidates,

Every decision that a survival machine takes is a gamble, and it is the business of genes to program brains in advance so that on average they take decisions that pay off. The currency used in the casino of evolution is survival, strictly gene survival, but for many purposes individual survival is a reasonable approximation.[33]

---

[30] Alterative analytic explanations also potentially exist; as one example, see S. M. Amadae and Daniel Lempert, "Long-term Viability of Team Reasoning," *Journal of Economic Methodology*, 2015 (published online May 11, 2015, DOI: 10.1080/1350178X.2015.1024880).

[31] This is John von Neumann and Oskar Morgenstern's original definition of a strategy set, *Theory of Games and Economic Behavior* (Princeton, NJ: Princeton University Press, 1944/1953), 31.

[32] Dawkins, *Selfish Gene*, 1976/2009, 99.

[33] Ibid., 55.

Game theory introduces a reward structure for decision outcomes reflecting fitness value. The same rules for survival apply at the genetic and individual levels. Dawkins applies his metaphor that is apt for neoliberal capitalism directly to units of DNA demarcated as genes. Within the evolutionary context, actors', in this case genes', preference rankings are externally supplied by the objective conditions characterizing the environment.

Selfish gene theory identifies the gene as the basic unit containing a strategy set determining action choices. The central idea is that the gene acts as if its fundamental mission were to procreate. Therefore, it as as though it aims to maximize an objective function directly correlating to its survival chances. Even though it may appear that the actor actively intends to survive, this view introduces an unwelcome teleology. Rather, it is appropriate to see that the genetic agents that did successfully survive and procreate had to have made choices that maximized their survival chances.[34] The application of noncooperative game theory to the gene, instead of to a group, introduces the assumption of methodological individualism. Noncooperative game theory vindicates selfish gene theory because it stipulates that elementary unitary actors optimize a function against one another; it does not permit joint maximization.[35] Noncooperative game theory is consistent with the widely accepted assumption that natural selection does not occur at the level of groups.[36] Individuals are responsible for their own survival chances in accordance with each individual's achieved fitness level and with more favorable odds that types of individuals with superior survival strategies encounter. This visualization of the mechanism of natural selection suggests that each individual actor must be viable in every round of engagement, much as how in neoliberal economic systems, all individuals are responsible for their self-care on a pay-as-you-go basis.

Once Dawkins has introduced the concept of the gene as a primary unit of action best thought of as a strategy set programming an organism's behavior so that it must act as though it is maximizing an objective function to exist, he quickly moves into the territory of game theory. He uses John Nash's equilibrium concept to argue that populations of individual members must live in this kind of equilibrium characterized by stable repeating patterns of mutual-best-replies.[37] Prior to evolutionary game theory, game theorists emphasized one-time games or games repeated over a fixed number of plays. In John von

---

[34] Philip Pettit makes clear how rational choice theory provides three types of explanation that permit its relevance to extend from biological programmed behavior to social norms derived from stable patterns of individuals' choices directly linked to their survival and to deliberate decisions made to conform to rational decision theory; "Three Aspects of Rational Explanation," in his *Rules, Reasons, and Norms*, 2002, 177–191.

[35] For an argument that joint maximization, also referred to as team reasoning, could be evolutionarily viable as an ESS, see Amadae and Lempert, "Long-Term Viability," 2015.

[36] For discussion, see Hargreaves Heap and Varoufakis, *Game Theory*, 2004, 248–250.

[37] For the discussion of Nash's equilibrium applied to evolutionary biology, see Hargreaves Heap and Varoukis, *Game Theory*, 2004, 211–244.

Neumann and Oskar Morgenstern's original two-person zero-sum games, indefinitely repeating games were contemplated as a mathematical artifice that rationalized mixed strategies, meaning those incorporating choice based on die rolls. However, within the context of evolution, theorists considered games played repeatedly over thousands or even millions of years. From this perspective, Nash's equilibrium concept of mutual-best-reply could be considered the outcome of a dynamic process in which agents compete with each other for objective value, and their phenotype is represented in the next generation according to their success relative to other phenotypes. In equilibrium, it does not behoove any individual to play a different strategy because, presumably, an agent representing that strategy would have tried it in the previous round. Within evolutionary games, the concept of an "evolutionary stable strategy" (ESS) was built on Nash's equilibrium to define a state in which even if a small number of actors deviated from playing this strategy, the original mutual-best-reply equilibrium still would be reasserted. An evolutionary stable strategy is more robust than a Nash mutual-best-reply equilibrium because it generates a favorable outcome for its wielder even if a new mutant strategy attempts to invade the population.[38]

John Maynard Smith introduces the concept of the evolutionary stable strategy as follows: "A 'strategy' is a behavioral phenotype; i.e., it is a specification of what an individual will do in any situation in which it may find itself. An ESS is a strategy such that, if all the members of a population adopt it, then no mutant strategy could invade the population under natural selection."[39] Because the ESS is a game theoretic artifact, its definition is formal and its applications are mathematical.[40] Even though models with variables can graphically make a point for any set of actual values, to be fully applicable to concrete circumstances of animals maximizing fitness, an objective means to appraise value is needed. One favored game for evolutionary game theorists is the Chicken game, aptly renamed Hawk-Dove. Dawkins supplies arbitrary values to make sense of applying game theory to evolutionary development: "Now as a purely arbitrary convention we allot contestants 'points.' Say 50 points for a win, 0 for losing, –100 for being seriously injured, and –10 for wasting time over a long contest." Crucially, in evolutionary game theory, the reward structure of the game is proposed to translate directly into fitness value much as the first generation of game theorists assumed that individuals maximize money. Dawkins continues,

---

[38] On the ESS concept, see Dawkins, *Selfish Gene*, 1976/2009, 69–87; John Maynard Smith, *Evolution and the Theory of Games* (Cambridge: Cambridge University Press, 1982), 10–11, 24; Hargreaves Heap and Varoufakis, *Game Theory*, 2004, 214–220.

[39] Maynard Smith, *Evolution*, 1982, 10, Maynard Smith's model assumes an infinite population. Note that populations may exist in equilibrium with two phenotypes playing different strategies; see Amadae and Lempert, "Long-Term Viability," 2015.

[40] See Maynard Smith, *Evolution*, 1982, 10–27; Hargreaves Heap and Varoufakis, *Game Theory*, 2004, 219.

These points can be thought of as being directly convertible into the currency of gene survival. An individual who scores high points, who has a high average 'pay-off', is an individual who leaves many genes behind in the gene pool.[41]

Nature plays the role of banker, so that the payoffs (or world states to which preferences refer in conventional game theory) are best thought of as objective survival criteria determined at a systemic level beyond an individual's control or individual evaluation. Survival requires meeting specific objective needs enforced by natural processes. In this biological economy of competition for scarce resources, game theorists propose that the entities being selected for must optimize against each other precisely as noncooperative game theory stipulates and do not explore the possibility that after a threshold level of fitness value is acquired, organisms may cease seeking to gain more.

Selfishness, or individualistic optimization, is alleged to be a condition of nature perfectly captured by game theory. In Dawkins's terms, "selfishness is to be expected in any entity that deserves the title of a basic unit of natural selection."[42] Game theory, originally applied to military strategy and subsequently to political economy, next proved a useful tool to apply to evolutionary biology. In turn, the apparent naturalness of noncooperative game theory to capture "the fierce competition for scarce resources in the relentless struggle to eat other survival machines, and to avoid being eaten" made it seem obvious that the game theoretic rational actor (here the selfish gene) exists prior to the theory discussing its behavior.[43]

Confident that nature hands out objective rewards for the fittest creatures, evolutionary game theorists are able to extrapolate to construct game theoretic payoff matrices to represent evolutionary interactions. The selfish gene, which must optimize tangible rewards as a condition of existence in competition with others, is primordial. In biological terms, just as all physics ultimately must be consistent with the laws governing the actions of fundamental participles, all of life must conform to the dictates regulating the selfish gene. The natural processes acting on genes force them to comply with the axioms of rational choice theory, at least on an "as if" basis. Purposive action, if not intended, is mimicked.[44] The pivotal point, on which the entire argument hinges, is that biological agents have objective needs structuring the payoffs of encounters. In Herbert Gintis's words, "fitness maximization is a precondition for evolutionary

[41] Dawkins, *Selfish Gene*, 1976/2009, 70; see also 151. Note that even though it is presumed that nature grants objective value, evolutionary game theorists are not able to measure this value; thus, it is postulated for demonstration purposes. In 1976, Dawkins writes, "Unfortunately, we know too little at present to assign realistic numbers to the costs and benefits of various outcomes in nature" (75), to which he responds in 2009 that one good example of values "comes from great golden digger wasps in North America" (283). He does not mention the values or present any other cases of specific cost-benefit value to be plugged into ESS models, nor does he consider that fitness and survival may more accurately reflect a threshold value.

[42] Ibid., 33.

[43] Quote from ibid., 47, see also 67.

[44] Ibid., 50–51, 196.

survival."[45] Acting in accordance with the axioms of rational choice theory is a condition of survival under the assumption that rewards are directly correlated to an actor's survival and reproduction prospects. Gintis further explains, "We can expect preferences to satisfy the completeness condition because an organism must be able to make a consistent choice in any situation it habitually faces or it will be outcompeted by another whose preference ordering can make such a choice."[46] It is important to keep in mind that it is the presumption that objective needs dictate the unfolding of long-term evolutionary processes that makes it possible to apply game theoretic payoff matrices. The payoff structure represents the nonnegotiable conditions hypothesized to be necessary to sustain life.

After identifying the selfish gene as the elementary unit on which natural selection functions, many treatments of evolutionary game theory move seamlessly between the animal world and the human world.[47] Dawkins is clear in his statement of the confluence of evolutionary biology and human behavior:

My own feeling is that a human society based simply on the gene's law of universal ruthless selfishness would be a very nasty society in which to live ... Be warned that if you wish, as I do, to build a society in which individuals cooperate generously and unselfishly toward a common good, you can expect little help from biological nature.[48]

Dawkins suggests that the idea of the state of nature populated by selfish genes serves as a ubiquitous model, no matter whether the gene is in a single-celled organism or a person. It is impossible to exit the state of nature, and noncooperative game theory specifying the laws of individualistic optimization best captures this inevitable competitive struggle for existence. Gratuitous altruism or universalistic selflessness is not only not favored by nature but also would be rapidly exploited and hence displaced from a population.[49] The fundamental premise is that any expression of human behavior must comport with the inveterate imprint of genetic selfishness. The logic of noncooperative game theory and of the selfish gene are one and the same. Individualistic optimization is the only mode of action available. Under the assumption that the value agents must maximize as a condition of their existence is externally and objectively determined, cooperation is a persistent Prisoner's Dilemma confronted by all organisms, and vying over territory is a Chicken game.[50]

---

[45] Herbert Gintis, *The Bounds of Reason: Game Theory and the Unification of the Behavioral Sciences* (Princeton, NJ: Princeton University Press, 2009), 7.

[46] Ibid., 8.

[47] See, for example, how Dawkins moves from worrying about being eaten by predators at the water-hole to gambling on the stock market; Dawkins, *Selfish Gene*, 1976/2009, 56

[48] Ibid., 3.

[49] For Dawkins's view that altruism such as it exists must be consistent with individualistic maximization see his *God Delusion* (New York: Bantam, 2008), 247–254.

[50] Note Maynard Smith's chapter on evolution and cooperation, titled "Honesty, Bargaining, and Commitment," which applies the same rationale to all life forms from the pre-intentional to *Homo sapiens*; *Evolution*, 1982, 147–173.

Culture, or learned behavior not programmed into individuals' genetic structure, is also presumed to obey the same rules of individualistic maximization. Dawkins coined the term "meme" to refer to "the new replicator ... the idea of a unit of cultural transmission, or unit of imitation."[51] Examples of memes are "tunes, ideas, catch-phrases, clothes, fashions, ways of making pots or building arches," yet memes are postulated to exist inside minds. In Dawkins's words, "memes propagate themselves in the meme pool by leaping from brain to brain via a process which, in the broad sense, can be called imitation."[52] Like genes, Dawkins's memes have the properties of longevity, fecundity, and copying fidelity.[53] Dawkins proposes that "just as we have found it convenient to think of genes as active agents, working purposefully for their own survival, perhaps it might be convenient to think of memes in the same way."[54] One example of a meme is the belief in God, understood to be a personal savior or almighty creator. Hinting toward his later work, *The God Delusion* (2006), Dawkins queries whether

the god meme ... [has] become associated with any other particular memes, and ... [if] this association assist[s] the survival of each of the participating memes? Perhaps we could regard an organized church, with its architecture, rituals, laws, music, art, and written tradition, as a co-adapted stable set of mutually-assisting memes.[55]

The concept of memes retains a fuzziness because unlike the phenomenon of DNA replication, which, arguably, has a specific site for identity and replication, cultural artifacts have a different ontology without a well-defined location for their existence. Whereas DNA exists in cells that replicate, memes, Dawkins suggests, exist as material entities in the world, yet propagate as a mind-virus "leaping from brain to brain."

Regardless of whether the analogy between gene and meme is either fully intelligible or sound, its underlying point is to suggest that human culture operates according to the same laws governing biological reproduction. Just as natural scientists hold that the laws of physics must obtain everywhere in the universe, so, too, does Dawkins reason that the fundamental laws of biology must hold everywhere throughout the realm of living organisms. This idea that biological science is unified by the principle of replication which obeys the laws of game theory is profound.[56] Dawkins ruminates,

What, after all, is so special about genes? The answer is that they are replicators. The laws of physics are supposed to be true all over the accessible universe. Are there any principles of biology that are likely to have similar universal validity? When astronauts

[51] Dawkins, *Selfish Gene*, 1976/2009, 192.
[52] Ibid., prior quote as well. Dawkins follows up in clarifying his meme concept in 2009, 322–323. He admits that ultimately a meme would have to exist in terms of molecular brain structure.
[53] Ibid., 194.
[54] Ibid., 196.
[55] Ibid., 197; see Dawkins, *The God Delusion*, 2008.
[56] This is the leading insight in Herbert Gintis's *The Bounds of Reason*, 2009.

voyage to distant planets and look for life, they can expect to find creatures too strange and unearthly to imagine. But is there anything that must be true of all life, wherever it is found, and whatever the basis of its chemistry? ... Obviously I do not know but, if I had to bet, I would put my money on one fundamental principle.[57]

To be sure that his readers have followed his message to this point, Dawkins emphasizes that this one fundamental principle is "the law that all life evolves by the differential survival of replicating entities."[58] The gene serves this function for Earth-based life.[59] It is best modeled by noncooperative game theory, which holds that it must conform to these behavior assumptions as a condition for its successful survival and propagation.[60]

Dawkins's original 1976 version of *The Selfish Gene* concludes that the game theoretic evolutionary stable strategy, or ESS, and noncooperative game theory require that individualistic actors must compete with one another on a momentary basis. As the next chapter discusses further, this vision of life holds that organisms must compete over scarce resources in a way best modeled as endless games of Prisoner's Dilemma. Solving games with conflict is the fundamental problem of life. Dawkins draws *The Selfish Gene* to its conclusion:

A simple replicator, whether gene or meme, cannot be expected to forgo short-term selfish advantage even if it would really pay, in the long run to do so. We saw this in the chapter on aggression [with the Hawk Dove game]. Even though a "conspiracy of doves" would be better for *every single individual* than the evolutionary stable strategy, natural selection is bound to favor the ESS.[61]

Relying on game theoretic analysis, Dawkins concludes that the entire domain of biological action – cellular, organismic, and social – is composed of individualistic entities that must optimize an objective function in competition with others on a momentary basis. In a Prisoner's Dilemma or a Hawk Dove (Chicken game), it would be better overall if agents peacefully cooperated. But this strategy is exploitable by defectors and aggressors who seek unilateral advantage in their apparently purposive struggle to survive and replicate.

SOCIAL IMPLICATIONS OF SELFISH GENE THEORY

Evolutionary game theorists identify a single organizing principle, the "differential survival of replicating entities," and apply it to all members of the animal kingdom. Human society is no different from a hive of bees with respect to meeting the requirement of maximizing fitness on an individual level. As is consistent throughout game theory, individuals vie with one another to maximize expected utility,

---

[57] Dawkins, *Selfish Gene*, 1976/2006, 191.
[58] Ibid., 192; Dawkins stands by this earlier claim in the 2009 edition of *Selfish Gene*, 322.
[59] Ibid., 192.
[60] The standard model does not permit joint maximization.
[61] Dawkins, *Selfish Gene*, 1976/2009, 200.

and patterns of stable regularity, or norms, emerge on this basis. Groups of individuals cannot act in concert because all individuals have the perpetual incentive to defect, and individual selection reinforces narrow self-interested maximization.[62] In evolutionary game theory, the currency of utility is fitness. This is not determined by subjective whim, but must reflect objective survival conditions structured by scarce resources and adaptive advantage in procuring them. Strategies of action are crucial. Whereas in classical game theory, the payoffs may be considered to be subjective values freely adopted by agents, in evolutionary game theory, the values structuring games express the material conditions necessary for reproduction. Preferences, or the payoffs of the games, are determined systemically in accordance with externally imposed success criteria. A far-reaching debate considers whether it is in fact the case that biological actors must evolve to be narrowly self-interested in accordance with objectively defined survival criteria.[63] However, it is clear that the axioms of rational choice theory, which require actors to act as if they had preferences over outcomes, are purported to pertain to life because biological reality is such that surviving requires acting as though one is maximizing fitness in accordance with objective, externally imposed standards. Behavioral norms, or typical behavior, arise and are sustained specifically because the actors exhibiting this behavior depend on it for their survival. Articulating this perspective, the versatile philosopher Philip Pettit observes, "A regularity will count as a norm for a system just in case the satisfaction of

---

[62] For an alternative account arguing for the viability of joint maximization, see Amadae and Lempert, "Long-Term Viability," 2015.

[63] Edward O. Wilson defends individualistic maximization in his *Sociobiology* (Cambridge, MA: Harvard University Press, 1975); for other arguments holding that behavior not conforming to narrow self-interested maximization will be eliminated; see D. M. Buss, "Evolutionary Psychology: A New Paradigm for Psychological Science," *Psychological Inquiry* (1995) 6, 1–30; T. Ketelaar and B. J. Ellis, "On the Natural Selection of Alternative Models: Evaluation of Explanations in Evolutionary Psychology," *Psychological Inquiry* (2000), 11, 56–68; D. T. Kenrick, N. P. Li, and J. Butner, "Dynamical Evolutionary Psychology: Individual Decision Rules and Emergent Social Norms," *Psychological Review* (2003) 110:1, 3–28; Ken Binmore, *Playing Fair* (Cambridge, MA: MIT Press, 1994); Binmore, *Just Playing* (Cambridge, MA: MIT Press, 1998); Binmore, *Natural Justice* (Oxford: Oxford University Press, 2005). Any putatively group-benefiting cooperative, altruistic behavior is explained as ultimately in the individual's self-interest; see Elliot Sober and David Sloan Wilson, *Unto Others* (Cambridge, MA: Harvard University Press, 1998), 31–35; Mark Van Vugt and Paul Van Lange, "The Altruism Puzzle," in *Evolution and Social Psychology*, ed. by Mark Schaller, Jeffrey A. Simpson, and Douglas T. Kenrick (New York: Psychology Press, 2006), 243–244; Herbert Gintis, Samuel Bowles, Robert Boyd, and Ernst Fehr, "Explaining Altruistic Behavior in Humans," *Evolution and Human Behavior* (2003), 24, 153–172, see 153–154. Cooperation among relatives is explained by the theory of inclusive fitness; see William Hamilton, "The Genetical Evolution of Social Behavior," *Journal of Theoretical Biology* (1964), 37, 1–52; however cooperation with non-kin is attributed to reciprocal altruism; Robert Trivers, "The Evolution of Reciprocal Altruism," *Quarterly Review of Biology* (1971), 46, 35–57. For a response and alternative view, see Marilynn Brewer and Linnda Caporael, "An Evolutionary Perspective on Social Identity: Revisiting Groups," in *Evolution and Social Psychology*, 2006.

that regularity is required for the system to succeed in the role for which it has been designed or selected."[64] The challenge of identifying markers of fitness that are independent of analytically specified survival criteria remains. The game theoretic standard is to symbolize the systemic constraints on successful action by payoffs reflecting objective value that must be obtained for effective survival and propagation. Evolutionary game theorists propose scores as the outcomes of games and presume that the highest scorers are represented in a higher proportion in the next round or generation of play.[65]

Contrary to evolutionary game theorists, most economists insist that the reward structure of games, reflecting expected utility, is the subjective preference of individuals.[66] Economists resist the idea that there are objective values that individuals must maximize as a condition of their survival and reproduction.[67] Even though Karl Marx may have been intrigued by and congratulatory of Dawkins's insights, a tension of worldviews exists between economic science and evolutionary game theory. Marx would agree that people confront a reward structure externally imposed on them that directly correlates to their likelihood of surviving and propagating. He would agree that differential access to the material conditions of life leads to differential success directly measurable in procreative success. Yet since the late nineteenth century, economic science has shied away from any language of objective needs, insisting that value only exists in the eyes of the beholder.[68] The game theoretic reliance on interpersonally transferable utility represents a middle ground that acknowledges a pre-social, ontologically significant source of value that denies interpersonal comparability of experiential satisfaction.

The final point of this chapter is to elucidate the implications of Dawkins's selfish gene argument for understanding social norms. The evolutionary stable strategy, or ESS, refers not only to the game theoretic equilibrium of John Nash's mutual-best-reply. It has the additional connotation that each enacted strategy exists in equilibrium with all the other enacted strategies such that no small perturbation would destabilize it. The continuity between classical game theory and evolutionary game theory lies in the specification that norms and

---

[64] Pettit, "Three Aspects of Rational Explanation," 2002, 183.

[65] Recall Dawkins's assignment of numeric payoffs and discussion about the difficulty of identifying actual rewards. See also Hargreaves Heap and Varoufakis's introduction of the ESS, *Game Theory*, 2004, 214–220. Evolutionary game theorists observe that the specific numbers of payoffs, although mathematically necessary to solve games, are not required to gain insights into the evolution of animal populations.

[66] Herbert Gintis proposes a "biological basis for expected utility" linked to survival; see his *Bounds of Reason*, 2009, 16.

[67] Partha Dasgupta is an exception here; see his *Inquiry into Well Being and Destitution* (Oxford: Oxford University Press, 1993), 324–333.

[68] Lionel Robbins, *Nature and Significance of Economic Science* (Auburn, AL: Ludwig von Mises Institute, [1932]/2007) provides one of the most authoritative statements of this now mainstream position. Amartya Sen's approach to social choice and poverty dutifully reflects this well-accepted sentiment. Dasgupta, in contrast, works to reintroduce objective needs in terms of caloric intake, *Inquiry into Well Being and Destitution*, 1993, 324–333.

customs are Nash equilibria. In Herbert Gintis's words, "at least since [Thomas] Schelling['s *Strategy of Conflict*, 1960] and [David] Lewis['s *Convention*, 1969], game theorists have interpreted social norms as Nash equilibria."[69] In game theoretic terms, norms are patterns of interconnected action that exist in equilibrium; no one would have any reason to change his or her action.[70] Mutual cooperation in a Prisoner's Dilemma or Chicken game is not self-sustaining by this analysis, then, because individuals are motivated to cheat or defect.[71]

Game theorists accept that strategic rationality stipulated by noncooperative game theory provides means to model and explain programmed behavior, norms or regularities of conduct reflecting survival imperatives, and deliberate choice.[72] Rational choice theory makes it possible to move seamlessly between these domains, as well as to the level of public policy and institutional design. Pettit acknowledges that "rational choice theory postulates that the things people generally do, whatever the basis on which they are chosen, are consistent with a major interest in economic gain and social acceptance."[73] The norms governing social conduct reflect individualistic competition. Where economists suggest that personal preferences structure the games from which norms arise in repeating contexts, evolutionary game theorists suggest that the underlying onto-logical real conditions determining fitness will give rise to norms. Great interest lies in discovering the process by which efficient norms, specifically those that do not represent suboptimal equilibrium such as mutual defection in the Prisoner's Dilemma, Stag Hunt, and Chicken games, arise and are maintained.

Therefore, regular patterns of behavior, social customs, and all laws are viewed as the product of games played by selfish genes or selfish individuals. No other explanation can gain purchase unless it can be distilled down to expected individualistic strategic utility maximization of factual finite fungi-ble resources.[74] This places the social science dictated by the fundamental assumptions of game theory on a collision course with other paradigms, including Jürgen Habermas's deliberative democracy and discursive ethics,

---

[69] Gintis, *Bounds of Reason*, 2009, 246.

[70] Philip Pettit has an in-depth discussion of this understanding of norms, "*Virtus Normativa*: Rational Choice Perspectives," in his *Rules, Reasons, and Norms*, 2002, 308–343.

[71] Pettit discusses how strategically rational individuals confront large-scale PD games, but he is vague about how the analysis spans two-person repeating PDs and large-scale anonymous games with a PD structure, see ibid., 320–321.

[72] Pettit clearly discusses this in "Three Aspects of Rational Explanation," 2002.

[73] Pettit, "*Virtus Normativa*," 2002, 340. Pettit encompasses seeking other's approval within the motive of self-interest; for commentary, see S. M. Amadae, "Utility, Universality, and Impartiality in Adam Smith's Jurisprudence," *Adam Smith Review* (2008), 4. Pettit works to introduce how individuals choose to interact with others based on a cost-benefit analysis of whether others will approve or disapprove of one's actions, "*Virtus Normativa*," 2002, 326–332.

[74] See, e.g., David Lewis, *Convention: A Philosophical Study* (Cambridge, MA: Harvard University Press, 1969); Thomas Schelling, *Strategy of Conflict* (Cambridge, MA: Harvard University Press, 1960); Pettit, "*Virtus Normativa*," 2002.

as well as all non-instrumental views of normativity.[75] Richard Dawkins's *The Selfish Gene*, with its preeminent acknowledgment of the evolutionary versions of Nash's game theoretic equilibrium, makes clear the far-reaching range of non-cooperative game theory to interpret all behavioral interactions in its terms.

CONCLUSION

We can surmise the following points:

1. According to Richard Dawkins's selfish gene theory, the primary evolutionary agents are genes that program behavior into their carriers and are perpetuated into the next generation if they aid the organism in accumulating more fitness value than competitors.
2. In evolutionary replicator dynamics modeling, actors must compete over an objective resource in accordance with the principles of noncooperative game theory to survive and propagate progeny in the next generation.
3. This demand on life forms extends throughout the universe in the same way that the laws of physics do.
4. Order, specifically regular patterns of interaction that conform to mutual-best-reply equilibrium, is the outcome of behavior that appears to be purposive in the same way that a missile moves toward its target or a thermostat maintains a constant temperature.
5. Norms, or regularized, repeating patterns of conduct, arise as individuals who are successful in obtaining objective value coexist in an equilibrium of mutual-best-reply; these are not necessarily optimal, but actors who benefit by participating in optimal equilibria will produce more offspring.
6. Humans have evolved with the logic of noncooperative game theory imprinted into their DNA and are likely to exhibit selfish behavior as a result.
7. The normalization inherent in neoliberal society consistent with strategic rationality is decidedly distinct from the modern normativity described by Michel Foucault in his analysis of Jeremy Bentham's Panoptic technology for behavioral modification.[76] In this system, which human scientists design to reform agents so that they will "internalize the gaze" of the authority and moderate their own behavior to become useful workers, norms are identified as the average conduct of membership populations. Experts study individuals to scientifically specify current average performance and to develop disciplinary exercises that transform the

---

[75] Jürgen Habermas, *Theory of Communicative Action*, vols. 1 and 2, trans. Thomas McCarthy (Cambridge, MA: MIT Press, 1985); Margaret Gilbert, *Joint Commitment: How We Make the Social World* (Oxford: Oxford University Press, 2013); Ludwig Wittgenstein, *Philosophical Investigations*, ed. P. M. S. Hacker and Joachim Schulte (New York: Wiley, 2009).

[76] Michel Foucault, *Discipline and Punish*, trans. Alan Sheridan (New York: Vintage, 1979).

performance of subjects, who then, over time, achieve a socially more satisfactory average. In game theory, norms are Nash equilibria to games in which actors' preferences and strategies are encoded to result in regularized patterns of play. The only ways to modify outcomes are to apply incentives or to shock the system such that possibly another stable mutual-best-reply equilibrium could emerge.

11

Tit for Tat

[Robert] Axelrod, like many political scientists, economists, mathematicians and psychologists, was fascinated by a simple gambling game called Prisoner's Dilemma. It is so simple that I have known clever men misunderstand it completely, thinking there must be something more to it! But its simplicity is deceptive. Whole shelves in libraries are devoted to the ramifications of this beguiling game. Many influential people think it holds the key to strategic defense planning, and that we should study it to prevent a third world war. As a biologist, I agree with Axelrod and [W. D.] Hamilton that many wild animals and plants are engaged in ceaseless games of Prisoner's Dilemma, played out in evolutionary time.

Richard Dawkins, 2009[1]

Since the Prisoner's Dilemma is so common in everything from personal relations to international relations, it would be useful to know how best to act when in this type of setting.

Robert Axelrod, 1984[2]

Noncooperative game theory has been applied to nuclear strategy, the social contract, public goods, and also evolutionary biology. Everywhere its logic is the same: optimization, or expected utility maximization, occurs on the level of individuals in a population who typically compete for scarce resources. The Prisoner's Dilemma game is emblematic of the perceived problem of cooperation: individuals seek propitious outcomes, but ultimately prefer their own gain, even at the expense of someone else's loss. Having generated the concept of an evolutionary stable strategy (ESS), evolutionary game theorists deduce that every population of actors must be impervious to a deviant member designed to exploit others (see Chapter 10). To protect against individuals' exploitation by other actors, natural selection eliminates acts of gratuitous altruism because they would undermine its perpetrators' survival chances, thereby eventually

[1] Richard Dawkins, *The Selfish Gene* (Oxford: Oxford University Press, [1989]/2009), 203.
[2] Robert Axelrod, *The Evolution of Cooperation* (New York: Basic Books, 1984), 17.

eliminating individuals disposed to this type of behavior. Even if group selection occurs among human societies, still every highly cooperative population must be impervious to individualistic exploiters. This, it is hypothesized, requires individualistic selfishness.

The Prisoner's Dilemma game has proven to be endlessly fascinating for representing the foil against which cooperative behavior must test its viability. Given evolutionary biologists' interest in the material conditions of survival, that is, caloric intake and wherewithal to procreate, the tendency of organisms to waste resources in the suboptimal Nash equilibrium of the Prisoner's Dilemma inspired copious research. It seems perplexing that animals competing for survival would repeatedly defect in Prisoner's Dilemmas, or in Stag Hunt and Chicken games as well, thereby achieving a mutually inferior outcome. This is the same problem surmised to plague the social contract and public goods, but in a state of nature there is no external authority to impose sanctions, through which actors could achieve cooperation.

The sanctions approach to solving a multi-agent, repeating, Prisoner's Dilemma game thought reflective of numerous social circumstances is limited for being costly and even infeasible or counterproductive to impose.[3] One difficulty is that the agents who are hired as enforcers likely have their own agendas and may be subject to corruption (breaking rules in their favor) and rent seeking (establishing rules in their favor). Evolutionary game theory offered two theoretical results to the ongoing and consequential study of cooperation. First, by modeling indefinitely repeated encounters, researchers added analytic complexity to their understanding of the profound bind organisms are hypothesized to be in: in their struggle for survival and competition for scarce resources, they necessarily optimize against one another on an individualistic and momentary basis. This individualistic competition is believed to be insurmountable because actors straying from this behavioral norm fail to survive and propagate, and it represents a fundamental principle of life best captured by noncooperative game theory. Second, theorists used the Nash equilibrium of mutual-best-reply and the evolutionary replicator dynamic method to show that in indefinitely repeated games, cooperation can emerge without the introduction of external sanctions through the evolutionary stable strategy of conditional cooperation. This not quite ESS is also called Tit for Tat, or reciprocal altruism. Tit for Tat lies at the heart of neoliberal theory, which argues that even under the most minimalist assumption of myopic strategic self-interest, cooperation can still emerge. This chapter presents the contours and derivation of this behavioral strategy.

Neoliberal theorists have unabated enthusiasm for the Tit for Tat strategy because it seems to be the only solution to the endless Prisoner's Dilemma encounters, perceived to characterize not just human society but life in general,

---

[3] See, e.g., Alan H. Goldman, "The Rationality of Complying with Rules: Paradox Resolved," *Ethics* (2006), 116, 453–470, specifically 457; see also Michael Taylor, *The Possibility of Cooperation* (Cambridge: Cambridge University Press, 1987), 125–179.

that does not require a strong state. Hence, conditional cooperation appears to be the silver bullet for Prisoner's Dilemmas. This is apparent in the United Kingdom's authoritative *Stern Review: The Economics of Climate Change* (2006). The London School of Economics economist Sir Nicholas Stern provides a comprehensive though succinct overview of state-of-the-art game theoretic findings on cooperation and public goods. His report observes,

Reciprocity plays a key role in situations where the players facing the prisoners' dilemma have the opportunity to play *repeated* games and remember the previous choices of the other player. In particular, many players adopt a strategy of *conditional cooperation*, in which they contribute more to the provision of the public good the more others contribute.[4]

This chapter introduces the Tit for Tat resolution of the repeated Prisoner's Dilemma game and the hopes, such as Stern's, pinned on it to solve the vexing problem of cooperation.

However, Tit for Tat more resembles an unstable physical equilibrium, like a ball balanced on the head of a pin, than it does a feasible approach to cooperation. This is because the conditions to actually tame a strategic rational actor by relying on conditional cooperation are so idealized that they cannot actually sponsor behavior leading to large-scale collaboration.[5] Nor is the approach of reciprocal altruism helpful for actors who either know the end point of their time frame for interaction or place higher emphasis on the present than the future. My points are three. First, conditional cooperators coexist in a purely rarefied theoretical realm presuming indefinitely repeating interactions of two individuals with perfect memory. Second, the efforts necessary to institutionalize the conditions turning Hobbes's Foole and Hume's knave into reciprocal altruists are not practically feasible. If reciprocal altruism is to enforce cooperation endogenously, then an elaborate practice of rewards and punishments must emerge to reflect the actors who prevailed in Axelrod's theoretical experiment, and the environment must perfectly resemble its iterated PD structure. Alternatively, if institutions are built to exogenously facilitate Tit for Tat behavior, then institutional designers must introduce mechanisms to ensure monitoring, transparency, and sanctioning conditions to nudge behavior into that resembling voluntary cooperation. And third, the hope of producing conditional cooperation through indefinite play and perfect transparency, on the premise that all life must obey the laws of individualistic maximization, ultimately fails to even realize that cultural artifacts and sociability could embody a different set of

---

[4] *Stern Review* (2006), "Part VI: International Collective Action," 452, available online. http://webarchive.nationalarchives.gov.uk/20100407172811/, http://www.hm-treasury.gov.uk/stern_review_report.htm, accessed July 15, 2015.

[5] Russell Hardin, "Individual Sanctions, Collective Benefits," in *Paradoxes of Rationality and Cooperation*, ed. by Richmond Campbell and Lanning Sowden (Vancouver: University of British Columbia Press, 1985), 339–354.

principles of conduct. These differences become evident in any attempt to use Tit for Tat to enforce a conduct of truth telling.[6]

ALMOST UNLOCKING THE REPEATING PRISONER'S DILEMMA

For people familiar with game theory, any introduction to the repeating Prisoner's Dilemma and its Tit for Tat solution will be remedial. Robert Axelrod's *The Evolution of Cooperation* (1984) is legendary for demonstrating that cooperation in Prisoner's Dilemma can emerge without any use of externally applied coercive force to induce it.[7] It is good to recall why theorists believe the Prisoner's Dilemma game is not only prevalent but also paradigmatic of the problem of achieving cooperation among dyads of individuals or in vast collective undertakings. Steven Kuhn's entry "Prisoner's Dilemma" in the *Stanford Encyclopedia of Philosophy* makes it clear: [Either] my temptation is to enjoy some benefits brought about by burdens shouldered by others ... [or my] temptation is to benefit myself by hurting others."[8] It is generally understood that in Prisoner's Dilemma scenarios, one benefits specifically because someone else is made worse off. Somebody else pays the price for one's success. Selfish gene theory, if accepted, requires that agents must adopt this behavioral trait consistent with noncooperative game theory as a condition of survival. Dawkins argues that we must accept the truth of "the gene's law of universal ruthless selfishness."[9] If we agree with this chain of argument, mandating the necessity of individualistic strategic optimization, then we must take the Prisoner's Dilemma seriously.[10]

Game theorists agree that in a series of Prisoner's Dilemma games played with exacting precision and a well-known end point, it is rational to defect on the first round of engagement, thereby missing out on the cooperative gains that could have been made.[11] Also, if the subsequent move's outcome is significantly less important than the current round's, then it will also pay to consistently defect in a repeating

---

[6] S. M. Amadae, "Normativity and Instrumentalism in David Lewis' Convention," *History of European Ideas* (2011), 37, 325–335; on verification and truth in verifying arms control among superpowers and the Tit for Tat strategy, see Steven J. Brams, *Superpower Games: Applying Game Theory to Superpower Conflict* (New Haven: Yale University Press, 1985), 138.

[7] Axelrod, *The Evolution of Cooperation*, 1984.

[8] Steven Kuhn, "Prisoner's Dilemma," *Stanford Encyclopedia of Philosophy*, 2007, plato.stanford .edu.entires/prisoner-dilemma/, accessed December 23, 2007, page 9 of 49.

[9] Dawkins, *Selfish Gene*, 1976/2009, 3.

[10] Philip Pettit makes clear the extent to which both rational choice and the Prisoner's Dilemma game continue to be viewed as the canonical form of rational behavior and the central problem for organisms and people to solve; see "*Virtus Normativa*: Rational Choice Perspectives," in his *Rules, Reasons, and Norms* (Oxford: Clarendon Press, 2002), 308–343.

[11] Axelrod, *The Evolution of Cooperation*, 1984, 10; see also Howard Luce and Duncan Raiffa, *Games and Decisions* (New York: Wiley, 1958), 94–102.

Prisoner's Dilemma.[12] Yet for game theorists, solving the Prisoner's Dilemma is pressing because they consider it both to exist everywhere and to entail a large-scale waste of resources that seems counterintuitive to accept.[13] Axelrod is emphatic about the significance of the Prisoner's Dilemma that is encountered by "not only people but also nations and bacteria."[14] His treatment of the repeating Prisoner's Dilemma is prescriptive. He tells us, "Since the Prisoner's Dilemma is so common in everything from personal relations to international relations, it would be useful to know how best to act when in this type of setting."[15]

Axelrod finds his research pertinent to the question of social order. He is dissatisfied with Thomas Hobbes's alleged solution of the Prisoner's Dilemma relying on a heavy-handed state. Thus, he poses the alternative question, "Under what conditions will cooperation emerge in a world of egoists without central authority?"[16] Axelrod exemplifies how by 1984, the time of *The Evolution of Cooperation*'s publication date, theorists widely accepted that all manner of interactions from the worlds of social relations and warfare to the animal kingdom are best described as Prisoner's Dilemma games. Axelrod's fame lies in presenting a solution to the Prisoner's Dilemma that does not rely on externally applied coercive force to achieve the cooperative outcome. Connecting the international relations problem of anarchy to the central thrust of his investigation, Axelrod observes,

Today nations interact without central authority. Therefore the requirements for the emergence of cooperation have relevance to many of the central issues of international politics. The most important problem is the security dilemma: nations often seek their own security through means which challenge the security of others.[17]

Axelrod finds the Prisoner's Dilemma in the Soviet's invasion of Afghanistan in 1979, because in his view, the Soviet Union was defecting while hoping the United States would cooperate. Inviting friends over to dinner is a Prisoner's Dilemma game because they may not reciprocate.[18] Journalists, business leaders, and congressional representatives are similarly locked into repeating Prisoner's Dilemma games. Each wants to gain from the other's cooperation without reciprocating but finds it necessary to reciprocate as a condition of the other's cooperation. Axelrod claims that the applicability of his result applies to political philosophy, international politics, economic and social exchange, and international political economy.[19] Axelrod refers to the "norm of reciprocity"

[12] For discussion of this see, Axelrod, *The Evolution of Cooperation*, 1984, 59. Note that the precise calculation depends on the specific reward structure of the game.
[13] Pettit, "*Virtus Normativa*," 2002, at 320, 327.
[14] Axelrod, *Evolution of Cooperation*, 1984, 18.
[15] Ibid., 27.
[16] Ibid., 3.
[17] Ibid., 4.
[18] Ibid., 4–5.
[19] Robert Axelrod, "The Emergence of Cooperation among Egoists," in *Paradoxes of Rationality and Cooperation*, 1985, 323–324.

and proceeds to deduce it from a tournament with combatants structured by a repeating Prisoner's Dilemma game in which the winning strategy was that of conditional cooperation, referred to as "Tit for Tat."[20]

Axelrod held an open tournament, inviting fellow academicians from any field to submit programs. These programs were instructions for play in every round of the 200-round supergame that specified which move the protagonist would take depending on the opponent's action and the outcome of the previous round. Axelrod designed a Prisoner's Dilemma game with the following rewards: suckering the other player yields 5 points, mutual cooperation yields 3 points each, mutual defection yields 1 point each, and being suckered yields 0 points. Note that as with the use of the models in evolutionary game theory, the rewards are fixed and objectively defined. Furthermore, the success criteria, in this case winning the tournament, are set at a systemic level beyond the subjective evaluation of participants. Axelrod's tournament went on for 200 rounds per dyad of players who could perfectly recollect the entire history of the game. The identity of the other agent is constant. Every entrant into the tournament "writes a program that embodies a rule to select the cooperative or noncooperative choice on each move."[21] The computer program uses this input and plays entrants' strategies against one another. All entrants fully understand the rules of noncooperative game theory and the logic of the Prisoner's Dilemma.

The result of this tournament stands as a eureka moment in the history of game theory. The winning strategy, submitted by Anatol Rapoport, was Tit for Tat. This strategy always cooperates first and then plays what the other player did in the subsequent round of play.[22] Here is what's interesting. Given Axelrod's payoff structure of 5, 3, 1, 0 for the 4 various Prisoner's Dilemma outcomes, the Tit for Tat strategy averaged 504 points, which is less than half of the amount, 1200, that would be shared among unconditional cooperators who always cooperated over 200 rounds of play. Axelrod's point is that even though it is possible to write programs that will steal points from cooperators, those who cooperate will gain more points overall than those who are prone to undermining others by defecting. Tit for Tat has the property of cooperating with other cooperators but defecting when paired with an exploitative strategy. Axelrod called this strategy "nice" because it always seeks cooperation first and only defects if the other player did so in the previous round. This strategy is considered benign because it does not try to get more than half of the spoils possible through cooperation.[23]

The Tit for Tat strategy applied to an indefinitely repeating Prisoner's Dilemma game may be evaluated for its mathematical robustness and for its

[20] Pettit shows how the PD and Tit for Tat have continued currency in mainstream philosophy and social science; Pettit, "*Virtus Normativa*," 2002, 320.
[21] Axelrod, *Evolution of Cooperation*, 1984, 30.
[22] Axelrod discusses this in ibid., 31.
[23] Ibid., 33.

social implications. From the perspective of evolutionary game theory, the main question is whether Tit for Tat represents an evolutionary stable strategy. This strategy does well when paired with itself, and both Axelrod and Dawkins stress that it is relatively impervious to challenge by mutant invaders.[24] This invader differs from the others and will be reflected by more of its phenotype in the next round of play if it gains more fitness points than other actors. However, despite the intrinsic allure that purely self-interested optimization could, in precisely replicating circumstances of dyadic play, demonstrably show the superiority of this nice strategy, Tit for Tat is not an evolutionary stable strategy.

To see this, assume that over many rounds of indefinite dyadic play, successful parents yield more offspring in the subsequent generation. Tit for Tat begins to become a predominant strategy, prevailing over its nasty cousins that are not able to reap rewards from cooperating. However, as Tit for Tat becomes more successful, even nicer strategies, specifically the Tit for Two Tats strategy proves even more successful. This nicer strategy is forgiving and only defects on another player after that player fails to cooperate in two rounds of play. Tit for Two Tats is then comparatively successful because Tit for Tat can easily become embroiled in a cycle of endless payback against another Tit for Tat strategy if the two players end up off-cycle because of erroneous play or imperfect memory recall. Tit for Two Tats is more forgiving and hence does not inadvertently spark an endless conflict. However, as Tit for Two Tats becomes predominant over Tit for Tat, it has the weakness that it opens the door again for the nasty strategy Probatory Retaliator that initiates play by first defecting and then cooperating. The moral is that even though a tendency toward cooperation is beneficial when many share the trait, it is susceptible to clever exploitation by a narrowly self-interested individually maximizing agent. Thus, even though Axelrod and Dawkins celebrate the outcome that the indefinitely repeated Prisoner's Dilemma could result in nice agents who prosper by enacting reciprocal altruism, this result is suggestive but not mathematically robust.[25] Thus, Tit for Tat sounds good on paper as a solution to a narrowly circumscribed PD encounter, but even in this idealized form, it fails to solidify cooperation as either necessary or likely.[26] Despite this weakness of Tit for Tat, even when considered solely between two players with good recall and perfectly repeating conditions, this strategy of conditional cooperation is still the mainstay of the neoliberal approach to explaining how cooperation can emerge from narrow self-interest and individualistic maximization.[27]

---

[24] Axelrod stresses the robustness of Tit for Tat in ibid., 96, as does Dawkins, *Selfish Gene*, 1989/ 2009, 216; however, Tit for Tat is not an evolutionary stable strategy.

[25] For a helpful clarifying discussion, see Steven Kuhn, "Prisoner's Dilemma," *Stanford Encyclopedia of Philosophy*, 2007.

[26] For discussion, see Shaun Hargreaves Heap and Yanis Varoufakis, *Game Theory: A Critical Text*, 2nd ed. (New York: Routledge, 2004), 199.

[27] For example, see Alexander Wendt, "Collective Identity Formation and the International State," *American Political Science Review* (1994), 88:2, 384–396.

Dawkins devotes the chapter "Nice Guys Finish First" to his treatment of Axelrod's tournament and the relative triumph of Tit for Tat in the 1989 edition of *The Selfish Gene*. Like Axelrod, he is optimistic about pure selfishness leading to cooperative conduct without introducing any extraneous motives or considerations. However, not only is Tit for Tat not evolutionarily stable, but it is also not clear that it can deliver on its promise to achieve social order in the Prisoner's Dilemma games widely believed by game theorists to characterize a significant class of interactions throughout society. The hope was that Tit for Tat, and the demonstration that cooperation can evolve in populations of egoists, could render order out of anarchy without the introduction of a central authority. This is especially true for the neoliberal school of international relations theorists.[28]

Both Dawkins and Axelrod move seamlessly in their discussions between actors who are simply replicating strategies and those who are human agents.[29] The keystone chapter in Axelrod's *Evolution of Cooperation* is "The Live-and-Let-Live System in Trench Warfare in World War I."[30] Like many other social interactions, trench warfare is determined by Axelrod to be a Prisoner's Dilemma game with the following payoff structure. Each agent chooses between either shooting to kill or shooting without aim. One's best outcome is to shoot to kill the other, while the other shoots aimlessly. Both prefer mutual aimlessness to mutually shooting to kill. Both least prefer to be the sucker who may be killed without attempting self-defense. Axelrod explains how by his analysis, trench warfare is "an iterated Prisoner's Dilemma in which conditional strategies are possible."[31] His idea is that a soldier would conditionally cooperate by withholding fire if the enemy did so in the prior round.

Yet counter to Axelrod's argument, Andrew Gelman argues in a recent paper that Axelrod's identification of a Prisoner's Dilemma payoff matrix underlying trench warfare is mistaken.[32] Gelman questions whether any single soldier cares whether or not he kills an opposing soldier. He further suggests that it is in no one's immediate self-interest to fire at all because this draws attention to oneself and makes it possible to identify one's location, thereby making oneself a more accessible target. Thus, again, even in an empirical application, it is not clear how relevant Tit for Tat conditional cooperation is to actors' choices.[33]

---

[28] Robert Axelrod and Robert O. Keohane, "Achieving Cooperation under Anarchy: Strategies and Institutions," in *Neorealism and Neoliberalism*, ed. by David A. Baldwin (New York: Columbia University Press, 1993), 85–115.
[29] See Dawkins's "Nice Guys Finish First," *Selfish Gene*, 1989/2009, 202–233; for a rationalization of this, see Philip Pettit, "Three Aspects of Rational Explanation," in his *Rules, Reasons, and Norms*, 2002, 177–191.
[30] Axelrod, *Evolution of Cooperation*, 1984, 73–87.
[31] Ibid., 1984, 77.
[32] Andrew Gelman, "Methodology as Ideology," *QA Revista dell'Associazione Rossi-Doria* (2014), 19:2, 167–175.
[33] In scenarios with multiple agents and different partners in each round of play, somehow the new interactant must punish defectors by singling them out; however, how this would be feasible remains unclear; see Pettit, "*Virtus Normativa*," 2002, 320.

Despite the fact that Tit for Tat is not an evolutionary stable strategy even under the refined conditions of Axelrod's tournament, neoliberals have pinned great hope on this apparent derivation of cooperation from the strict assumptions of myopic self-interest. Here is a typical explanation of this hope in the domain of international relations theory:

Most game theoretic studies of international cooperation [including Axelrod (1984)] and regimes have focused on the Prisoner's Dilemma (PD). PD is attractive since it can produce cooperative behavior under "realist" conditions. If play is repeated, the costs of defecting on any single move must be calculated not only with reference to the immediate payoff, but with reference to the opportunity costs associated with future interactions. Yet under assumptions of complex interdependence, the "dilemma" of PD diminishes. The very existence of a network of regimes and transnational relations among the advanced industrial states facilitates communication, enhances the importance of reputation, and lengthens the "shadow of the future." In its heuristic use PD indicates why these institutions deter suboptimal outcomes.[34]

Axelrod's suggestive demonstration that cooperation can spontaneously evolve without the imposition of a reward structure by a centralized authority led to enthusiasm for building institutions that created conditions of transparency. The idea is that if agents' actions are visible, then the alchemy of Axelrod's experimental setting would lead the actors to adopt the nice Tit for Tat strategy. Hume's knaves and Hobbes's Foole are transmuted, like lead into gold, to behave as enlightened individuals. This optimism about institutional design deserves to be considered an iteration of liberalism because it looks to structural constraints to induce cooperation among rational egoists. Institutions that work through the principle of transparency are more attractive than those working through surveillance, monitoring, policing, and sanctioning.

However, the hopes for institutions that could emulate Axelrod's tournaments to induce the intentional adoption of Tit for Tat are overstated. In the replicator dynamics used to identify evolutionary stable strategies, behavior is understood to be programmed into actors. Thus, nasty strategies are eliminated in successive generations of play. At first, they succeed rather well. However, as the easy prey is starved out of existence, and the competition becomes increasingly nasty, nastiness cannot pay off against itself. At this point, Tit for Tat prevails. However, as discussed earlier, Tit for Tat does not play as well as the more forgiving strategy Tit for Two Tats among less myopically self-interested actors. Biological evolution does not presume that actors may adopt strategies at will or whim, depending on a specific environment. Therefore, even if a clever institutional designer were to perfectly build institutions on the premise of indefinitely repeated perfectly transparent rounds of dyadic PD play, predatory opportunism cannot be effectively removed from the rational actor's motivation set. As soon as the end of play is in sight, or the next round is not almost as

[34] Stephen Haggard and Beth A. Simmons, "Theories of International Regimes," *International Organization* (1987), 41:3, 491–517, 505–506, internal citations omitted.

important as the present one, or players interact with other players only a few times, the hopes for Tit for Tat fade.[35]

Both Russell Hardin and Ken Binmore, two ardent devotees of the parsimony of the rational choice approach to politics, recognize this reality. Hardin, although not critical of the implications of Axelrod's study for ongoing two-person interactions, levies heavy criticism on the ability of Tit for Tat to achieve cooperation in an anarchic situation with many actors.[36] Hardin, as discussed in Chapter 9, views basic exchange as a Prisoner's Dilemma.[37] He emphasizes a conclusion he did not budge from over the course of his career: "my failure to cooperate in a large-number collective action ... is most likely to be a genuinely self-serving action that makes me better off now and in the longer run."[38] Hardin argues that individualistic optimization is the course determined by rational choice. He directly counters Axelrod's hopeful counsel for institutional designers, noting that "an understanding of dyadic iterated exchange relations does not yield us an analogous understanding of large-number iterated collective actions."[39]

Ken Binmore, a mathematical game theorist who migrated into political philosophy, similarly has a fondness for the Tit for Tat strategy.[40] However, by the end of his naturalized approach to justice recognizing solely rationally self-interested actors, he too must accept that dyadic conditional cooperation is insufficient to achieve large-scale social cooperation. Contemporary society, which gave rise to the Enron scandal and then worse lapses in collective encounters during the 2007 financial crisis, is far beyond repeated dyadic encounters or even small groups of individuals. Even though Binmore stresses the value of decentralization, institutions must still be designed to prevent the failure of collective action predetermined by rational agency. In his conclusion "Designing a Social Mechanism," he laments,

To achieve an efficient outcome, it is necessary to decentralize decision-making, but the agents to whom decisions are decentralized won't usually share the objectives the designer is seeking to achieve. Some of them will want to embezzle the funds entrusted to their care ... To some extent, the agents' behavior can be controlled by straightforward regulation that forbids certain practices, and makes others mandatory.[41]

Even though the liberal hope is to avoid a Leviathan state, and neoliberalism strives to design institutions purely on the premise of transparency, even Binmore acknowledges that ultimately behavior must be policed to enforce compliance. He states, "But only behavior that can be effectively monitored can be controlled in this way. Deviants can then be detected and punished by the external enforcement

---

[35] Luce and Raiffa, *Games and Decisions*, 1958, 109.
[36] Hardin, "Individual Sanctions, Collective Benefits," 1985, 339–354.
[37] Ibid., 340.
[38] Ibid., 350.
[39] Ibid.
[40] See his section on Tit for Tat; Ken Binmore, *Natural Justice* (Oxford University Press, 2005), 77–92.
[41] Ibid., 193.

agency whose existence is taken for granted in classical mechanism design."[42] Despite the optimistic reception of Tit for Tat, it remains unclear how this remedy for two-person indefinitely repeating PDs can possibly be extended to situations in which a new actor is encountered in almost every round of play.[43]

TIT FOR TAT AND RULE FOLLOWING

The axioms of rational choice theory are rules that could be programmed into actors, could exist as behavioral norms reinforced as survival criteria by environmental constraints, or could inform deliberate conduct. Game theory requires that agents have a set of strategies, based on their evaluation of the expected utility gain from every possible outcome, that tells the agent how to act in every conceivable circumstance.[44] These rules defining instrumental agency are deemed to be nonnegotiable. Thus, when game theorists identify Prisoner's Dilemma scenarios throughout political economy, international relations, and civil society, the message is that agents must defect. This is because defecting in a PD has implications directly relating to survival; although mutual cooperation is better than mutual defection, sole cooperation has a cost, and unilateral defection a reward. In evolution, natural selection chooses behavioral traits conducive to survival. According to rational choice theory, normal behavior among humans must have the similar property that it provides gain to actors.[45] Thus, it is unclear that one can teach a knave or a fool when to be a conditional cooperator, or that the rulebook would, unlike the Golden Rule, have more rules than exceptions.[46] In Axelrod's words,

Perhaps the most widely accepted moral standard is the Golden Rule: Do unto others as you would have them do unto you. In the context of the Prisoner's Dilemma, the Golden Rule would seem to imply that you should always cooperate, since cooperation is what you want from the other player.[47]

Axelrod suggests that unconditional altruism may be the best interpretation of the Golden Rule, but he believes that reciprocal altruism that accepts equity in the distribution of shares is ultimately a sufficient, and even better, approximation.[48]

---

[42] Ibid.

[43] Pettit seems to only consider PD games in which any single actor's defection in a multi-person game leads to other actor's subsequent defection in the subsequent round, but it is hard to see at which point a cooperative outcome could be captured; "*Virtus Normativa*," 2002, 320; see also Hardin, "Individual Sanctions," 1985.

[44] John von Neumann and Oskar Morgenstern, *Theory of Games and Economic Behavior* (Princeton, Princeton University Press, 1944/1953/2004), 31.

[45] Pettit, "*Virtus Normativa*," 2002.

[46] For a similar question about the guidelines to action recommended in David Lewis's *Convention: A Philosophical Study* (Oxford: Blackwell, 1969); see also Margaret Gilbert, "Game Theory and Convention," *Synthese* (1981), 46:1, 41–93.

[47] Axelrod, *Evolution of Cooperation*, 1984, 136.

[48] Ibid., 136–137; see also Binmore's appropriation of the Golden Rule, title of his chap. 9, *Natural Justice*, 2005, 129–145.

In biological evolution, organisms are programmed with strategies that game theorists typically argue must be individualistically optimizing. However, in human societies, people can adopt strategies at will. Axelrod's *Evolution of Cooperation* makes an unabashedly open case for action predicated on the principle of Tit for Tat. However, given interactions that, although defined by a PD payoff matrix reflecting tangible rewards, are not limited to two individuals and indefinite interaction, it is not at all clear how to implement the Tit for Tat rule of conduct. Not only is the individualistic maximizer free to try all strategies, including Probatory Retaliator, but in the end, the only safe strategy to avoid being suckered and to accrue the rewards from unilateral defection is to decline to cooperate unless incentivized to do so by institutional design. Neoliberal institutions must attempt to achieve conditions that make individuals subject to being punished for any round of play in which they defected in a prior round. This takes strong institutional infrastructures with mechanisms for monitoring, keeping dossiers on actors' earlier moves, and providing sanctioning devices to steer behavior into a mutually cooperative mold. However, the price paid for treating individuals in accordance with the assumption that strategic rationality is the only logic for their choice may end up driving out the type of voluntary cooperation consistent with logics of appropriateness, team reasoning, or other-regarding considerations.[49]

CONCLUSION

In conclusion, the following points are salient:

1. Tit for Tat is not an evolutionary stable strategy of indefinitely repeated two-actor Prisoner's Dilemmas. Notwithstanding this fact, evolutionary game theorists view its relative success among other strategies as a promising explanation for how cooperation could emerge among actors whose conduct is limited to the prescriptions of strategic rationality, and who must compete over finite resources as a condition of their survival.

2. Evolutionary game theorists describe Tit for Tat as a nice strategy because its actors only seek at most half of the total environmentally available resources. Its relative success in the iterated PD game suggests that if individuals confront circumstances resembling this model, then they may do well to emulate it, and moreover it may be possible to act in accordance with this benign strategy and still obey the competitive logic of noncooperative game theory.

[49] Elinor Ostrom, "Policies That Crowd out Collective Action," in *Moral Sentiments and Material Interests*, ed. by Herbert Gintis, Samuel Bowles, Robert Boyd, and Ernst Fehr (Cambridge, MA: MIT Press, 2005); Dan Kahan, "The Logic of Reciprocity," in ibid.; Dan Kahan, "Social Influence, Social Meaning, and Deterrence," *Virginia Law Review* (1997) 83, 349–396; for an overview, see Bruno Frey, "Does Monitoring Increase Work Effort? The Rivalry with Trust and Loyalty," *Economic Inquiry* (1983) 31, 663–670.

3. However, in evolutionary games, behavioral traits, such as always cooperating, never cooperating, or conditionally cooperating in a repeating PD with the same opponent, are programmed into agents rather than freely adopted on consideration and will.

4. Hence, for both the reasons that the repeating PD models such a rarified context that it does not pertain to those circumstances with many actors, distant futures, or few encounters, and that the behavioral norms derived as Nash equilibria of games emerge behind the back of agents without their reflection or conscious compliance, it remains unclear how valuable Tit for Tat is for achieving either the rule of law, private market exchange, or stability among nations.

CONCLUSION

12

Pax Americana

> I have, therefore, chosen this time and place to discuss a topic on which ignorance too often abounds and the truth too rarely perceived. And that is the most important topic on earth: peace. What kind of peace do I mean and what kind of a peace do we seek? Not a *Pax Americana* enforced on the world by American weapons of war. Not the peace of the grave or the security of the slave. I am talking about genuine peace, the kind of peace that makes life on earth worth living, and the kind that enables men and nations to grow, and to hope, and build a better life for their children – not merely peace for Americans but peace for all men and women, not merely peace in our time but peace in all time.
>
> John Fitzgerald Kennedy[1]

Throughout the twentieth century, despite the shadow cast by US capitalism and dominion, Americans stood confident that their country was grounded by values and commitments that rejected tyranny and oppression in favor of the impartial rule of law and individual self-determination. Progress toward realizing these values was uneven and halting. But for American citizens, hopeful immigrants, and foreigners inspired by these ideals, the US Constitution and Declaration of Independence perpetuated the Enlightenment ethos of self-emancipation. A *Pax Americana* could be envisioned to be inseparable from the American Dream of inclusive wealth creation "with opportunity for each according to his ability or achievement."[2]

It goes without saying that one person's claim to freedom cannot, without contradiction, involve denying another's equal claim. However nostalgically one may remember the moral high ground of World War II or the prolific power of mid-century "made in USA" industry, the twenty-first century social contract is blatantly antithetical to an inclusive vision for prosperity. Strategic

---

[1] J. F. Kennedy, "Commencement Address American University," June 10, 1963, Americanrhetoric .com. Accessed July 29, 2014.
[2] James Thruslow Adams, *The Epic of America* (Little, Brown, 1931), 214–215.

rationality and the Prisoner's Dilemma model for social relations that is its central logical paradox recommend both political domination and economic chicanery.

Game theory is ubiquitous. It dominates academic curricula and economic models, and its application spans market practices, institutional design, and public policy implementation. Strategic rationality is so much part of our contemporary worldview that it is intertwined with the evolving meaning of the word "cynical." Game theory, like late twentieth-century cynicism, has three moments. The first captures the initial meaning of ancient Greek cynicism: using reason to understand the world to live in accordance with its true nature and thus strive to achieve happiness, or *eudemonia*. The second captures the latent anxiety that the world may not be forgiving, and that human nature and thermodynamics may inevitably reflect endless competition and the gradual heat death of the universe.[3] At this stage in the 1970s, neoliberal theorists understood their challenge to be to accept the worst but still try to find a way to reconstruct the positive system of classical liberalism in the hope of building a mutually prosperous basis for capitalism and democracy. The eventual end of the Cold War seemed to offer this prospect. All that was needed was for the baby boom generation to rise to the occasion of this optimistic future symbolized by the Apollo 11 lunar landing on July 20, 1969. Yet instead, the optimism soon faded. President Richard Milhous Nixon mired the country in impeachment proceedings as a result of his paranoid excesses, destroying the public's faith in government. The "three strikes" penal reform initiative resulted in a mass incarceration state. The peace dividend that was supposed to follow the Cold War never appeared, giving way instead to successive wars and mounting debt. As a result, the third wave of post–World War II neoliberal cynicism exudes a pessimism that the world could ever be otherwise: limited resources and the selfish gene are the realities with which we are confronted, and satisfying desire as best as we can, in the present moment of this life, is all that remains of the existential significance of human subjectivity and life experience.

From this perspective, returning to classical liberalism is not an option. Its ethical compass of voluntary self-restraint in respecting others' right to exist is contrary to one's prerogative to survive and propagate. The commons and the public are antiquated relics beset with crisis and tragedy. The Enlightenment equipoise between a private sphere for expressing self-identity in faith or without and a public realm that encompasses both commons and all private possessions, with the full acknowledgment of the zero-sum quality of physical property yet the belief in the positive-sum benefit of security in ownership, has vanished. In its place stands a single metric of value that treats all human experience as commensurable and reducible to an individual's willingness and ability to pay for goods and outcomes, regardless of how close a person lives to

---

[3] Randall Schweller, *Maxwell's Demon and the Golden Apple: Global Discord in the New Millennium* (Johns Hopkins University Press, 2014).

the level of subsistence, or even below. The rule of law, as well as individuals' equality before it, is a nostalgic artifact with mythopoeic dimensions of an Atlantian era transcending the brutal reality that in a neoliberal regime, laws and their enforcement are open to the highest bidder: "to each according to his threat advantage."[4] The modern Third Estate, informing public debate and effective citizenry, is in free fall because information is valued as a strategic commodity and not an indispensable public good.

This book has focused on the pivotal role of the invention of nuclear weapons in the neoliberal turn. Missiles and bombs, no matter how deadly, provide no security unless they are integrated into a war plan that coherently advances and effectively executes the goals of a sovereign nation. Rational deterrence theory made it possible to wield threats of unfathomable terror as a preeminent means to project national power. The first rational actor was the thermonuclear-armed national security state. The nuclear security dilemma and arms race became prime examples of the Prisoner's Dilemma game. Mutual assured destruction (MAD) offered the means to secure American nuclear sovereignty by relying on unassailable counter-strike capability. However, even Thomas Schelling, its pioneer, was not able to marshal an effective defense of it in light of the perceived immorality and incredibility of executing the threatened retaliation if deterrence failed. As a result, nuclear utilization targeting selection (NUTS) was developed to demonstrate the US intention and capability to wage and prevail in warfare at any level of conflict. This way, destruction served the constructive purpose of furthering American security goals rather than submitting to defeat in a final act of cataclysmic ruination.

Rational choice provided a means of performing national security via commanding and controlling a vast nuclear arsenal in accordance with the principles of rationality. It helped provide a seemingly rational order to the irrational arms race. This exercise of sovereignty soon permeated all levels of social order, as game theory became the preferred means to describe, enact, and explain purposive agency. Analysts embraced strategies and models used to address a worst-case scenario requiring urgent and painstaking attention as the best means to understand civil society, markets, government, and even interpersonal relations. The prerogatives of national security were integrated into the most mundane of interactions. A social contract based on consent and voluntary compliance yielded to the neoliberal regime of de facto possession, coercive bargaining, *ex post* consent, the reduction of individual worth to willingness and ability to pay, and single criterion accounting that reduces all phenomena and entities to one finite commensurable source of value.

---

[4] Edward F. McClennen, "The Tragedy of Sovereignty," in *Nuclear Weapons and the Future of Humanity*, ed. by Avner Cohen and Steven Lee (Totowa, NJ: Rowman and Allanheld, 1986), 391–405 at 399.

NEOLIBERAL POLITICAL PHILOSOPHY

If there is a contentious element to the argument presented throughout *Prisoners of Reason*, that neoliberal subjectivity and governance are uniquely conceived by noncooperative game theory, the point of controversy is not the wide-reaching application of rational choice but the claim that it offers a distinctive and unprecedented understanding of action. Game theory is used to descriptively model action, its guidelines for action may be programmed into mechanical decision-making devices, and it can also offer prescriptive advice for individual decision makers, institutional designers, and public policy analysts. However the prescriptive validity of expected utility theory on which game theory relies for its concept of solutions is not subject to empirical validation.[5] The economist and Nobel Memorial Prize winner Roger C. Myerson observes that "one can only ask whether a person who understands the model would feel that he would be making a mistake if he did not make decisions according to the model."[6] At the same time, game theory was presented as a comprehensive theory informing individuals "how to behave in every situation which may conceivably arise."[7] Yet realistically, to be operationalizable, the decisions individuals make must take on the simplicity of parlor games, with rewards identified in terms of a salient tangible property such as cash value.[8] Processes, means, and side constraints cannot factor into choice. Actors must pursue gain independently of and despite others.

Each of the chapters of *Prisoners of Reason* is designed to impart a counter-intuitive self-contained insight that either clarifies the distinctive quality of neoliberal politics and economics from its modern classical liberal predecessor or sheds light on a misunderstanding theorists may have inadvertently perpetuated to maintain the continuity between classical liberalism and neoliberalism. Chapter 1, "Neoliberalism," argues that the assumptions underlying strategic rationality rule out non-consequentialist considerations for judgment. Game theory does not recognize side-constraints on action consistent with fair play, commitment, or the Archimedean classical liberal perfect duty not to harm other individuals.

Chapter 2, "Prisoner's Dilemma," presents how the standard assumptions used to operationalize game theory in most contexts – only outcomes factor into choice, value is fungible and precedes social context, individuals act independently despite others, and gratuitous generosity is counter to instrumental rationality – are precisely those on which the Prisoner's Dilemma game depends to generate its troubling outcome. Neoliberal subjectivity results from internalizing these

---

[5] Roger B. Myerson, *Game Theory* (Cambridge, MA: Harvard University Press, 1991), 22; on the analytic structure of game theory, see also Philip Green, *Deadly Logic* (Columbus: Ohio State University Press, 1966).

[6] Ibid., 22.

[7] John von Neumann and Oskar Morgenstern, *Theory of Games and Economic Behavior* (Princeton, NJ: Princeton University Press, 1994/2014), 31.

[8] Anatol Rapoport, *Fights, Games, and Debates* (Ann Arbor: University of Michigan Press, 1960), 107–109, 126–127; Myerson, *Game Theory*, 1991, 3.

guidelines for action. Clarity over the assumptions locking actors into a Prisoner's Dilemma enables us to exit at will by disregarding a fundamental assumption of strategic rationality: we may consider means and act with commitment, explore non-finite sources of value derived from particular relations and circumstances, work as a team, and give to others without any expectations of gain to ourselves.

In Part I, "War," Chapter 3, "Assurance," Thomas Schelling derives a Prisoner's Dilemma from an Assurance Dilemma to model the nuclear security dilemma. He initiated what would become the neoliberal institutionalist approach to international relations: accepting the most pessimistic motives of actors yet showing how they may still achieve a mutually beneficial cooperative outcome. Thus, Schelling used the PD model to defend mutual assured destruction, which mimicked the classical liberal security stance of cooperating when others do but meeting aggression with aggression. Yet, in attempting to make his argument valid in the worst-case scenario, Schelling conceded the comprehensive logic of game theory, and that benign cooperators must adopt the characteristics of aggressors in self-defense. Thus, in striving to seek mutual assurance through reciprocal deterrent threats, he laid the groundwork for shifting from the liberal framework of commitment to achieve cooperation when possible to the neoliberal framework of maintaining security through coercive bargaining and asymmetric deterrent threats.

Chapter 4, "Deterrence," argues that the new neoliberal approach to agency could not salvage the classical liberal approach to mutual security based on assurance and legitimacy recognizing people's right to exist. The chapter analyzes how the strategists tasked with rendering deterrence rational, credible, and effective could not and did not decide the nuclear security debate between MAD and NUTS in favor of the former. Thus, in its highest expression of sovereignty, the ability to command and threaten others with nuclear devastation, the United States shifted its basis for engaging other nations to the sole platform of strategic rationality. Notwithstanding the existential certainty of MAD, no contemporary philosophy relying on strategic rationality could provide a sound argument to President Carter to counter his defense analysts' conclusion that the only way to avoid nuclear combat was to prepare to fight a prolonged nuclear war.

Part II, "Government," follows rational choice theorists' conclusion that anarchy, like the intense nuclear security dilemma, whether among nations or among individuals, is best modeled by the Prisoner's Dilemma game. Chapter 5, "Hobbesian Anarchy," argues that game theorists' assertion that *Leviathan* provided a solution to the Prisoner's Dilemma game underlying Hobbes's state of nature misses that Hobbes's actors are better regarded to be in an Assurance Game than a Prisoner's Dilemma. Moreover, strategic rationality proposes the opposite to Hobbes's solution: whereas game theory assumes that actors calculate what to do based on knowledge of value, probabilities of outcomes, and common knowledge of others' preferences, Hobbes assumes that radical uncertainty characterizes anarchy. Actors must stop acting out of the hope for immediate gain and must in every action seek peace by desisting from injuring others, obeying the commands of the sovereign, and upholding agreements personally made.

Chapter 6, "Social Contract," analyzes James M. Buchanan's construction of the social contract from a Prisoner's Dilemma payoff matrix. Buchanan's innovation, the core of neoliberal market exchange and governance, renders explicit the paradigmatic shift from classical liberalism to neoliberalism. The neoliberal social contract, or Prisoner's Dilemma model of governance, displaces normative bargaining predicated on the no-harm principle and voluntary compliance in favor of coercive bargaining and a strong security state. Actors can seek more gain by credibly threatening harm to others, and the strong can exploit the weak's pursuit of subsistence by lowering their expectations should no social contract be achieved.

Chapter 7, "Unanimity," investigates Buchanan's transformation of this principle from a foolproof means to validate legitimate governance to a momentary expression of aligned preferences. Consistent with strategic rationality, Buchanan cuts the principle of unanimity free from either a foundation in an interpersonally evident basis of preference satisfaction or the expression of consent validated by commitment to uphold agreements. Hence, this formerly powerful means to sanction market exchange and validate governance loses the normative purchase to legitimate any collective outcome. Even if mutual cooperation is unanimously preferred to unanimous defection in considering a social contract versus anarchy, if an individual could achieve a greater division through a new agreement, then that person may find it worthwhile to treat the prior mutually cooperative outcome as indistinguishable from mutual defection. This holdout tactic could force others to cooperate for less if they are more desperate for an agreement.

Chapter 8, "Consent," examines how rational choice theory suggests that because individuals' preferences over outcomes are path independent and stand outside of time, *ex ante* agreement and *ex post* agreement to terms are indistinguishable. Thus, actors with a greater willingness coupled with the ability to pay for rights or capability to generate greater income streams with those rights can legally acquire others' property without their *ex ante* consent simply by offering *ex post* compensation. Richard Posner's concept of justice as wealth maximization permits the state to reassign rights if a reallocation enables those acquiring the entitlement to generate more income from the property. Moreover, the Kaldor-Hicks compensation principle Posner relies on recognizes that wealth is generated in the case that winners could compensate the losers of an entitlement transfer, *even if no compensation is paid.*

Chapter 9, "Collective Action," discusses rational choice theorists' analysis of the challenge of achieving collective action in terms of the multiple-actor Prisoner's Dilemma game. It compares this analysis to that of Mancur Olson, who argued that very large groups are bound to have a difficult time securing cooperation, even if those actors are motivated by side constraints or concerns for others' well-being, because individuals realize that their efforts make no causal impact on whether the collective undertaking fails or succeeds. Thus, we must distinguish between the problem of collective action in large groups that results from the failure of instrumental action through the worry of causal negligibility, and the Prisoner's Dilemma logic in which actors defect because what they gain by suckering others is the causal result of displacing costs onto their partner in interaction. As under perfect competition, strategic rationality has no purchase in a large-scale collective action.

Part III, "Evolution," draws attention to the extension of noncooperative game theory from international relations and civil society to evolutionary biology because this theory works at multiple levels from programming behavior, to modeling how behavioral norms arise from repeated patterns of individuals' expected utility maximization, to guiding individuals' decision making and public policies. Chapter 10, "Selfish Gene," explains how Richard Dawkins relied on noncooperative game theory and the concept of an evolutionary stable strategy to develop his theory that individuals are necessarily programmed to be individually and momentarily self-seeking in the cosmic and competitive game of life and death. Evolutionary game theorists propose that humans evolved to maximize instrumentally salient properties of the world, and that game theory provides a means to unify the social and behavioral sciences. Preferences are exogenously determined, since agents' preference rankings must reflect environmentally dictated success criteria.

Chapter 11, "Tit for Tat," assesses the neoliberals' silver bullet for explaining how cooperation can emerge among strategically rational actors. Although assuming consequentialism, realism, and individualism, the Tit for Tat behavioral rule mimics the Golden Rule by programming actors to first choose cooperation rather than defection in indefinitely repeated, two-actor Prisoner's Dilemma games. However, Tit for Tat is not an evolutionary stable strategy, and the settings in which it could prove an effective decision rule are so limited that it is better regarded as contributing to a precarious equilibrium much like balancing a ball on the pinnacle of a cone.

RETROSPECTIVE AND PROSPECTIVE

The classical liberals built a world celebrating individual freedom to pursue one's own goals in an organized civil society that relies on individuals' reciprocal respect. Neoliberals seek to approximate this same world but on the assumption that individuals satisfy their preferences despite others. Human civilization differs little from that of other creatures on earth or even throughout the cosmos. Behavioral traits are programmed into organisms as a condition of their survival and propagation. Alternatively, habitual action conforming to Nash's mutual-best-reply equilibrium can emerge and become mutually sustaining over cycles of repeating encounter. On this view, language, communication, and truth-telling emerge as behavioral patterns that serve life forms in their individual maximization of a salient environmental property necessary for survival.[9]

As an analytic paradigm, game theory cannot be falsified, nor can it be empirically demonstrated to be prescriptively valid.[10] In its standard form, it is limited to the logic of consequences and cannot acknowledge gratuitous generosity. It thus negates the classical liberal perfect and imperfect duties. Furthermore, it typically only accepts noncooperative individualistic maximization and limits value to scarce sources resembling energy in physics. As a result actors cannot work together as teams, and they compete for limited sources of value as in the parlor games John von Neumann originally studied.

Neoliberal explanations will gravitate toward the analytic and empirical centrality of the Prisoner's Dilemma game.[11] Neoliberal institutions will treat actors as though they were strategically rational and solely motivated by incentives.[12] Acknowledging that the Prisoner's Dilemma impasse lies at the core of neoliberal political economy, the political theorist Philip Pettit observes,

> The norms that have been at the focus of concern in the rational choice literature are those such that conformity to them enables people to resolve free-rider problems, in particular problems that are also many-party prisoner's dilemmas. In a prisoner's dilemma each party faces options of cooperating or defecting in some way and the following two conditions are fulfilled: universal cooperation is Pareto-superior to universal defection, being better for some – perhaps for all – and worse for none; but defecting is the dominant option, being better for each regardless of what others do.[13]

Pettit finds that reliably telling the truth, keeping promises, refraining from theft or violence, serving public interest in the capacity of an officer of state, and reliably contributing to public interest that serves everyone's goals has the structure of a multi-actor Prisoner's Dilemma. He states, "Arguably,

---

[9] David Lewis sets out this view of language and truth in *Convention* (Cambridge, MA: Harvard University Press, 1969); this is discussed by S. M. Amadae, "Normativity and Instrumentalism in David Lewis' Convention," *History of European Ideas* (2011), 37, 325–335; see Philip Pettit for discussion of how Tit for Tat could account for the norm of truth telling: "*Virtus Normativa*: Rational Choice Perspectives," in his *Rules, Reasons, and Norms* (Oxford: Clarendon Press, 2002), 321; Brian Skyrms, *The Stag Hunt and the Evolution of Social Structure* (Cambridge: Cambridge University Press, 2004), 45–81.

[10] Ken Binmore, *Playing Fair* (Cambridge, MA: MIT Press), 95; Myerson, *Game Theory*, 1991, 22.

[11] Pettit, "*Virtus Normativa*," 2002, 319.

[12] These incentives are typically tangible, although Pettit argues that people can also be motivated by the cost-benefit analysis of the intangible incentive of approval or disapproval; ibid., 326–337; see also Pettit and Geoffrey Brennan, "Hands Invisible and Intangible," *Synthese* (1993), 94, 191–225. Joseph Heath models deontic constraints as though they function like preferences over outcomes as an additional consideration, similar to Pettit's cost-benefit analysis of approval but instead focused on perceived rightness of an action; see also Joseph Heath, *Following the Rules* (Oxford: Oxford University Press, 2011), 65–98.

[13] Pettit, "*Virtus Normativa*," 2002, at 319.

conforming to norms like ... [this preceding list] is equivalent to cooperating in a many-party prisoner's dilemma, so that universal ... [or] fairly general conformity ... represents an escape from the predicament."[14] Behavioral norms are the solutions to games but, as the Nash mutual-best-reply equilibrium indicates, may not exhibit patterns of optimal resource use. Behavioral tactics including either strong governance or endogenous punishment through vigilant shaming of defectors are necessary to prevent every actor from pursuing the temptation of free riding on others.

Analyzing and modeling "social dilemmas" with the PD game and identifying some incentive structure, either through endogenous punishment or exogenous sanctioning, have characterized much of the neoliberal enterprise. However, we know from experiments that situations that researchers frame as PDs are often not interpreted as such by subjects.[15] Moreover research reveals that treating individuals as strategic rational actors who are moved only by incentives can crowd out ethical and other-regarding conduct.[16] Analytic modeling suggests that opportunists, or the standard rational actors Pettit describes who have the preference to defect while others cooperate, can also negatively impact their partners in interaction who respect alternative logics or values for action.[17] Hence, noncooperative game theory may best be viewed as a transformative paradigm that may be used for descriptive modeling, but it is also presented as a tool for individual decision making, in addition to public policy and institutional design.[18]

Cultivating neoliberal subjectivity conforming to strategic rational action then can be achieved through learning its method, inhabiting institutions structured in accordance with its logic, and being exposed to neoliberal actors. *Prisoners of Reason* opened with the contrast between the theoretical underpinnings of classical versus neoliberal agency and political economy to demonstrate at a minimum the unique and idiosyncratic character of strategic rationality, and more ambitiously to offer its Western predecessor as an alternative understanding of action familiar to many within a generation's reach of

---

[14] Ibid.

[15] Anatol Rapoport and Albert M. Chammah, *Prisoner's Dilemma* (Ann Arbor: University of Michigan Press, 1970); Jeremy Cone and David G. Rand, "Time Pressure Increases Cooperation in Competitively Framed Social Dilemmas," *PLoS ONE* (2014) 9:12, e115756. doi:10.1371/journal.pone.0115756.

[16] Elinor Ostrom, "Policies That Crowd out Collective Action," *Moral Sentiments and Material Interests*, ed. by Herbert Gintis, Samuel Bowles, Robert Boyd, and Ernst Fehr (Cambridge, MA: MIT Press, 2005); Dan Kahan, "The Logic of Reciprocity," in ibid.; Dan Kahan, "Social Influence, Social Meaning, and Deterrence," *Virginia Law Review* (1997), 83, 349–396; Bruno Frey and Reto Jegen, "Motivation Crowding Theory," *Journal of Economic Surveys* (2001), 15:5, 589–511.

[17] Marie-Laure Cabon-Dhersin, "Opportunism, Trust and Cooperation: A Game Theoretic Approach With Heterogeneous Agents," *Rationality and Society* (2007), 19:2, 203–228.

[18] Myerson, *Game Theory*, 1991, 22.

the present.[19] An individual who wishes to resist neoliberal subjectivity and imperatives of action faces the challenge of going against the grain of a cultural era in which noncooperative strategic rationality is the academically sanctioned norm at elite universities and leading institutions and polices are designed to implement its logic.[20]

The demands of expected utility theory are great: every conceivable world state must be ranked ahead of time and for all time on a single consistent scale including lotteries of potential world states, and this utility function renders superfluous the actual moment or act of judgment.[21] *Ex ante* and *ex post* consent are, in this worldview, irrelevant distinctions. Consent is gathered directly from preferences. The operationalization of game theory relies on simplifications, the most prominent being the introduction of monetary denominations to reflect individuals' preference rankings. However, the single metric criterion, which must operate as a limited resource, reveals that, ultimately neoliberalism deploys a philosophy of value that more resembles mercantilism than classical liberalism. The single criterion scale requires that preferences for an outcome be stated as a willingness and ability to pay for an outcome with resources already on hand. All value then is constrained by the finite scale that must have a well-defined cap to be useful. This method is perfectly transparent if, as von Neumann and Morgenstern originally observed, expected utility functions such as temperature in physics and is responsive to an underlying objective energy state of the system.

RESISTING NEOLIBERAL SUBJECTIVITY

There are three ways to respond to the standardized application of game theory. First, one could accept the paradigm and strive to identify means within it to construct institutions that will best approximate the classical liberal vision of mutual prosperity under limited government. However, the theoretical concessions necessary to live within the confines of strategic rationality make this impossible. The NUTS resolution of the nuclear security debate, via escalation dominance and coercive bargaining, will characterize other relations bounded by strategic action. By contrast, the liberal aspiration was to secure freedom by living under principles and agreements ratified by consent in opposition to laws

[19] Robert Nozick, *Anarchy, State, and Utopia*, 2nd ed. (New York: Basic Books, 2013); and John Rawls, *A Theory of Justice* (Cambridge, MA: Belknap Press, 1971) are both classical liberal texts. Nozick invokes side constraints, and Rawls relies on fair play.

[20] Gary Becker, *Economic Approach to Human Behavior* (Chicago: University of Chicago Press, 1978); Douglass C. North, *Institutions, Institutional Change, and Economic Performance* (Cambridge: Cambridge University Press, 1990); and Nicolas Stern, *Stern Review on the Economics of Climate Change*, available online, http://webarchive.nationalarchives.gov.uk/20100407172811/http://www.hm-treasury.gov.uk/stern_review_report.htm, accessed July 21, 2015.

[21] David Lewis, *Convention*, 1969; von Neumann and Morgenstern, *Theory of Games*, 1944/2004, 31.

imposed de facto as a *fait accompli*. Perhaps, as some rational choice theorists have surmised, the rationality paradigm cannot surpass the Hobbesian Leviathan state and in fact must be even more invasive than Hobbes's early modern commonwealth to achieve the necessary means to leverage compliance.

Second, those who are dissatisfied with the implications of strategic rationality may expand the bases of rational choice to permit sufficient leeway to exit the Prisoner's Dilemma gracefully. This can be accomplished in several ways. The least demanding is to recognize that individuals can act as members of a team. From this perspective, individuals identify with a group and play their role to maximize a shared expected utility function. Thus, the members of the group could be the two or more individuals in any social dilemma classified as a Prisoner's Dilemma. Construing a choice problem in this way saves instrumental rationality, although it denies the premise of individual maximization.[22]

Alternatively, an individual could superimpose a personal subjective evaluation on a tangible resource dilemma so that mutual cooperation registers a higher value than unilateral gain. This is a less tractable resolution because instrumental rationality typically tracks actors' obtainment of tangible outcomes. A new value metric would need to be devised to reflect payoffs in terms other than cash rewards, or any other such interpersonally transferable scarce resource. However, regardless of how game theory may be augmented to accommodate an individual's subjective predilection for mutually cooperative outcomes over unilateral success, the litmus test for determining if one is unwillingly and unnecessarily trapped in a Prisoner's Dilemma game is what one would do if one were assured of the other's cooperation in a tangible resource dilemma. If one defects, then one's choice conforms to Prisoner's Dilemma logic of seeking self-advancement despite others. If one cooperates, then the Prisoner's Dilemma payoff matrix does not apply to one's choice in the first place. Thus, this individual is not and never was in a Prisoner's Dilemma, but was instead in an Assurance Dilemma or Assurance Game. Individuals cannot be trapped in a Prisoner's Dilemma against their will. Although alleviating an Assurance Dilemma may not be trivial, it is logically surmountable unlike the PD (in which actors hope to sucker others) and is amenable to concrete solutions.[23]

Third, and finally, actors may conclude that strategic rationality is too constraining to offer guidance in all, many, or even just some choice situations. Students of rational choice may be surprised by the realization that strategic rationality contradicts the classical liberal rule of law and free market systems that regulates private affairs through self-binding agreements. Game theory

---

[22] Robert Sugden, "Thinking as a Team: Towards an Explanation of Nonselfish Behavior," *Social Science and Policy* (1993) 10:1, 69–89, S.M. Amadae and Daniel Lempert, "Long-Term Viability of Team Reasoning," *Journal of Economic Methodology*, 2015.

[23] Randall L. Schweller argues this point within international relations theory in "Neorealism's Status-Quo Bias: What Security Dilemma?," *Security Studies*, (1996) 5:3, 90–121.

encourages substituting coercive bargaining, or pressing one's threat advantage, for normative bargaining that respects the no-harm principle because the former is consistent with noncooperative behavior. Game theory's strict denial of the meaningfulness of self-determined normativity replaces self-governance by participation and consent with governance by incentives. Neoliberal institutions reframe trust, loyalty, and commitment as practices that must be formalized in systems of rewards and sanctions that rely on institutional means to monitor and record individuals' actions.

The following considerations offer ways to resist neoliberal subjectivity:

1. Abraham Maslow's hierarchy of needs – physiological, safety, love and belonging, esteem, and self-actualization – reminds us that many, if not most, human aspirations are inherently unbounded in their potential for satisfaction. Even elementary physiological needs for food, clothing, shelter, transportation and sanitation may be more constructively envisioned in line with the Epicurean ethic of *ataraxia*, or tranquility in sufficiency and affinity, rather than with bottomless consumption and ceaseless competition.[24]

2. Psychologists have revealed the transformative power of the belief in volition and free will against the view of biological destiny or physical determinism.[25]

3. We may consider that democracy is participatory and relational, as opposed to perfunctory and irrational, drawing on experiences from the developing world.[26] Against the moral of narrow self-interest, we may perceive generosity and sharing abundance to be the hallmarks of independence and freedom from constraint.[27]

4. We can subscribe to Joseph S. Nye Jr.'s foreign policy of cosmopolitan realism, embracing a classical liberal platform for mutual security. The ethical commitments underlying the American Dream and *Pax Americana* pave a surer path to global security than do fighting terror with escalating threats of violence.[28]

---

[24] Abraham Maslow, *Motivation and Personality* (New York: Harper, 1954).

[25] John Tierney, "Do You Have Free Will? Yes, it's the only Choice," *New York Times*, March 21, 2011, D; see also Thomas Nagel, *Mind and Cosmos: Why the Materialist Neo-Darwinian Conception of Nature Is Almost Certainly False* (New York: Oxford University Press, 2012).

[26] Read the United States pragmatic political theorist who is credited with influencing the Arab Spring 2010–2011: Gene Sharp, *From Dictatorship to Democracy: A Conceptual Framework for Revolution*, 4rth edition (Boston, MA: Albert Einstein Institution, 2010).

[27] Seth Godin, *Lynchpin: Are You Indispensible*, (New York: Portfolio, 2011); see also references to the virtue of magnanimity in Adam Smith, *Theory of Moral Sentiments*, ed. by D.D. Raphael and A. L. Macfie (Indianapolis: Liberty Fund, 1982).

[28] Joseph S. Nye, Jr., *Nuclear Ethics*, (New York: Free Press, 1986), see also Michael Lind, *The American Way of Strategy: U.S. Foreign Policy and the American Way of Life* (New York: Oxford University Press, 2008).

Bibliography

"The Prisoner's Dilemma." In *Introduction to Philosophy: Classical and Contemporary Readings*, edited by Michael E. Bratman and John Perry. New York: Oxford University Press, 1993.

"Economic Focus: Never the Twain Shall Meet." *Economist*, February 2, 2002.

"Robert Aumann's and Thomas Schelling's Contributions to Game Theory: Analyses of Conflict and Cooperation." In *The Royal Swedish Academy of Sciences*, 2005.

Achen, Christopher H. and Duncan Snidal. "Rational Deterrence Theory and Comparative Case Studies." *World Politics* 41, no. 2 (1989): 143–169.

Adams, James Thurslow. *Epic of America*. Boston: Little, Brown, 1931.

Alexander, Larry and Emily Sherwin. *The Rule of Rules: Morality, Rules, and the Dilemmas of Law*. Durham: Duke University Press, 2001.

Alt, James E., Margaret Levi, and Elinor Ostrom. *Competition and Cooperation: Conversations with Nobelists about Economics and Political Science*. New York: Russell Sage Foundation, 1999.

Amadae, S. M. *Rationalizing Capitalist Democracy: The Cold War Origins of Rational Choice Liberalism*. Chicago: University of Chicago Press, 2003.

Amadae, S. M. "Impartiality, Utility and Induction in Adam Smith's Jurisprudence." In *The Adam Smith Review*, edited by Vivienne Brown, 4, 238–246. London: Routledge, 2008.

Amadae, S. M. and Daniel Lempert. "Long-Term Viability of Tem Reasoning." *Journal of Economic Methodology* (2015).

Ariely, Dan and Nina Mazar. "Dishonesty in Everyday Life and Its Policy Implications." *Journal of Public Policy and Marketing* 25, no. 1 (2006): 1–21.

Arrow, Kenneth J. *Social Choice and Individual Values*. New York; London: Wiley; Chapman & Hall, 1951.

Arrow, Kenneth J. *Social Choice and Individual Values*. 2nd ed. New Haven: Yale University Press, 1963.

Attanasio, John. "Aggregate Autonomy, the Difference Principle, and the Calabresian Approach to Products Liability," in *Philosophical Foundations of Tort Law*, edited by David G. Owen (Oxford: Clarendon Press, 1995), 229–320.

Avanzadi, Javier. *Liberalism against Liberalism: Theoretical Analysis of the Writings of Ludwig Von Mises and Gary Becker*. London: Routledge, 2006.

Axelrod, Robert M. *Conflict of Interest: A Theory of Divergent Goals with Applications to Politics*, Markham Political Science Series. Chicago: Markham, 1970.

Axelrod, Robert M. *The Evolution of Cooperation*. New York: Basic Books, 1984.

Axelrod, Robert M. "The Emergence of Cooperation among Egoists." In *Paradoxes of Rationality and Cooperation: Prisoner's Dilemma and Newcomb's Problem*, edited by Richmond Campbell and Lanning Sowden, 320–338. Vancouver: The University of British Columbia Press, 1985.

Axelrod, Robert and Lisa D'Ambrosio. "Announcement for Bibliography on the Evolution of Cooperation." *Journal of Conflict Resolution* 39 (1995): 190.

Axelrod, Robert and Robert O. Keohane. "Achieving Cooperation under Anarchy: Strategies and Institutions." In *Neorealism and Neoliberalism*, edited by David A. Baldwin, 85–115. New York: Columbia University Press, 1993.

Bacharach, Michael. "Interactive Team Reasoning: A Contribution to the Theory of Co-Operation." *Research in Economics* 53 (1999): 117–147.

Bacharach, Michael. *Beyond Individual Choice*. Princeton, NJ: Princeton University Press, 2006.

Bailyn, Bernard. "The Ideological Origins of the American Revolution." (1967).

Baldwin, David A., ed. *Neorealism and Neoliberalism: The Contemporary Debate*. New Directions in World Politics. New York: Columbia University Press, 1993.

Baliga, Sandeep and Tomas Sjöström. "Arms Races and Negotiations." *Review of Economic Studies* 71 (2004): 351–369.

Ball, Desmond and Jeffrey Richelson. *Strategic Nuclear Targeting*. Cornell Studies in Security Affairs. Ithaca: Cornell University Press, 1986.

Barker, Al. "Where the Police Go Military." *The New York Times*, December 4, 2011, SR6.

Barry, Brian and Russell Hardin. *Rational Man and Irrational Society?: An Introduction and Sourcebook*. Beverly Hills: Sage Publications, 1982.

Becker, Gary. *The Economic Approach to Human Behavior*. Chicago: University of Chicago Press, 1978.

Beer, Francis A. "Games and Metaphors: Review Article." *The Journal of Conflict Resolution* 30, no. 1 (1986): 171–191.

Beitz, Charles R. *Political Theory and International Relations*. Princeton, NJ: Princeton University Press, 1979.

Bell, Duncan. "What Is Liberalism?" *Political Theory* 42, no. 6 (2014), 682–715.

Bellah, Robert et al. *Habits of the Heart: Individualism and Commitment in American Life*. Berkeley: University of California Press, 1985.

Berkes, Fikret. "Social Systems, Ecological Systems, and Property Rights." In *Rights to Nature*, edited by Susan Hanna, Carl Folke, and Karl-Göran Mäler, 87–110. Washington, DC: Island Press, 1996.

Berlin, Isaiah, Henry Hardy, and Ian Harris. *Liberty: Incorporating Four Essays on Liberty*. Oxford: Oxford University Press, 2002.

Bernard, Jessie. "Some Current Conceptualizations in the Field of Conflict." *The American Journal of Sociology* 70, no. 4 (1965): 442–454.

Bernholz, Peter. "Is a Paretian Liberal Really Impossible?" *Public Choice* 20 (1974): 99–107.

Binmore, Ken. "Bargaining and Morality." In *Rationality, Justice and the Social Contract: Themes from Morals by Agreement*, edited by David Gauthier and Robert Sugden, 131–156. Ann Arbor: The University of Michigan Press, 1993.

Binmore, Ken. *Game Theory and the Social Contract*, Vol 1: *Playing Fair*. Cambridge, MA: MIT Press, 1994.

Binmore, Ken. *Game Theory and the Social Contract,* Vol. 2: *Just Playing.* London: The MIT Press, 1998.

Binmore, Ken. *Natural Justice.* New York: Oxford University Press, 2005.

Binmore, Ken. "Why Do People Cooperate?" *Politics, Philosophy & Economics* 5, no. 1 (2006): 91–96.

Binmore, Ken. *Game Theory: A Very Short Introduction.* New York: Oxford University Press, 2007.

Binmore, Ken. *Rational Decisions.* The Gorman Lectures in Economics. Princeton, NJ: Princeton University Press, 2009.

Binmore, Ken and M. J. Herrero. "Security Equilibrium." *The Review of Economic Studies* (1988): 33–48.

Bohnenblust, H. F. et al. "Mathematical Theory of Zero-Sum Two-Person Games with a Finite Number or a Continuum of Strategies." Santa Monica, CA: The RAND Corporation, September 3 1948.

Bourdieu, Pierre. *Acts of Resistance: Against the Tyranny of the Market.* New York: New Press, 1999.

Bowles, Samuel and Herbert Gintis. *Democracy and Capitalism: Property, Community, and the Contradictions of Modern Social Thought.* New York: Basic Books, 1986.

Brams, Steven J. *Superpower Games: Applying Game Theory to Superpower Conflict.* New Haven: Yale University Press, 1985.

Brams, Steven J. *Biblical Games: Game Theory and the Hebrew Bible.* Cambridge, MA: MIT Press, 2003.

Brams, Steven J. *Game Theory and the Humanities: Bridging Two Worlds* (Cambridge, MA: MIT Press, 2012).

Brams, Steven J., Morton D. Davis, and Philip D. Straffin. "The Geometry of the Arms Race." *International Studies Quarterly* 23, no. 4 (1979): 567–588.

Brams, Steven J. and D. Marc Kilgour. *Game Theory and National Security.* New York: Basil Blackwell, 1988.

Bratman, Michael E. "Toxin, Temptation, and Stability of Intention." In *Rational Commitment and Social Justice,* edited by Jules L. Coleman and Christopher W. Morris, 59–83. Cambridge: Cambridge University Press, 1998.

Bratman, Michael E. "Intention, Practical Rationality, and Self-Governance." *Ethics* (2009): 411–443.

Brennan, Geoffrey and Philip Pettit. "Hands Invisible and Intangible." *Synthese* 94 (1993): 191–225.

Brewer, Marilynn B. and Linnda R. Caporael. "Reviewing Evolutionary Psychology: Biology Meets Society." *Journal of Social Issues* 47, no. 3 (1991): 187–195.

Brewer, Marilynn B. and Linnda R. Caporael. "The Quest for Human Nature: Social and Scientific Issues in Evolutionary Psychology." *Journal of Social Issues* 47, no. 3 (1991): 1–9.

Brewer, Marilynn B. and Linnda R. Caporael. "An Evolutionary Perspective on Social Identity: Revisiting Groups." In *Evolution and Social Psychology,* edited by M. Schaller, J. A. Simpson, and D. T. Kendrick. New York: Psychology Press, 2006.

Brewer, Marilynn B. and Roderick Kramer. "Choice Behavior in Social Dilemmas." *Journal of Personality and Social Psychology* 50, no. 3 (1986): 543–549.

Brewer, Marilynn B. and Sherry Schneider. "Social Identity and Social Dilemmas." In *Social Identity Theory: Constructive and Critical Advances,* edited by Dominic Abrams and Michael A. Hoggs. London: Harvester-Wheatsheaf, 1990.

Brodie, Bernard. "Why Were We So (Strategically) Wrong?" *Foreign Policy*, no. 5 (1972): 151–161.

Brodie, Bernard. "The Development of Nuclear Strategy." *International Security* 2, no. 4 (1978): 65–83.

Brodie, Bernard and Rand Corporation. *Strategy in the Missile Age*. Princeton, NJ: Princeton University Press, 1959.

Brody, Richard A. "Some Systemic Effects of the Spread of Nuclear Weapons Technology: A Study through Simulation of a Multi-Nuclear Future." *The Journal of Conflict Resolution* 7, no. 4 (1963): 663–754.

Brown, Wendy. *Undoing the Demos: Neoliberalism's Stealth Revolution*. Brooklyn, NY: Zone Books, 2015.

Brown, Wendy. "American Nightmare: Neoliberalism, Neoconservatism, and De-Democratization." *Political Theory* 34, no. 6 (2006): 690–714.

Brunner, Karl. *Economics & Social Institutions: Insights from the Conferences on Analysis & Ideology: [Selections]*. Rochester Studies in Economics and Policy Issues. Boston: Martinus Nijhoff, 1979.

Brzezinski, Zbigniew. *Power and Principle: Memoirs of the National Security Adviser, 1977–1981*. New York: Farrar, Straus, Giroux, 1983.

Brzezinski, Zbigniew. *The Choice: Global Domination or Global Leadership*. New York: Basic Books, 2004.

Buchanan, James M. *The Limits of Liberty: Between Anarchy and Leviathan*. Chicago: University of Chicago Press, 1975.

Buchanan, James M. *The Economics and the Ethics of Constitutional Order*. Ann Arbor: University of Michigan Press, 1991.

Buchanan, James M. and Gordon Tullock. *The Calculus of Consent*. Ann Arbor: University of Michigan Press, 1965.

Bull, Hedley. "Strategic Studies and Its Critics." *World Politics* 20, no. 4 (1986): 593–605.

Burgin, Angus. *The Great Persuasion: Reinventing Free Markets since the Depression*. Cambridge: Harvard University Press, 2012.

Burke, Edmund. *A Letter to a Member of the National Assembly Revolution and Romanticism, 1789–1834*. Oxford; New York: Woodstock Books, 1990.

Buss, D. M. "Evolutionary Psychology: A New Paradigm for Psychological Science." *Psychological Inquiry* 6, (1995): 1–30.

Cahn, Anne Hessing. "Team B: The Trillion-Dollar Experiment." *Bulletin of the Atomic Scientists* 1993, 22, 24–27.

Callahan, David. *The Cheating Culture: Why More Americans Are Doing Wrong to Get Ahead*. New York: Mariner Books, 2004.

Camerer, Colin. *Behavioral Game Theory: Experiments in Strategic Interaction*. The Roundtable Series in Behavioral Economics. New York; Princeton, NJ: Russell Sage Foundation; Princeton University Press, 2003.

Camerer, Colin. *Behavioral Game Theory*. Princeton, NJ: Princeton University Press, 2003.

Camerer, C. F. "When Does 'Economic Man' Dominate Social Behavior?" *Science*, 311, (2006): 47–52.

Campbell, Richmond. "Background for the Uninitiated." In *Paradoxes of Rationality and Cooperation: Prisoner's Dilemma and Newcomb's Problem*, edited by Richmond Campbell and Lanning Sowden, 1–41. Vancouver: University of British Columbia Press, 1985.

Campbell, Richmond and Lanning Snowden. *Paradoxes of Rationality and Cooperation: Prisoner's Dilemma and Newcomb's Problem*. Vancouver: University of British Columbia Press, 1985.

Caporael, Linnda R., Robyn M. Dawes, John Orbell, and Alphons van de Kragt. "Selfishness Examined." *Behavioral Sciences* 12 (1989): 683–699.

Carter, Jimmy. *White House Diary*. New York: Farrar, Straus and Giroux, 2010.

Caspary, William R. "Richardson's Model of Arms Races: Description, Critique, and an Alternative Model." *International Studies Quarterly* 11, no. 1 (1967): 63–88.

Cassidy, John. *How Markets Fail: The Logic of Economic Calamities*. New York: Farrar, Straus and Giroux, 2009.

Cesarani, David. *Becoming Eichmann: Rethinking the Life, Crimes, and Trial of a "Desk Murderer."* 1st Da Capo Press pbk. ed. Cambridge, MA: Da Capo Press, 2007.

Churchill, Winston. *Mr. Churchill's Speech in the House of Commons*. Stockholm: I. Haeggström, 1944.

Clausewitz, Carl von, F. N. Maude, and Anatol Rapoport. *On War*. New & revised ed. Pelican Classics. Harmondsworth: Penguin, 1968.

Coase, Ronald H. "The Problem of Social Cost." *The Journal of Law & Economics* 3, 1–44. Chicago: University of Chicago Law School, 1960.

Cohen, Avner and Steven Lee. *Nuclear Weapons and the Future of Humanity: The Fundamental Questions*. Philosophy and Society Series. Totowa, NJ: Rowman & Allanheld, 1986.

Cohen, G. A. *Why Not Socialism?* Princeton, NJ: Princeton University Press, 2009.

Cohen-Cole, Jamie. *The Open Mind: Cold War Politics and the Sciences of Human Nature*. Chicago: Chicago University Press, 2014.

Coleman, Jules L. *Markets, Morals, and the Law*. Cambridge; New York: Cambridge University Press, 1988.

Coleman, Jules L. and John Ferejohn. "Democracy and Social Choice." *Ethics* 97, no. 1 (1986): 6–25.

Coleman, Jules L., Christopher W. Morris, and Gregory S. Kavka. *Rational Commitment and Social Justice: Essays for Gregory Kavka*. Cambridge; New York: Cambridge University Press, 1998.

Collins, Francis S. *The Language of God: A Scientist Presents Evidence for Belief*. New York: Free Press, 2006.

Conybeare, John A. C. "Public Goods, Prisoners' Dilemmas and the International Political Economy." *International Studies Quarterly* 28, no. 1 (1984): 5–22.

Cook, Karen S., Russell Hardin, and Margaret Levi. *Cooperation without Trust?* The Russell Sage Foundation Series on Trust v. 9. New York: Russell Sage Foundation, 2005.

Cooter, Robert. "Coase Theorem." *The New Palgrave Dictionary of Economics*, vol. 1. London: Palgrave MacMillan, 1987.

Copeland, Dale C. "The Constructivist Challenge to Structural Realism: A Review Essay." *International Security* 25, no. 2 (2000): 187–212.

Copeland, Dale. *The Origins of Major War*. Ithaca: Cornell University Press, 2000.

Crowther-Heyck, Hunter. *Herbert A. Simon: The Bounds of Reason in Modern America*. Baltimore: Johns Hopkins University Press, 2005.

Cullity, Garrett. "Review: Ethics Done Right: Practical Reasoning as a Foundation for Moral Theory." *Ethics* (2009): 581–585.

Cuneo, Terence. "Review: The Nature of Normativity." *Ethics* (2009): 397–402.

Curry, Lynne and Winnebago County (WI). Dept. of Social Services. *The Deshaney Case: Child Abuse, Family Rights, and the Dilemma of State Intervention.* Landmark Law Cases & American Society. Lawrence: University Press of Kansas, 2007.

Dasgupta, Partha. *An Inquiry into Well-Being and Destitution.* Oxford; New York: Clarendon Press; Oxford University Press, 1993.

David, Anthony. "The Apprentice: From Albert Wohlstetter to Paul Wolfowitz to Lewis Libby, Apocaplyptic Neoconservatism Was Passed Down through Generations – to Become Official U.S. Policy." *The American Prospect* 18, no. 6 (2007): 38–41.

Davidson, Donald. "The Folly of Trying to Define Truth." *Journal of Philosophy* 93 (1996): 263–278.

Davidson, Donald. "Truth and Meaning," in *Inquiries into Truth and Interpretation* (Oxford: Clarendon, 1984), 17–36.

Davis, Lawrence H. "Prisoners, Paradox and Rationality." In *Paradoxes of Rationality and Cooperation: Prisoner's Dilemma and Newcomb's Problem,* edited by Richmond Campbell and Lanning Sowden, 45–58. Vancouver: The University of British Columbia Press, 1985.

Davis, Lawrence H. "Is the Symmetry Argument Valid." In *Paradoxes of Rationality and Cooperation,* edited by Richmond Campbell and Lanning Sowden, 255–263. Vancouver: University of British Columbia Press, 1985.

Dawes, Robyn M. and Joachim I. Krueger. *Rationality and Social Responsibility: Essays in Honor of Robyn Mason Dawes.* Modern Pioneers in Psychological Science. New York: Psychology Press, 2008.

Dawkins, Richard. *The Selfish Gene.* Oxford: Oxford University Press, 1979.

Dawkins, Richard. *The Selfish Gene.* New ed. Oxford; New York: Oxford University Press, 1989.

Dawkins, Richard. *The God Delusion.* Boston: Houghton Mifflin, 2006.

Dawkins, Richard. *The Selfish Gene.* 30th anniversary ed. Oxford; New York: Oxford University Press, 2006.

deLeon, Peter. "Review: Freeze: The Literature of the Nuclear Weapons Debate." *The Journal of Conflict Resolution* 27, no. 1 (1983): 181–189.

DeNardo, James. *The Amateur Strategist: Intuitive Deterrence Theories and the Politics of the Nuclear Arms Race.* Cambridge Studies in Political Psychology and Public Opinion. Cambridge; New York: Cambridge University Press, 1995.

Deutsch, Morton. "Trust and Suspicion." *The Journal of Conflict Resolution* 2, no. 4 (1958): 265–279.

Dewey, John. *Public and Its Problems.* Athens, OH: Swallow Press, 1954.

Diesing, Paul. *Science & Ideology in the Policy Sciences.* New York: Aldine, 1982.

Donniger, Christian. "Is It Always Efficient to Be Nice?" In *Paradoxical Effects of Social Behavior,* edited by Anatol Rapoport, Andreas Diekmann, and Peter Mitter, 123–134. Heidelberg: Physica Verlag, 1986.

Downs, George W. "The Rational Deterrence Debate." *World Politics* 41, no. 2 (1989): 225–237.

Downs, George W., David M. Rocke, and Peter N. Barsoom. "Is the Good News About Compliance Good News About Cooperation?" *International Organization* 50, no. 3 (1996): 379–406.

Downs, George W., David M. Rocke, and Randolph M. Siverson. "Arms Races and Cooperation." *World Politics* 38, no. 1 (1985): 118–146.

Doyle, Michael W. *Ways of War and Peace.* New York: W. W. Norton, 1997.

Doyle, Michael W. "Kant, Liberal Legacies, and Foreign Affairs." *Philosophy and Public Affairs* 12:3 (1983), 205–235.

Dresher, Melvin. *Mathematical Theory of Zero-Sum Two-Person Games with a Finite Number or a Continuum of Strategies.* Santa Monica, CA: Rand Corp., 1949.

Eatwell, John, Murray Milgate, Peter Newman, and Robert Harry Inglis Palgrave. *The New Palgrave: A Dictionary of Economics.* 4 vols. London; New York; Tokyo: Macmillan; Stockton Press; Maruzen, 1987.

Eden, Lynn and Steven E. Miller. *Nuclear Arguments: Understanding the Strategic Nuclear Arms and Arms Control Debates.* Cornell Studies in Security Affairs. Ithaca: Cornell University Press, 1989.

Ehrenberg, John. *Civil Society: The Critical History of an Idea.* New York: New York University Press, 1999.

Elloitt, Sober and David Sloan Wilson. *Unto Others: The Evolution and Psychology of Unselfish Behavior.* Cambridge, MA: Harvard University Press, 1998.

Ellsberg, Daniel. "Theory of the Reluctant Duelist." *The American Economic Review* 46, no. 5 (1956): 909–923.

Ellsberg, Daniel. *Risk, Ambiguity, and Decision.* Studies in Philosophy. New York: Garland, 2001.

Elster, Jon. *Sour Grapes: Studies in the Subversion of Rationality.* New York: Cambridge University Press, 1985.

Elster, Jon. *Ulysses and the Sirens: Studies in Rationality and Irrationality.* Cambridge: Cambridge University Press, 1988.

Epstein, Richard. *Takings: Private Property and the Power of Eminent Domain.* Cambridge, MA: Harvard University Press, 1985.

Erickson, Paul. *The World the Game Theorists Made.* Chicago: University of Chicago Press, 2015.

Erickson, Paul et al. *How Reason Almost Lost Its Mind.* Chicago: University of Chicago Press, 2013.

Ernst, Zachary. "Explaining the Social Contract." *British Journal of Philosophical Science* 52, (2001): 1–24.

Evensky, Jerry. *Adam Smith's Moral Philosophy: A Historical and Contemporary Perspective on Markets, Law, Ethics, and Culture.* Cambridge; New York: Cambridge University Press, 2005.

Fairbanks Jr., Charles H. "Mad and U.S. Strategy." In *Getting MAD: Nuclear Mutual Assured Destruction, Its Origins and Practice,* edited by Henry D. Sokolski, 137–147. Darby, PA: Diane Publishing 2004.

Farrell, Daniel M. "Hobbes as Moralist." *Philosophical Studies* 48, (1985): 257–283.

Farrell, Daniel M. "A New Paradox of Deterrence." In *Rational Commitment and Social Justice,* edited by Jules L. Coleman and Christopher W. Morris, 22–46. Cambridge: Cambridge University Press, 1998.

Fearon, James D. "Bargaining, Enforcement, and International Cooperation." *International Organization* 52, no. 2 (1998): 269–305.

Fehr, E. and Urs Fischenbacher. "The Nature of Human Altruism." *Nature* 425, (2003): 785–791.

Ferguson, Charles H. *Predator Nation: Corporate Criminals, Political Corruption, and the Hijacking of America.* New York: Crown Business, 2012.

Forst, Brian and Judith Lucianovic. "Prisoner's Dilemma: Theory and Reality." *Journal of Criminal Justice* 5 (1977): 55–64.

Foster, William C. "Prospects for Arms Control." *Foreign Affairs* 47, no. 3 (1969): 413–448.

Foucault, Michel. *Discipline and Punish*. Translated by Alan Sheridan. New York: Vintage, 1979.

Foucault, Michel. "Governmentality." In *The Foucault Effect: Studies in Governmentality*, ed. by Graham Burchell, Colin Gordon, and Peter Miller. Chicago: University of Chicago Press, 1991, 87–104.

Foucault, Michel. *Birth of Biopolitics: Lectures at the College De France*. New York: Picador, 2010.

Frank, Robert, Thomas Gilovich, and Donald Regan. "Does Studying Economics Inhibit Cooperation?" *Journal of Economic Perspectives* 7 (1993): 159–171.

Freedman, Lawrence. *The Evolution of Nuclear Strategy*. New York: St. Martin's Press, 1981.

Freedman, Lawrence. *Deterrence*. Cambridge, UK; Malden, MA: Polity Press, 2004.

Frey, Bruno and Reto Jegen. "Motivation Crowding Theory." *Journal of Economic Surveys* 15, no. 5 (2001): 589–611.

Friedman, James W. *Game Theory with Applications to Economics*. New York: Oxford University Press, 1986.

Friedman, Milton. *Capitalism and Freedom, 20th Anniversary Edition*. Chicago: University of Chicago Press, 2002.

Friedman, Thomas L. "This Column Is Not Sponsored by Anyone." *New York Times*, May 13, 2012, SR13-13.

Fukuyama, Francis. *The End of History and the Last Man*. New York; Toronto: Free Press; Maxwell Macmillan Canada; Maxwell Macmillan International, 1992.

Gaddis, John. *The Cold War: A New History*. London: Penguin, 2006.

Gafni, Amiram. "Willingness-to-Pay as a Measure of Benefits: Relevant Questions in the Context of Public Decisionmaking about Health Care Programs." *Medical Care* 29, no. 12 (1991), 1246–1252.

Gardenfors, Peter. "Rights, Games and Social Choice." *Nous* 15, no. 3 (1981): 341–356.

Gaus, Gerald "The Diversity of Comprehensive Liberalisms." In *Handbook of Political Theory*, edited by Gerald F. and Chandran Kukathas Gaus. London: Sage, 2004, 100–114.

Gaus, Gerald F. *The Order of Public Reason: A Theory of Freedom and Morality in a Diverse and Bounded World*. New York: Cambridge University Press, 2011.

Gaus, Gerald and Shane Courtland. "Liberalism." In *Stanford Encyclopedia of Philosophy*, 2014. http://plato.stanford.edu/entries/liberalism/, accessed July 1, 2015.

Gauthier, David P. *The Logic of Leviathan: The Moral and Political Theory of Thomas Hobbes*. Oxford: Clarendon Press, 1969.

Gauthier, David. "Deterrence, Maximization, and Rationality." *Ethics* 94, no. 3 (1984), 474–495.

Gauthier, David P. *Morals by Agreement*. Oxford; New York: Clarendon Press; Oxford University Press, 1987.

Gauthier, David. "Rethinking the Toxin Puzzle." In *Rational Commitment and Social Justice*, edited by Jules L. Coleman and Christopher W. Morris, 47–58. Cambridge: Cambridge University Press, 1998.

Gelman, Andrew. "Methodology as Ideology: Mathematical Modeling of Trench Warfare." 2007. SSRN: http://ssrn.com/abstract=1010642 or http://dx.doi.org/10.2139/ssrn.1010642.

George, Alexander L. and Richard Smoke. *Deterrence in American Foreign Policy: Theory and Practice*. New York: Columbia University Press, 1974.

George, Alexander L. and Richard Smoke. "Deterrence and Foreign Policy." *World Politics* 41, no. 2 (1989): 170–182.

Ghamari-Tabrizi, Sharon. *The Worlds of Herman Kahn: The Intuitive Science of Thermonuclear War*. Cambridge, MA: Harvard University Press, 2005.

Gilbert, Margaret. *A Theory of Political Obligation: Membership, Commitment, and the Bonds of Society*. Oxford; New York: Clarendon Press; Oxford University Press, 2006.

Gilbert, Margaret. *Joint Commitment: How We Make the Social World*. New York: Oxford University Press, 2013.

Gintis, Herbert. *Moral Sentiments and Material Interests: The Foundations of Cooperation in Economic Life*. Economic Learning and Social Evolution. Cambridge, MA: MIT Press, 2004.

Gintis, Herbert. "Behavioral Ethics Meets Natural Justice." *Politics, Philosophy & Economics* 5, no. 1 (2005): 5–32.

Gintis, Herbert. *The Bounds of Reason: Game Theory and the Unification of the Behavioral Sciences*. Princeton, NJ: Princeton University Press, 2009.

Gintis, Herbert and Samuel Bowles. "The Evolution of Altruistic Punishment." In *The Origin and Evolution of Cultures*, edited by Robert Boyd and Peter Richerson. New York: Oxford University Press 2005.

Gintis, Herbert, Samuel Bowles, Robert Boyd and Ernst Fehr. "Explaining Altruistic Behavior in Humans." *Evolution of Human Behavior* 24, (2003): 153–172.

Giocoli, Nicola. "Nash Equilibrium." *History of Political Economy* 36, no. 4 (2004): 649–666.

Giocoli, Nicola. "Do Prudent Agents Play Lotteries: Von Neumann's Contributions to the Theory of Rational Behavior." *Journal for the History of Economic Thought* 28, no. 1 (2006): 95–109.

Giroux, Henry A. *The Terror of Neoliberalism: Authoritarianism and the Eclipse of Democracy*. London: Paradigm Publishers, 2004.

Glaser, Charles L. "Why Do Strategists Disagree about the Requirements of Strategic Nuclear Deterence?" In *Nuclear Arguments*, edited by Lynn Eden and Steven E. Miller, 109–171. London: Cornell University Press, 1989.

Glaser, Charles L. "The Security Dilemma Revisited." *World Politics* 50, no. 1 (1997): 171–201.

Gödel, Kurt. *On Formally Undecidable Propositions of* Principia Mathematica *and Related Systems*. New York: Basic Books, 1962.

Gold, Natalie and Robert Sugden. "Collective Intentions and Team Agency." *The Journal of Philosophy*, no. 3 (2007): 109–137.

Goldman, Alan H. *Practical Rules: When We Need Them and When We Don't*. Cambridge Studies in Philosophy. Cambridge; New York: Cambridge University Press, 2001.

Goodin, Robert E. "Equal Rationality and Initial Endowments." In *Rationality, Justice and the Social Contract: Themes from Morals by Agreement*, edited by David Gauthier and Robert Sugden, 116–130. Ann Arbor: University of Michigan Press, 1993.

Goodwin, Craufurd D. W. *Economics and National Security: A History of Their Interaction*. Annual Supplement to History of Political Economy. Durham; London: Duke University Press, 1991.

Gowa, Joanne. "Review: Anarchy, Egoism, and Third Images: The Evolution of Cooperation and International Relations." *International Organization* 40, no. 1 (1986): 167–186.

Gralnick, Alexander. "Deterrence, Realism, and Nuclear Omnicide." *Political Psychology* 9, no. 1 (1988): 175–188.

Gray, Colin S. "What RAND Hath Wrought." *Foreign Policy*, no. 4 (1971): 111–129.
Gray, Colin S. "The Arms Race Is About Politics." *Foreign Policy*, no. 9 (1973): 117–129.
Gray, Colin S. "Foreign Policy – There Is No Choice." *Foreign Policy*, no. 24 (1976): 114–127.
Gray, Colin S. "Nuclear Strategy: The Case for a Theory of Victory." *International Security* 4, no. 1 (1979): 54–87.
Gray, Colin S. *Strategic Studies and Public Policy: The American Experience.* Lexington: University Press of Kentucky, 1982.
Gray, Colin S. and Jeffrey G. Barlow. "Inexcusable Restraint: The Decline of American Military Power in the 1970s." *International Security* 10, no. 2 (1985): 27–69.
Gray, Colin S. and Michael Howard. "Perspectives on Fighting Nuclear War." *International Security* 6, no. 1 (1981): 185–187.
Gray, Colin S. and Keith Payne. "Victory Is Possible." *Foreign Policy*, no. 39 (1980): 14–27.
Green, Donald and Ian Shapiro. *Pathologies of Rational Choice.* New Haven: Yale University Press, 1994.
Green, Philip. *Deadly Logic: The Theory of Nuclear Deterrence.* Columbus: Ohio State University Press, 1966.
Green, T. H. *Lectures on the Principles of Political Obligation.* Kitchener, ONT: Batoche Books, 1999.
Grieco, Joseph M. "Anarchy and the Limits of Cooperation: A Realist Critique of the Newest Liberal Institutionalism." In *Neorealism and Neoliberalism: The Contemporary Debate*, edited by David A. Baldwin, 116–140. New York: Columbia University Press, 1993.
Grotius, Hugo. *The Rights of War and Peace.* vols. 1–3, edited by Richard Tuck. Indianapolis: Liberty Fund, 2005.
Guetzkow, Harold and Lloyd Jensen. "Research Activities on Simulated International Processes." *Background* 9, no. 4 (1966): 261–274.
Habermas, Jürgen. *Legitimation Crisis.* Boston: Beacon Press, 1975.
Habermas, Jürgen. *The Theory of Communicative Action*, vols., 1 and 2. Trans. by Thomas McCarthy. Cambridge: MIT Press, 1984 and 1987.
Habermas, Jürgen. *Between Facts and Norms: Contributions to a Discourse Theory of Law and Democracy.* Studies in Contemporary German Social Thought. Cambridge, MA: MIT Press, 1996.
Haggard, Stephan and Beth A. Simmons. "Theories of International Regimes." *International Organization* 41, no. 3 (1987): 491–517.
Halberstam, David. *The Best and the Brightest.* New York: Ballantine Books, 1993.
Hamilton, William. "The Genetical Evolution of Social Behavior." *Journal of Theoretical Biology* 37 (1964): 1–52.
Hampton, Jean. *Hobbes and the Social Contract Tradition.* Cambridge; New York: Cambridge University Press, 1986.
Hampton, Jean. *The Authority of Reason.* Cambridge; New York: Cambridge University Press, 1998.
Han, Sam. "American Cultural Theory." In *Routledge Handbook of Social and Cultural Theory*, edited by Anthony Elliot, 239–256. New York: Routledge, 2013.
Hanna, Susan and Carl Jentoft. "Human Use of the Natural Environment: An Overview of Social and Economic Dimensions." In *Rights to Nature*, edited by Susan Hanna, Carl Folke, and Karl-Göran Mäler, 35–56. Washington, DC: Island Press, 1996.

Hardin, Garrett James, Richard D. Lyons, and Edward Edelson. *Tragedy of the Commons Revisited*. Sound recording. Washington, DC: American Association for the Advancement of Science, 1973.

Hardin, Russell. *Collective Action*. Baltimore: Published for Resources for the Future by the Johns Hopkins University Press, 1982.

Hardin, Russell. *Social Science Information and Decision Making*. vol. 2 FID Studies in Social Science Information and Documentation. Budapest, Hungary: Published on behalf of FID by the Economic Information Unit, Hungarian Academy of Science, 1984.

Hardin, Russell, ed. "Individual Sanctions, Collective Benefits." In *Paradoxes of Rationality and Cooperation: Prisoner's Dilemma and Newcomb's Problem*, edited by Richmond Campbell and Lanning Sowden, 339–354. Vancouver: The University of British Columbia Press, 1985.

Hardin, Russell, ed. *Nuclear Deterrence: Ethics and Strategy*. Chicago: University of Chicago Press, 1985.

Hardin, Russell. "Risking Armageddon." In *Nuclear Weapons and the Future of Humanity*, edited by Avner Cohen and Steven Lee, 201–223. Totowa: Rowman & Allanheld, 1986.

Hardin, Russell. "The Utilitarian Logic of Liberalism." *Ethics* 97, no. 1 (1986): 47–74.

Hardin, Russell. "Does Might Make Right?" In *Authority Revisited*, edited by J. Roland Pennock and John William Chapman, 201–217. New York: New York University Press, 1987.

Hardin, Russell. *Morality within the Limits of Reason*. Chicago: University of Chicago Press, 1988.

Hardin, Russell. "Contractarianism: Wistful Thinking." *Constitutional Political Economy* 1, no. 2 (1990), 35–52.

Hardin, Russell. *Liberalism, Constitutionalism, and Democracy*. Oxford; New York: Oxford University Press, 1999.

Hardin, Russell. *Trust and Trustworthiness*. The Russell Sage Foundation Series on Trust v. 4. New York: Russell Sage Foundation, 2002.

Hardin, Russell. "The Free Rider Problem." In *Stanford Encyclopedia of Philosophy*, edited by Edward N. Zalta. Stanford, 2003. http://plato.stanford.edu/entries/free -rider/.

Hartz, Louis. *The Liberal Tradition in America; an Interpretation of American Political Thought since the Revolution*. New York: Harcourt, 1955.

Harvey, David. *A Brief History of Neoliberalism*. New York: Oxford University Press, 2007.

Hausman, Daniel M. and Michael S. McPherson. *Economic Analysis, Moral Philosophy, and Public Policy*. 2nd ed. New York: Cambridge University Press, 2006.

Hayek, Friedrich A. von. *The Constitution of Liberty*. Chicago: University of Chicago Press, 1960.

Hayek, Friedrich A. von. *The Constitution of Liberty: The Definitive Edition*. The Collected Works of F. A Hayek, edited by Ronald Hamowy. Chicago: University of Chicago Press, 2011.

Heap, Shaun Hargreaves and Yanis Varoufakis. *Game Theory: A Critical Text*. 2nd ed. London; New York: Routledge, 2004.

Heath, Joseph. *Communicative Action and Rational Choice*. Studies in Contemporary German Social Thought. Cambridge, MA: MIT Press, 2001.

Heath, Joseph. *Following the Rules: Practical Reasoning and Deontic Constraints.* New York: Oxford University Press, 2011.

Heinrich, Joseph et al. "Economic Man in Cross-Cultural Perspective." *Behavioral and Brain Sciences* 28 (2005): 795–855.

Heinrich, Joseph et al. "Costly Punishment across Human Perspective." *Science* 312, (2006): 1767–1770.

Held, Virginia. "Rationality and Social Value in Game-Theoretical Analyses." *Ethics* 76, no. 3 (1966): 215–220.

Herf, Jeffrey. *Reactionary Modernism: Technology, Culture, and Politics in Weimar and the Third Reich.* Cambridge; New York: Cambridge University Press, 1984.

Herken, Gregg. *Counsels of War.* New York: Knopf; distributed by Random House, 1985.

Hess, Charlotte and Elinor Ostrom. *Understanding Knowledge as a Commons: From Theory to Practice.* Cambridge, MA: MIT Press, 2007.

Hessing Cahn, Anne. "Team B: The Trillion-Dollar Experiment." *Bulletin of the Atomic Scientists* 49, no. 3 (1993): 22, 24–27.

Hilgers, Mathieu. "The Three Anthropological Approaches to Neoliberalism." *International Social Science Journal* 61, no. 202 (2010): 351–364.

Hilgers, Mathieu. "The Historicity of the Neoliberal State." *Social Anthropology* 20, no. 1 (2012): 80–94.

Hobbes, Thomas. *Leviathan,* edited by Richard Tuck. Cambridge: Cambridge University Press, [1651] 1996.

Hobbes, Thomas and Richard Tuck. *Leviathan.* Cambridge Texts in the History of Political Thought. Cambridge; New York: Cambridge University Press, 1991.

Hollis, Martin. *The Cunning of Reason.* Cambridge; New York: Cambridge University Press, 1987.

Hollis, Martin. "The Agriculture of the Mind." In *Rationality, Justice and the Social Contract: Themes from Morals by Agreement,* edited by David Gauthier and Robert Sugden, 40–52. Ann Arbor: University of Michigan Press, 1993.

Hollis, Martin. *Trust within Reason.* Cambridge; New York: Cambridge University Press, 1998.

Hollis, Martin and Steve Smith. *Explaining and Understanding International Relations.* Oxford; New York: Clarendon Press; Oxford University Press, 1990.

Hont, Istvan and Michael Ignatiev, eds. *Wealth and Virtue: The Shaping of Political Economy in the Scottish Enlightenment.* Cambridge; New York: Cambridge University Press, 1983.

Hont, Istvan and Michael Ignatiev. "Needs and Justice in the Wealth of Nations." In *Wealth and Virtue: The Shaping of the Political Economy in the Scottish Enlightenment,* edited by Istvan Hont and Michael Ignatiev, 1–44. Cambridge: Cambridge University Press, 1983.

Horkheimer, Max and Theodor W. Adorno. *Dialectic of Enlightenment.* Trans by John Cumming. New York: Herder and Herder, 1972.

Hubin, Donald C. "The Groundless Normativity of Instrumental Rationality." *The Journal of Philosophy* 98, no. 9 (2001): 445–465.

Hume, David. *A Treatise of Human Nature.* Pelican Classics. Baltimore: Penguin Books, 1969.

Ikenberry, John G. *Institutions, Strategic Restraint, and the Rebuilding of Order after Major Wars.* Princeton, NJ: Princeton University Press, 2001.

Innocenti, Alessandro. "Linking Strategic Interaction and Bargaining Theory: The Harsanyi-Schelling Debate on the Axiom of Symmetry." *History of Political Economy* 40, no. 1 (2008): 111–132.

Jasanoff, Sheila. *States of Knowledge: The Co-Production of Science and the Social Order.* International Library of Sociology. London; New York: Routledge, 2004.

Jervis, Robert. *Perception and Misperception in International Politics.* Princeton, NJ: Princeton University Press, 1976.

Jervis, Robert. "Cooperation under the Security Dilemma." *World Politics* 30, no. 2 (1978): 167–214.

Jervis, Robert. "Review: Deterrence Theory Revisited." *World Politics* 31, no. 2 (1979): 298–324.

Jervis, Robert. *The Illogic of American Nuclear Strategy.* Cornell Studies in Security Affairs. Ithaca: Cornell University Press, 1984.

Jervis, Robert. "Realism, Game Theory, and Cooperation." *World Politics* 40, no. 3 (1988): 317–349.

Jervis, Robert. "Rational Deterrence: Theory and Evidence." *World Politics* 41, no. 2 (1989): 183–207.

Jervis, Robert. *The Meaning of the Nuclear Revolution: Statecraft and the Prospect of Armageddon.* Cornell Studies in Security Affairs. Ithaca: Cornell University Press, 1989.

Jervis, Robert. "Realism, Neoliberalism, and Cooperation: Understanding the Debate." *International Security* 24, no. 1 (1999): 42–63.

Jervis, Robert. "Was the Cold War a Security Dilemma?" *Journal of Cold War Studies* 3, no. 1 (2001): 36–60.

Jervis, Robert. "Mutual Assured Destruction." *Foreign Policy*, no. 133 (2002): 40–42.

Kahan, Dan. "Social Influence, Social Meaning, and Deterrence." *Virginia Law Review* 83, (1997): 349–395.

Kahan, Dan. "The Logic of Reciprocity." In *Sentiments and Material Interests*, edited by Herbert Gintis, Samuel Bowles, Robert Boyd and Ernst Fehr. Cambridge, MA: MIT Press, 2005.

Kahn, Herman. *On Thermonuclear War.* Princeton, NJ: Princeton University Press, 1960.

Kahn, Herman. *The Nature and Feasibility of War and Deterrence, a Study.* 86th Cong., 2d Sess. Senate. Document. Washington: US Government Printing Office, 1960.

Kahn, Herman. *Thinking about the Unthinkable.* New York: Horizon Press, 1962.

Kahn, Herman. *On Escalation: Metaphors and Scenarios.* New York: Praeger, 1965.

Kant, Immanuel. *Foundations of the Metaphysics of Morals, and What Is Enlightenment?* Translated and introduced by Lewis White Beck. The Library of Liberal Arts. New York: Liberal Arts Press, 1959.

Kant, Immanuel, *Groundwork of the Metaphysic of Morals.* Edited and translated by John Ladd. New York: Harper and Row, 1964.

Kant, Immanuel and John Ladd. *Metaphysical Elements of Justice: Part I of the Metaphysics of Morals.* 2nd ed. Indianapolis: Hackett, 1999.

Kaplan, Fred M. *The Wizards of Armageddon.* New York: Simon and Schuster, 1983.

Kaplin, Morton A. "World Politics." *World Politics* 11, no. 1 (1958): 20–43.

Kapstein, Ethan B. "Two Dismal Sciences Are Better Than One – Economics and the Study of National Security: A Review Essay." *International Security* 27, no. 3 (2001): 158–187.

Kattenburg, Paul M. "MAD Is Still the Moral Position." In *After the Cold War: Questioning the Morality of Nuclear Deterrence*, edited by Charles W. Kegley Jr. and Kenneth L. Schwab, 111–120. Boulder, CO: Westview Press, 1991.

Kavka, Gregory S. "Deterrence, Utility, and Rational Choice." *Theory and Decision* 12, no. 1 (1980): 41–60.

Kavka, Gregory S. "Some Paradoxes of Deterrence." *The Journal of Philosophy* 75, no. 6 (1978): 285–302.

Kavka, Gregory S. "The Toxin Puzzle." *Analysis* 43, no. 1 (1983): 33–36.

Kavka, Gregory S. *Hobbesian Moral and Political Theory*. Studies in Moral, Political, and Legal Philosophy. Princeton, NJ: Princeton University Press, 1986.

Kavka, Gregory S. "Morality and Nuclear Politics: Lessons of the Missile Crisis." In *Nuclear Weapons and the Future of Humanity*, edited by Avner Cohen and Steven Lee, 233–254. Totowa NJ: Rowman & Allanheld, 1986.

Kavka, Gregory S. *Moral Paradoxes of Nuclear Deterrence*. Cambridge; New York: Cambridge University Press, 1987.

Kegley, Charles W. and Kenneth L. Schwab. *After the Cold War: Questioning the Morality of Nuclear Deterrence*. Boulder, CO: Westview Press, 1991.

Kenrick, D. T., N. P. Li, and J. Butner. "Dynamical Evolutionary Psychology: Individual Decision Rules and Emergent Social Norms." *Psychological Review* 110, no. 1 (2003): 3–28.

Keohane, Robert O. *After Hegemony: Cooperation and Discord in the World Political Economy*. Princeton, NJ: Princeton University Press, 1984.

Keohane, Robert O. *Neorealism and Its Critics*. The Political Economy of International Change. New York: Columbia University Press, 1986.

Ketelaar, T. and B. J. Ellis. "Are Evolutionary Explanations Unfalsifiable? Evolutionary Psychology and the Lakatosian Philosophy of Science." *Psychological Inquiry* 11, (2000), 1–21.

Ketelaar, T., and B. J. Ellis. "On the Natural Selection of Alternative Models: Evaluation of Explanations in Evolutionary Psychology." *Psychological Inquiry*, 11, (2000), 56–68.

King, Gary, Robert O. Keohane, and Stanley Verba. *Designing Social Inquiry: Scientific Inference in Qualitative Research*. Princeton, NJ: Princeton University Press, 1994.

Kloppenberg, James T. *The Virtues of Liberalism*. New York: Oxford University Press, 1998.

Klosko, George. *Political Obligations*. Oxford; New York: Oxford University Press, 2005.

Kolbert, E. "The Calculator: How Kenneth Feinberg Determines the Value of Three Thousand Lives." *The New Yorker Magazine*, November 25, 2002.

Kollock, Peter. "Social Dilemmas: The Anatomy of Cooperation." *Annual Review of Sociology* 24, (1998): 183–214.

Kornhauser, Lewis. "Economic Analysis of the Law." *Stanford Encyclopedia of Philosophy*, published November 26, 2001, revised August 12, 2011, available online, http://plato.stanford.edu/entries/legal-econanalysis/, accessed July 15, 2015.

Krasner, Stephen D. *Sovereignty: Organized Hypocrisy*. Princeton, NJ: Princeton University Press, 1999.

Krause, Keith. "Rationality and Deterrence in Theory and Practice." In *Contemporary Security and Strategy*, edited by Craig A. Snyder, 120–149. New York: Routledge, 1999.

Krueger, Joachim I. and Mellisa Acevedo. "A Game-Theoretic View of Voting." *Journal of Social Issues* 64, no. 3 (2008): 467–485.

Kuhn, Steven. "Prisoner's Dilemma." *Stanford Encyclopedia of Philosophy* (2007, 2014). http://plato.stanford.edu/entries/prisoner-dilemma/.

Kuklick, Bruce. *Blind Oracles: Intellectuals and War from Kennan to Kissinger*. Princeton, NJ: Princeton University Press, 2006.

Kull, Steven. "Nuclear Nonsense." *Foreign Policy*, no. 58 (1985): 28–52.

Kydd, Andrew. "Game Theory and the Spiral Model." *World Politics* 49 (1997): 371–400.

Kydd, Andrew H. *Trust and Mistrust in International Relations.* Princeton, NJ: Princeton University Press, 2005.

Lackey, Douglas. "Missiles and Morals: A Utilitarian Look at Nuclear Deterrence." *Philosophy and Public Affairs* 11, no. 3 (1982): 189–231.

Lackey, Douglas. "Disarmament Revisited: A Reply to Kavka and Hardin." *Philosophy and Public Affairs* 12, no. 3 (1983): 261–256.

Lackey, Douglas. "The American Debate on Nuclear Weapons Policy: A Review of the Literature 1945–85." *Analyse and Kritik* 9 (1987): 7–46.

Langlois, Jean-Pierre P. "Modeling Deterrence and International Crises." *Journal of Conflict Resolution* 33, no. 1 (1989): 67–83.

Lasanga, Louis. "Oath of Lasanga." Undated. in Box 10, Folder 3, D.302, University of Rochester Archive.

Lasker, Emanuel. *Kampf.* New York: Lasker's, 1907.

Lawler, Peter Augustine. "Fukuyama Versus the End of History." In *After History?: Francis Fukuyama and His Critics*, edited by Timothy Burns, 63–79. London: Rowman & Littlefield, 1994.

Lawrence, Philip K. "Strategy, Hegemony and Ideology: The Role of Intellectuals." *Political Studies* XLIV (1996): 44–59.

Lebow, Richard Ned and Janice Gross Stein. "Rational Deterrence Theory: I Think, Therefore I Deter." *World Politics* 41, no. 2 (1989): 208–224.

Lemann, Nicholas. "The Quiet Man; Dick Cheney's Discrete Rise to Unprecedented Power." *The New Yorker*, May 7, 2001.

Leonard, Robert. *Von Neumann, Morgenstern, and the Creation of Game Theory: From Chess to Social Science, 1900–1960.* Historical Perspectives on Modern Economics. New York: Cambridge University Press, 2010.

Lewin, Leonard C. *Report from Iron Mountain: On the Possibility and Desirability of Peace.* New York: The Free Press, 1996.

Lewis, David K. *Convention: A Philosophical Study.* Cambridge, MA: Harvard University Press, 1969.

Lewis, David K. "Prisoners' Dilemma Is a Newcomb Problem." In *Paradoxes of Rationality and Cooperation*, edited by Richmond Campbell and Lanning Sowden, 251–255. Vancouver: University of British Columbia Press, 1985.

Lewis, David K. "Finite Counterforce." In *Nuclear Deterrence and Moral Restraint*. Ed. by Henry Shue (Cambridge: Cambridge University Press, 1989).

Lewis, David K. *Papers in Ethics and Social Philosophy*, vol. 3. Cambridge Studies in Philosophy. Cambridge; New York: Cambridge University Press, 2000.

Lewis, David K. *Convention: A Philosophical Study*, reissue. Oxford: Wiley-Blackwell, 2002.

Licklider, Roy E. *The Private Nuclear Strategists.* Columbus: Ohio State University Press, 1971.

Locke, John. *The Second Treatise of Government*, edited by C. B. Macpherson. Indianapolis: Hackett, 1980.

Lohmann, Susanne. "The Poverty of Green and Shapiro." *Critical Review* 9 (1995): 127–154.

Luce, R. Duncan and Howard Raiffa. *Games and Decisions; Introduction and Critical Survey.* New York: Wiley, 1957.

Lynn-Jones, Sean M. "Preface." In *Rational Choice and Security Studies: Stephen Walt and His Critics*, edited by Michael E. Brown, Owen R. Cote Jr., Sean M. Lynn-Jones, and Steven E. Miller, ix–xix. Cambridge, MA: The MIT Press, 2000.

MacGilvray, Eric. *Market Freedom*. New York: Cambridge University Press, 2010.

Mack, Eric and Gerald Gaus. "Classic Liberalism and Libertarianism." In *Handbook of Political Theory*, edited by Gerald F. and Chandran Kukathas Gaus. London: Sage, 2004.

MacLean, Douglas. *The Security Gamble: Deterrence Dilemmas in the Nuclear Age*. Maryland Studies in Public Philosophy. Totowa, NJ: Rowman & Allanheld, 1984.

Majeski, Stephen J. "Technological Innovation and Cooperation in Arms Races." *International Studies Quarterly* 30, no. 2 (1986): 175–191.

Maloy, J. S. *The Colonial American Origins of Modern Democratic Thought*. New York: Cambridge University Press, 2008.

Malthus, T. R. *An Essay on the Principle of Population*. London: J. Johnson, 1798.

Mariotti, Marco. "Fair Bargains: Distributive Justice and Nash Bargaining Theory." *The Review of Economic Studies* 66, no. 3 (1999): 733–741.

Martin, Lisa L. and Beth A. Simmons. "Theories and Empirical Studies of International Institutions." *International Organization* 52, no. 4 (1998): 729–757.

Maslow, Abraham. *Motivation and Personality*. New York: Harper, 1954.

Maynard Smith, John. "The Theory of Games and the Evolution of Animal Conflicts." *Journal of Theoretical Biology* 47, no. 1 (1974): 209–221.

Maynard Smith, John. "Evolution and the Theory of Games: In Situations Characterized by Conflict of Interest, the Best Strategy to Adopt Depends on What Others Are Doing." *American Scientist* 64, no.1 (1976): 41–45.

Maynard Smith, John. "Optimization Theory in Evolution." *Annual Review of Ecology and Systematics* 9 (1978): 31–56.

Maynard Smith, John. *Evolution and the Theory of Games*. Cambridge; New York: Cambridge University Press, 1982.

McCabe, Donald, Kenneth Butterfield, and Linda Klebe Trevino. "Academic Dishonesty in Graduate Business Programs." *Academy of Management Learning and Education* 5, no. 3 (2006): 294–305.

McCabe, Donald and Linda Klebe Trevino. "Academic Dishonesty." *Journal of Higher Education* 64, no. 5 (1993): 522–538.

McCabe, Donald and Linda Klebe Trevino. "Cheating among Business Students." *Journal of Management Education* 19, no. 2 (1995): 205–218.

McCay, Bonnie J. "Common and Private Concerns." In *Rights to Nature*, edited by Susan Hanna, Carl Folke, and Karl-Göran Mäler, 111–126. Washington, DC: Island Press, 1996.

McClennen, Edward F. "The Tragedy of National Sovereignty." In *Nuclear Weapons and the Future of Humanity*, edited by Avner Cohen and Steven Lee, 391–405. Totowa, NJ: Rowman & Allanheld, 1986.

McCumber, John. *Time in the Ditch: American Philosophy and the McCarthy Era*. Evanston, IL: Northwestern University Press, 2001.

McCumber, John. *Philosophical Excavations: Reason, Truth and Politics in the Early Cold War*. Chicago: University of Chicago Press, forthcoming.

McKean, Margaret A. "Common-Property Regimes as a Solution to Problems of Scale." In *Rights to Nature*, edited by Susan Hanna, Carl Folke, and Karl-Göran Mäler. Washington, DC: Island Press, 1996.

McMahon, Christopher. *Collective Rationality and Collective Reasoning.* Cambridge; New York: Cambridge University Press, 2001.

McNamara, Robert S. *Blundering into Disaster: Surviving the First Century of the Nuclear Age.* New York: Pantheon Books, 1986.

Mearsheimer, John J. *The Tragedy of Great Power Politics.* New York: W. W. Norton, 2001.

Mensch, A. and NATO Science Committee. *Theory of Games; Techniques and Applications. Proceedings of a Conference under the Aegis of the NATO Scientific Affairs Committee, Toulon, 29th June–3rd July 1964.* New York: American Elsevier, 1966.

Mesquita, Bruce Bueno de and William H Riker. "An Assessment of the Merits of Selective Nuclear Proliferation." *The Journal of Conflict Resolution* 26, no. 2 (1982): 283–306.

Mill, John Stuart. *On Liberty and Other Writings,* edited by Stephan Collini. Cambridge: Cambridge University Press, 1989.

Miller, Dale. "The Norm of Self-Interest." *American Psychologist* 54, no. 12 (1999), 1053–1060.

Milner, Helen. "Review: International Theories of Cooperation among Nations: Strengths and Weaknesses." *World Politics* 44, no. 3 (1992): 466–496.

Mirowski, Philip. *More Heat Than Light: Economics as Social Physics, Physics as Nature's Economics.* Historical Perspectives on Modern Economics. Cambridge; New York: Cambridge University Press, 1989.

Mirowski, Philip. *Economics and National Security: A History of Their Interaction. Annual Supplement to History of Political Economy,* edited by Craufurd D. W. Goodwin. Durham; London: Duke University Press, 1991.

Mirowski, Philip. *Machine Dreams: Economics Becomes a Cyborg Science.* Cambridge; New York: Cambridge University Press, 2002.

Mirowski, Philip and Dieter Plehwe. *The Road from Mont PèLerin: The Making of the Neoliberal Thought Collective.* Cambridge, MA: Harvard University Press, 2009.

Moehler, Michael. "Why Hobbes' State of Nature Is Best Modeled by an Assurance Game." *Utilitas* 21, no. 3 (2009): 297–326.

Moglewer, S. "A Resource Allocation Model for Tactical Air War." In *Theory of Games: Techniques and Applications,* edited by A. Mensch, 219–235. New York: American Elsevier, 1966.

Montgomery, Evan Braden. "Breaking out of the Security Dilemma: Realism, Reassurance, and the Problem of Uncertainty." *International Security* 31, no. 2 (2006): 151–185.

Morgan, Mary S. "The Curious Case of the Prisoner's Dilemma: Model Situation? Exemplary Narrative?" In *Science without Laws,* 157–188. Durham: Duke University Press, 2007.

Morgan, Patrick M. Deterrence Now. Cambridge: Cambridge University Press, 2003.

Morgan, Patrick M. "New Directions in Deterrence." In *Nuclear Weapons and the Future of Humanity,* edited by Avner Cohen and Steven Lee, 169–189. Totowa, NJ: Rowman & Allanheld, 1986.

Morgenstern, Oskar. *The Question of National Defense.* New York: Random House, 1959.

Morris, Christopher W. "A Contractarian Defense of Nuclear Deterrence." *Ethics* 95 (1985): 479–496.

Moyn, Samuel. *The Last Utopia: Human Rights in History.* Cambridge, MA: Belknap Press of Harvard University Press, 2010.

Mueller, Dennis C. *Public Choice III*. Cambridge; New York: Cambridge University Press, 2003.

Mueller, John E. *Capitalism, Democracy, and Ralph's Pretty Good Grocery*. Princeton, NJ: Princeton University Press, 1999.

Mueller, John E. *Atomic Obsession: Nuclear Alarmism from Hiroshima to Al Qaeda*. New York: Oxford University Press, 2009.

Myerson, Roger B. *Game Theory: Analysis of Conflict*. Cambridge, MA: Harvard University Press, 1991.

Nash, John F. "The Bargaining Problem." *Econometrica* 18, no. 2 (1950): 155–162.

Nash, John. "Non-Cooperative Games." *The Annals of Mathematics* 54, no. 2 (1951): 54, 286–295.

Nash, John F. *Essays on Game Theory*. Cheltenham, UK; Brookfield, VT: E. Elgar, 1996.

Neal, Fred Warner and Bruce D. Hamlett. "The Never-Never Land of International Relations." *International Studies Quarterly* 13, no. 3 (1969): 281–305.

Neal, Patrick. "Hobbes and Rational Choice Theory." *The Western Political Quarterly* 41, no. 4 (1988): 635–652.

Neblo, Michael A. *Deliberative Democracy: Between Theory and Practice*. New York: Cambridge University Press, 2015.

Von Neumann, John and Oskar Morgenstern. *Theory of Games and Economic Behavior*. 60th anniversary ed. Princeton, NJ; Woodstock: Princeton University Press, 2004.

Nitze, Paul H. "Atoms, Strategy and Policy." *Foreign Affairs* 34, no. 2 (1956): 187–198.

Nitze, Paul H. "The Strategic Balance between Hope and Skepticism." *Foreign Policy*, no. 17 (1975): 136–156.

Nitze, Paul H. "Assuring Strategic Stability in an Era of Détente." *Foreign Affairs* 54, no. 2 (1976): 207–231.

Nitze, Paul H. "Deterring Our Deterrent." *Foreign Policy*, no. 25 (1977): 195–210.

Nitze, Paul H. "Strategy in the Decade of the 1980s." *Foreign Affairs* 59, no. 1 (1980): 82–101.

North, Douglass Cecil. *Institutions, Institutional Change, and Economic Performance*. The Political Economy of Institutions and Decisions. Cambridge; New York: Cambridge University Press, 1990.

Nozick, Robert. "The Construction of the Good." In *Philosophy, Science, and Method*, edited by Sidney Morgenbesser, Patrick Suppes, and Morton White, 440–472. New York: St. Martin's Press, 1969.

Nozick, Robert. *Anarchy, State, and Utopia*. Oxford: Blackwell, 1975.

Nozick, Robert. *Nature of Rationality*. Princeton, NJ: Princeton University Press, 1994.

Nye, Joseph S., Jr. *Nuclear Ethics*. New York: Free Press, 1986.

Nye, Joseph S., Jr. *Soft Power: The Means to Success in World Politics*. New York: Public Affairs, 2005.

Nye, Joseph S. and Sean M. Lynn-Jones. "International Security Studies: A Report of a Conference on the State of the Field." *International Security* 12, no. 4 (1988): 5–27.

Olson, Mancur. *The Logic of Collective Action; Public Goods and the Theory of Groups*. Harvard Economic Studies, v. 124. Cambridge, MA: Harvard University Press, 1965.

Ostrom, Elinor. *Governing the Commons: The Evolution of Institutions for Collective Action*. The Political Economy of Institutions and Decisions. Cambridge; New York: Cambridge University Press, 1990.

Ostrom, Elinor. "Policies That Crowd out Collective Action." In *Moral Sentiments and Material Interests*, edited by Herbert Gintis, Samuel Bowles, Robert Boyd, and Ernst Fehr. Cambridge, MA: MIT Press, 2005.

Ostrom, Elinor and Edella Schlager. "The Formation of Property Rights." In *Rights to Nature*, edited by Susan Hanna, Carl Folke, and Karl-Göran Mäler, 127–156. Washington, DC: Island Press, 1996.

Oye, Kenneth A. "Explaining Cooperation under Anarchy: Hypotheses and Strategies." *World Politics* 38, no. 1 (1985): 1–24.

Parrington, Col. Alan J. "Mutually Assured Destruction Revisited. Strategic Doctrine in Question." *Airpower Journal* (1997): 4–18.

Pettit, Philip. "The Prisoner's Dilemma and Social Theory: An Overview of Some Issues." *Australian Journal of Political Science* 20, no. 1 (1985): 1–11.

Pettit, Philip. "Free Riding and Foul Dealing." *Journal of Philosophy* 83, no. 7 (1986), 361–379.

Pettit, Philip. "Preserving the Prisoner's Dilemma." *Synthese* 68, no. 1 (1986): 181–184.

Pettit, Philip. *Republicanism: A Theory of Freedom and Government*. Oxford Political Theory. Oxford; New York: Clarendon Press; Oxford University Press, 1997.

Pettit, Philip. *Rules, Reasons, and Norms: Selected Essays*. Oxford; New York: Oxford University Press; Clarendon Press, 2002.

Pettit, Philip. "*Virtus Normativa*: Rational Choice Perspectives." In *Rules, Reasons, and Norms*, edited by Philip Pettit. Oxford: Clarendon Press, 2002.

Pettit, Philip and Geoffrey Brennan. "Restrictive Consequentialism." *Australasian Journal of Philosophy* 64, no. 4 (1986): 438–455.

Pipes, Richard. *Survival Is Not Enough: Soviet Realities and America's Future*. New York: Simon and Schuster, 1984.

Pocock, J. G. A. *The Machiavellian Moment: Florentine Political Thought and the Atlantic Republican Tradition*. 2nd ed. Princeton, NJ: Princeton University Press, 2003.

Posner, Richard A. *The Economics of Justice*. Cambridge, MA: Harvard University Press, 1981.

Posner, Richard A. "Wealth Maximization and Judicial Decision-Making." *International Review of Law and Economics* 4 (1984): 131–135.

Posner, Richard A. "Wealth Maximization Revisited." *Notre Dame Journal of Law, Ethics, and Public Policy* 2 (1985): 85–106.

Poundstone, William. *Prisoner's Dilemma*. New York: Anchor Books, 1993.

Quackenbush, Stephen. "The Rationality of Rational Choice Theory." *International Interactions* 30, no. 2 (2004): 87–107.

Quester, George H. "Some Possible Surprises in Our Nuclear Future." *Small Wars & Insurgencies* 11, no. 2 (2000): 38–51.

Rand, David G., Joshua D. Greene, and Martin A. Nowak. "Spontaneous Giving and Calculated Greed." *Nature* 489, no. 11457 (2012), doi:10.1038/nature11467; www.nature.com/nature/journal/v489/n7416/abs/nature11467.html, accessed July 1, 2015.

Rapoport, Anatol. *Fights, Games, and Debates*. Ann Arbor: University of Michigan Press, 1960.

Rapoport, Anatol. *Strategy and Conscience*. New York: Harper & Row, 1964.

Rapoport, Anatol. "The Role of Game Theory in Uncovering Non-Strategic Principles of Decision." In *Theory of Games: Techniques and Applications*, edited by A. Mensch, 410–431. New York: American Elsevier Publishing Company, Inc., 1966.

Rapoport, Anatol. *The Big Two; Soviet-American Perceptions of Foreign Policy*. American Involvement in the World. New York: Pegasus, 1971.

Rapoport, Anatol and Albert M. Chammah. *Prisoner's Dilemma; a Study in Conflict and Cooperation*. Ann Arbor: University of Michigan Press, 1965.

Rawls, John. *A Theory of Justice*. Cambridge, MA: Belknap Press of Harvard University Press, 1971.

Rawls, John. "Justice as Fairness: Political Not Metaphysical." *Philosophy and Public Affairs* 14, no. 3 (1985): 223–251.

Rawls, John. *A Theory of Justice*. Original ed. Cambridge, MA: Belknap Press, 2005.

Riker, William H. and Peter C. Ordeshook. *An Introduction to Positive Political Theory*. Prentice-Hall Contemporary Political Theory Series. Englewood Cliffs, NJ: Prentice-Hall, 1973.

Robbins, Lionel Robbins. *An Essay on the Nature & Significance of Economic Science*. London: Macmillan, 1932.

Robbins, Lionel Robbins. *An Essay on the Nature and Significance of Economic Science*. 3rd ed. New York: New York University Press, 1984.

Rodgers, Daniel T. *Age of Fracture*. Cambridge, MA: Belknap Press, 2012.

Roemer, John. "The Mismarriage of Bargaining Theory and Distributive Justice." *Ethics* 97, no. 1 (1986): 88–110.

Roese, Neal J. and James M. Olson. *What Might Have Been: The Social Psychology of Counterfactual Thinking*. Mahwah, NJ: Erlbaum, 1995.

Ross, Don. "Evolutionary Game Theory and the Normative Theory of Institutional Design: Binmore and Behavioral Economics." *Politics, Philosophy & Economics* 5, no. 1 (2006): 51–79.

Ross, Don. "Game Theory." *Stanford Encyclopedia of Philosophy*, 2006, revised 2014. http://plato.stanford.edu/entries/game-theory/.

Rousseau, Jean-Jacques. *The Social Contract*. Translated and Introduced by Maurice Cranston. New York: Penguin, [1762] 1968.

Ryan, Alan. *The Making of Modern Liberalism*. Princeton, NJ: Princeton University Press, 2014.

Ryan, Matthew J. "Mathematicians as Great Economists: John Forbes Nash Jr." *Agenda* 9, no. 2 (2002): 121–134.

Saladoff, Susan. *Hot Coffee*. Docurama Films: New Video, 2011.

Sandel, Michael. *Liberalism and the Limits of Justice*. 2nd ed. Cambridge; New York: Cambridge University Press, 1998.

Sandel, Michael. *What Money Can't Buy: The Moral Limits of Markets*. New York: Farrar, Straus, Giroux, 2013.

Savage, Leonard J. *The Foundations of Statistics*. New York: Wiley, 1954.

Scarry, Elaine. *Thermonuclear Monarchy: Choosing between Democracy and Doom*. New York: W. W. Norton, 2014.

Schelling, Thomas C. "The Strategy of Conflict Prospectus for a Reorientation of Game Theory." *The Journal of Conflict Resolution* 2, no. 3 (1958): 203–264.

Schelling, Thomas C. "The Retarded Science of International Strategy." *Midwest Journal of Political Science* 4, no. 2 (1960): 107–137.

Schelling, Thomas C. *The Strategy of Conflict*. Cambridge, MA: Harvard University Press, 1960.

Schelling, Thomas C. "Review: War without Pain, and Other Models." *World Politics* 15, no. 3 (1963): 465–487.

Schelling, Thomas C. "Review: *Strategy and Conscience* by Anatol Rapoport." *The American Economic Review* 54, no. 6 (1964): 1082–1088.

Schelling, Thomas C. "Strategy, Tactics, and Non-Zero-Sum Theory." In *Theory of Games: Techniques and Applications*, edited by A. Mensch, 469–480. New York: American Elsevier, 1966.

Schelling, Thomas C. "Hockey Helmets, Concealed Weapons, and Daylight Saving: A Study of Binary Choices with Externalities." *The Journal of Conflict Resolution* 17, no. 3 (1973): 381–428.

Schelling, Thomas. "A Framework for the Analysis of Arms-Control Proposals." *Daedalus* 104, no. 3 (1975): 187–200.

Schelling, Thomas C. "The Role of War Games and Exercises." In *Managing Nuclear Operations*, edited by Ashton B. Carter, John D. Steinbruner, and Charles A. Zraket, 426–444. Washington, DC: The Brookings Institution 1987.

Schelling, Thomas C. "An Astonishing Sixty Years: The Legacy of Hiroshima." In *Nobel Prize Lecture*, 2005. Available online at www.nobelprize.org/nobel_prizes/economic-sciences/laureates/2005/schelling-lecture.html.

Schelling, Thomas C. and Morton H. Halperin. *Strategy and Arms Control*. New York: Twentieth Century Fund, 1961.

Schelling, Thomas C. and Harvard University. Center for International Affairs. *Arms and Influence*. New Haven: Yale University Press, 1966.

Scheper-Hughes, Nancy. *The Last Commodity: Post-Human Ethics, Global (in)Justice, and the Traffic in Organs*. Dissenting Knowledges Pamphlet Series. Penang: Multiversity & Citizens International, 2008.

Scherer, F. M. "Review: Analysis for Military Decisions by E. S. Quade." *The American Economic Review* 55, no. 5 (1965): 1191–1192.

Schilling, Warner S. "US Strategic Nuclear Concepts in the 1970s: The Search for Sufficiently Equivalent Countervailing Parity." *International Security* 6, no. 2 (1981).

Schlesinger, James R. "European Security and the Nuclear Threat since 1945." RAND Report P3574, April 1967.

Schlesinger, James R. "Systems Analysis and the Political Process" RAND Report P3464, June 1967.

Schlesinger, James R. "Uses and Abuses of Analysis." *Survival: Global Politics and Strategy* 10, no. 10 (1968): 334–342.

Schlesinger, James R. Interview, Carter Presidency Project, Miller Center of Public Affairs, available at Jimmy Carter Presidential Library, 1984.

Schlosser, Eric. *Command and Control: Nuclear Weapons, the Damascus Accident, and the Illusion of Safety*. London: Penguin, 2014.

Schmidtz, David. "Pettit's Free Riding and Foul Dealing." *Australasian Journal of Philosophy* 66, no. 2 (1988): 230–233.

Schrecker, Ellen. *Cold War Triumphalism: The Misuse of History after the Fall of Communism*. New York: New Press, distributed by W. W. Norton, 2004.

Schweller, Randall L. "Bandwagoning for Profit: Bringing the Revisionist State Back In." *International Security* 19, no. 1 (1994): 72–107.

Schweller, Randall L. "Neorealism's Status-Quo Bias: What Security Dilemma?" *Security Studies* 5, no. 3 (1996): 90–121.

Schweller, Randall L. *Deadly Imbalances: Tripolarity and Hitler's Strategy of World Conquest*. New York: Columbia University Press, 1998.

Schweller, Randall L. "Realism and the Present Great Power System: Growth and Positional Conflict over Scarce Resources." In *Unipolar Politics: Realism and State Strategies after the Cold War*, edited by Ethan B. Kapstein and Michael Mastanduno, 28–68. New York: Columbia University Press, 1999.

Schweller, Randall L. "Unanswered Threats." *International Security* 29, no. 2 (2004): 159–201.

Schweller, Randall L. "Neoclassical Realism and State Mobilization: Expansionist Ideology in the Age of Mass Politics." In *Neoclassical Realism, the State, and Foreign Policy*, edited by Steven E. Lobell, Norrin M. RIpsman, and Jeffrey W. Taliaferro, 227–250. Cambridge: Cambridge University Press, 2009.

Schweller, Randall L. *Maxwell's Demon and the Golden Apple: Global Discord and the New Millennium*. Baltimore: JHUP, 2014.

Scodel, Alvin. "Induced Collaboration in Some Non-Zero-Sum Games." *The Journal of Conflict Resolution* 6, no. 4 (1962): 335–340.

Seabright, Paul. "Social Choice and Social Theories." *Philosophy and Public Affairs* 18, no. 4 (1989): 365–387.

Seabright, Paul. "The Evolution of Fairness Norms: An Essay on Ken Binmore's Natural Justice." *Politics, Philosophy & Economics* 5, no. 1 (2006): 33–50.

Searle, John R. *The Construction of Social Reality*. New York: Free Press, 1995.

Sen, Amartya. *Collective Choice and Social Welfare*. Mathematical Economics Texts. San Francisco: Holden-Day, 1970.

Sen, Amartya. "The Impossibility of a Paretian Liberal." *The Journal of Political Economy* 78, no. 1 (1970): 152–157.

Sen, Amartya. "Rational Fools: A Critique of the Behavioral Foundations of Economic Theory." *Philosophy and Public Affairs* 6, no. 4 (1977): 317–344.

Sen, Amartya. "Social Choice Theory: A Re-Examination." *Econometrica* 45, no. 1 (1977): 53–89.

Sen, Amartya. *Choice, Welfare and Measurement*. Cambridge, MA: Harvard University Press, 1982.

Sen, Amartya. "Liberty and Social Choice." *The Journal of Philosophy* 80, no. 1 (1983): 5–28.

Sen, Amartya. "Minimal Liberty." *Economica* 59, no. 234 (1992): 139–159.

Sen, Amartya. "The Possibility of Social Choice." *The American Economic Review* 89, no. 3 (1999): 349–378.

Sen, Amartya. *Rationality and Freedom*. Cambridge, MA: Belknap Press, 2002.

Sen, Amartya. "Elements of a Theory of Human Rights." *Philosophy and Public Affairs* 32, no. 4 (2005): 315–356.

Sent, Esther-Mirjam. "Some Like It Cold: Thomas Schelling as a Cold Warrior." *The Journal of Economic Methodology* 14, no. 4 (2007): 455–471.

Shapiro, Ian and Russell Hardin. *Political Order Nomos XXXVIII*. New York: New York University Press, 1996.

Shapiro, Ian and Alexander Wendt. "The Difference That Realism Makes: Social Science and the Politics of Consent." *Politics and Society* 20, no. 2 (1992): 197–223.

Shapley, Deborah. *Promise and Power: The Life and Times of Robert McNamara*. Boston: Little, Brown, 1993.

Shimshoni, Jonathan. *Israel and Conventional Deterrence: Border Warfare from 1953 to 1970*. Ithaca: Cornell University Press, 1988.

Shubik, Martin. "On the Study of Disarmament and Escalation." *The Journal of Conflict Resolution* 12, no. 1 (1968): 83–101.

Shue, Henry. *Nuclear Deterrence and Moral Restraint: Critical Choices for American Strategy*. Cambridge; New York: Cambridge University Press, 1989.

Sigal, Leon V. "Rethinking the Unthinkable." *Foreign Policy*, no. 34 (1979): 35–51.

Simmel, Georg. *The Philosophy of Money*. London; Boston: Routledge & Kegan Paul, 1978.

Simmel, Georg and David Frisby. *The Philosophy of Money*. 2nd enl. ed. London; New York: Routledge, 1990.

Simon, Herbert A. *Models of Bounded Rationality: Economic Analysis and Public Policy*. Cambridge: MIT Press, 1984.

Singer, Peter. "Famine, Affluence, and Morality," *Philosophy and Public Affairs* 1, no. 1 (1972): 229–243.

Skinner, Quentin. *The Foundations of Modern Political Thought*. 2 vols. Cambridge; New York: Cambridge University Press, 1978.

Skyrms, Brian. "The Shadow of the Future." In *Rational Commitment and Social Justice*, edited by Jules L. Coleman and Christopher W. Morris, 12–21. Cambridge: Cambridge University Press, 1998.

Skyrms, Brian. "The Stag Hunt and the Evolution of Social Structure." *Economics and Philosophy* 22 (2004): 441–468.

Skyrms, Brian. *The Stag Hunt and the Evolution of Social Structure*. Cambridge: Cambridge University Press, 2004.

Skyrms, Brian. "Trust, Risk, and the Social Contract." *Synthese* 160 (2006): 21–25.

Slocombe, Walter. "The Countervailing Strategy." *International Security* 5, no. 4 (1981): 18–27.

Smith, Adam. *Wealth of Nations*. vols. 1 and 2, edited by R. H. Campbell and A. S. Skinner. Indianapolis: Liberty Fund, [1776] 1976.

Smith, Adam. *The Theory of Moral Sentiments*, edited by D. D. Raphael and A. L. Macfie. Indianapolis: Liberty Fund, [1759] 1982.

Smith, Maynard. *Evolution and the Theory of Games*. New York: Cambridge University Press, 1982.

Snidal, Duncan. "Cooperation vs. Prisoners' Dilemma." *Journal of Personality and Social Psychology* 79 (1985): 923–942.

Snidal, Duncan. "Coordination versus Prisoners' Dilemma: Implications for International Cooperation and Regimes." *The American Political Science Review* 79, no. 4 (1985): 923–942.

Snidal, Duncan. "The Game Theory of International Politics." *World Politics* 38, no. 1 (1985): 25–57.

Snidal, Duncan. "The Limits of Hegemonic Stability Theory." *International Organization* 39, no. 4 (1985): 579–614.

Snyder, Glenn Herald. *Deterrence and Defense; toward a Theory of National Security*. Princeton, NJ: Princeton University Press, 1961.

Snyder, Glenn Herald. "'Prisoner's Dilemma' and 'Chicken' Models in International Politics." *International Studies Quarterly* 15, no. 1 (1971): 66–103.

Snyder, Glenn Herald. "The Security Dilemma in Alliance Politics." *World Politics* 36, no. 1 (1984): 461–495.

Snyder, Glenn Herald and Paul Diesing. *Conflict among Nations: Bargaining, Decision Making, and System Structure in International Crises*. Princeton, NJ: Princeton University Press, 1977.

Snyder, Jack L. *The Soviet Strategic Culture: Implications for Limited Nuclear Options*. Santa Monica, CA: RAND Corporation, September 1977.

Soll, Jacob. *The Reckoning: Financial Accountability and the Rise and Fall of Nations*. New York: Basic Books.

Solovey, Mark. *Shaky Foundations: The Politics-Patronage-Social Science Nexus in Cold War America*. New Brunswick, NJ: Rutgers University Press, 2013.

Southwood, Nicholas. "Vindicating the Normativity of Rationality." *Ethics* (October 2008): 9–30.

Spencer, Metta. "Rapoport at Ninety." *Connections* 24, no. 3 (2001): 104–107.

Starobin, Joseph R. "Origins of the Cold War: The Communist Dimension." *Foreign Affairs*, no. 47 (1969): 681–696.

Steger, Manfred B. and Ravi K. Roy. *Neoliberalism: A Very Short Introduction*. New York: Oxford University Press, 2010.

Stein, Arthur. "Coordination and Collaboration: Regimes in an Anarchic World." *International Organization* 36, no. 2 (1982): 299–324.

Stein, Arthur. "Coordination and Collaboration: Regimes in an Anarchic World." In *Neorealism and Neoliberalism: The Contemporary Debate*, edited by David A. Baldwin, 3–29. New York: Columbia University Press, 1993.

Steinbruner, John. "Beyond Rational Deterrence: The Struggle for a New Concept." *World Politics* 38, no. 2 (1976), 223–245.

Stern, N. H. and Great Britain Treasury. *Stern Review: The Economics of Climate Change*. London: HM Treasury, 2006.

Stokey, Edith and Richard Zeckhauser. *A Primer for Policy Analysis*. New York: W. W. Norton, 1978.

Sugden, Robert. "Rationality and Impartiality: Is the Contractarian Enterprise Possible?" In *Rationality, Justice and the Social Contract: Themes from Morals by Agreement*, edited by David Gauthier and Robert Sugden, 157–175. Ann Arbor: University of Michigan Press, 1993.

Sugden, Robert. "The Contractarian Enterprise." In *Rationality, Justice and the Social Contract: Themes from Morals by Agreement*, edited by David Gauthier and Robert Sugden, 1–23. Ann Arbor: University of Michigan Press, 1993.

Sugden, Robert. "What We Desire, What We Have Reason to Desire, Whatever We Might Desire: Mill and Sen on the Value of Opportunity." *Utilitas* 18, no. 1 (2006): 33–51.

Sunstein, Cass and Richard Thaler. *Nudge: Improving Decisions about Health, Wealth, and Happiness*. New York: Penguin, 2009.

Taliaferro, Jeffrew W. "Security Seeking under Anarchy: Defensive Realism Revisited." *International Security* 25, no. 3 (2000): 128–161.

Tang, Shiping. *A Theory of Security Strategy for Our Time: Defensive Realism*. New York: Palgrave Macmillan, 2010.

Taylor, Michael. *Anarchy and Cooperation*. London; New York: Wiley, 1976.

Taylor, Michael. *The Possibility of Cooperation*. Studies in Rationality and Social Change. Cambridge; New York: Cambridge University Press, 1987.

Taylor, Michael. *Rationality and the Ideology of Disconnection*. Contemporary Political Theory. Cambridge; New York: Cambridge University Press, 2006.

Thomas, Clayton J. "Some Past Applications of Game Theory to Problems of the United States Air Force." In *Theory of Games: Techniques and Applications*, edited by A. Mensch, 205–267. New York: American Elsevier, 1966.

Thomas, William. *Rational Action: The Sciences of Policy in Britain and America, 1940–1960*. Cambridge, MA: MIT Press, 2015.

Thompson, Nicholas. *The Hawk and the Dove: Paul Nitze, George Kennan, and the History of the Cold War*. New York: Henry Holt, 2009.

Tierney, John. "Do You Have Free Will? Yes, it's the only Choice." *New York Times*, March 21, 2011, D.

Tonelson, Alan. "Nitze's World." *Foreign Policy*, no. 35 (1979): 74–90.

Trivers, Robert. "The Evolution of Reciprocal Altrusim," *Quarterly Review of Biology* 46, (1971): 35–37.

Tuck, Richard. *Hobbes*. Past Masters. Oxford; New York: Oxford University Press, 1989.

Tuck, Richard. *The Rights of War and Peace: Political Thought and the International Order from Grotius to Kant*. Oxford: Oxford University Press, 2001.

Tuck, Richard. *Hobbes: A Very Short Introduction*. Oxford: Oxford University Press, 2002.

Tuck, Richard. *Free Riding*. Cambridge, MA: Harvard University Press, 2008.

Twing, Stephen W. *Myths, Models & U.S. Foreign Policy: The Cultural Shaping of Three Cold Warriors*. Boulder, CO: Lynne Rienner, 1998.

Ungar, Sheldon. "Moral Panics, the Military-Industrial Complex, and the Arms Race." *The Sociological Quarterly* 31, no. 2 (1990): 165–185.

United States, John Dunlap, Peter Force, David Ridgely, and Printed Ephemera Collection (Library of Congress). In *Congress, July 4, 1776, a Declaration by the Representatives of the United States of America, in General Congress Assembled*. Philadelphia: Printed by John Dunlap, 1776.

University of Southern California School of Philosophy. "Pacific Philosophical Quarterly." In *Hobbes on Reason*, edited by Bernard Gert, 82, 243–257. Los Angeles: School of Philosophy, University of Southern California, 1980.

Van Vugt, Mark and Paul Van Lange. "The Altrusim Puzzle." In *Evolution and Psychology*, edited by M. Schaller, J. A. Simpson, and D. T. Kenrick. New York: Psychology Press, 2006.

Veblen, Thorstein. "Why Is Economics Not an Evolutionary Science." In *The Place of Science in Modern Civilization*, edited by Thorstein Veblen. New York: Echo, 1919/2012.

Venugopal, Rajesh. "Neoliberalism as a Concept." *Economy and Society* 44, no. 2 (2015): 165–187.

Volmar, Daniel. "The Power of the Atom: US Nuclear Command, Control, and Communications, 1945–1965." PhD diss., Harvard University, 2016.

Wagner, Harrison. "The Theory of Games and the Problem of International Cooperation." *American Political Science Review* 70 (1983): 330–346.

Wald, Abraham. "Generalization of a Theorem by v. Neumann Concerning Zero Sum Two Person Games" *Annals of Mathematics*, 46, no. 2 (1945): 281–286.

Wald, Abraham. "Statistical Functions Which Minimize the Maximum Risk." *Annals of Mathematics*, 46 (1945): 265–280.

Wald, Abraham. *Statistical Decision Functions*. New York: Wiley, 1950.

Wallace, Njorn et al. "Heritability of Ultimatum Game Respondent Behavior." *Proceedings of the National Academy of Sciences* 104, no. 40 (2007): 15631–15634.

Walt, Stephen M. "Rigor or Rigor Mortis: Rational Choice and Security Studies." *International Security* 23, no. 4 (1999): 5–48.

Waltz, Kenneth. *Man, the State, and War: A Theoretical Analysis*. New York: Columbia University Press, 1959.

Waltz, Kenneth N. "Structural Realism after the Cold War." *International Security* 25, no. 1 (2000): 5–42.

Warnke, Paul C. "Apes on an Treadmill." *Foreign Policy*, no. 18 (1975): 12–29.

Weale, Albert. "Justice, Social Union and the Separateness of Persons." In *Rationality, Justice and the Social Contract: Themes from Morals by Agreement*, edited by David

Gauthier and Robert Sugden, 75–94. Ann Arbor: University of Michigan Press, 1993.

Weber, Max. *Economy and Society: An Outline of Interpretive Sociology.* Berkeley: University of California Press, 1978.

Weber, Max. *Protestant Ethic and the Spirit of Capitalism.* London: Routledge, 1985.

Wendt, Alexander. "Collective Identity Formation and the International State." *American Political Science Review*, 88, No. 2 (1994), 384–396.

Wendt, Alexander. *Social Theory of International Politics.* Cambridge Studies in International Relations. Cambridge; New York: Cambridge University Press, 1999.

Wicksell, Knut. "A New Principle of Just Taxation." In *Classics in the Theory of Public Finance*, edited by Richard A. and Alan T. Peacock Musgrave, 72–119. London: MacMillan, 1958.

Wicksell, Knut. *Lectures on Political Economy.* 2 vols. Reprints of Economic Classics. New York: A. M. Kelley, 1967.

Williams, Michael C. "What Is the National Interest? The Neoconservative Challenge in IR Theory." *European Journal of International Relations* 11, no. 3 (2005): 307–337.

Williamson, John. "Democracy and the 'Washington Consensus.'" *World Development* 21, no. 8 (1991): 1329–1336.

Wilson, Edward O. *Sociobiology.* Cambridge, MA: Harvard University Press, 1975.

Wohlstetter, Albert. "Sin and Games in America." In *Game Theory and Related Approaches to Social Behavior*, edited by Martin Shubik, 209–229. New York: John Wiley, 1964.

Wohlstetter, Albert. "Analysis and Design of Conflict Systems." In *Analysis for Military Decisions*, edited by E. S. Quade, 103–148. Chicago: Rand McNally, 1964.

Wohlstetter, Albert. "Theory and Opposed-Systems Design." *The Journal of Conflict Resolution* 12, no. 3 (1968): 302–331.

Wohlstetter, Albert. "Is There a Strategic Arms Race?" *Foreign Policy*, no. 15 (1974): 3–20.

Wohlstetter, Albert. "Optimal Ways to Confuse Ourselves." *Foreign Policy*, no. 20 (1975): 170–196.

Wohlstetter, Albert, Paul H. Nitze, Joseph Alsop, Morton H. Halperin, and Jeremy J. "Is There a Strategic Arms Race?: Rivals but No 'Race.'" *Foreign Policy*, no. 16 (1974): 48–92.

Wolfowitz, Paul D. "The New Defense Strategy." In *Rethinking America's Security: Beyond Cold War to New World Order*, edited by Graham Allison and Gregory F. Treverton, 176–195. New York: W. W. Norton, 1992.

Wood, Gordon S. and Institute of Early American History and Culture (Williamsburg, VA). *The Creation of the American Republic, 1776–1787.* Chapel Hill: Published for the Institute of Early American History and Culture at Williamsburg, VA, 1969.

Woodward, P. A. "The 'Game' of Nuclear Strategy: Kavkas on Strategic Defense." *Ethics* 99, no. 3 (1989): 563–571.

Yamagishi, Toshio. "The Provision of a Sanctioning System as a Public Good." *Journal of Personality and Social Psychology* 51, no. 1 (1986): 110–116.

Zagare, Frank C. "Classical Deterrence Theory: A Critical Assessment." *International Interactions* 21, no. 4 (1996): 365–387.

Zagare, Frank C. and D. Marc Kilgour. *Perfect Deterrence*. Cambridge Studies in International Relations 72. Cambridge; New York: Cambridge University Press, 2000.

Zhang, Dongmo. "A Logical Model of Nash Bargaining Solutions." In *International Joint Conference On Artificial Intelligence*, 983–988. Edinburgh: Morgan Kaufmann, 2005.

Index

Lightning Source UK Ltd.
Milton Keynes UK
UKHW022251250920
370556UK00009B/78